Sustainable Landscape Planning and Natural Resources Management

IEREK Interdisciplinary Series for Sustainable Development

Sustainable Landscape Planning and Natural Resources Management (SLNR) is a peer-reviewed book series that presents groundbreaking research through the intersection of two main areas that of which are 'Sustainable Landscape Planning' and 'Natural Resources Management'. The series initially covers a wide range of academic disciplines in sustainability, including landscape ecology, green urbanism, urban regeneration, land restoration and renaturation, energy management, and nature conservation. Moreover, it also incorporates the latest developments and research in engineering topics that contribute to more sustainable cities, such as Geosciences, infrastructure, civil engineering, geology, mining, water, and natural resources conservation and management.

By showcasing methods for discovering, managing, and employing natural resources, the series aims to provide approaches that enable the responsible use of these resources. Ultimately, the SLNR series is committed to advancing the United Nations' Sustainable Development Goals for 2030.

The wide range of topics covered in this series are primarily related to sustainable development, involving durable infrastructure projects, civil engineering, urban regeneration, the built, sustainability, climate change mitigation, energy management, geosciences, geology, and the future of cities, all of which depend on the availability of natural resources and effective wealth management. The series also features conceptual and empirical contributions from various interrelated fields of earth science, with an emphasis on the era of new discovery. It focuses on providing practical solutions to ensure sustainable development and provides new case studies that offer concrete examples of how to manage natural resources effectively.

The series encompasses the best research papers from various international conferences organized by IEREK conferences, also accepts proposals from other international conferences, including interdisciplinary contributions. It exhibits innovative approaches and highlights how they can best support both economic and sustainable development through the responsible use of natural resources and global, local, and individual activities for the welfare of all societies on Earth.

The SLNR series is a valuable source for experts in government, international organizations, consultancies, industry, business sectors, and academics. Published in cooperation with IEREK, the Springer SLNR series displays important new research in various sectors, delivering valuable insights and practical solutions to the complex challenges of sustainable development for the readers.

Nitish Kumar
Editor

Biotechnology in Plant Conservation

Relevance to Sustainable Management of Natural Plant Resources

Springer

Editor
Nitish Kumar
Department of Biotechnology
Central University of South Bihar
Gaya, Bihar, India

ISSN 2948-1910 ISSN 2948-1929 (electronic)
Sustainable Landscape Planning and Natural Resources Management
ISBN 978-3-032-02604-0 ISBN 978-3-032-02605-7 (eBook)
https://doi.org/10.1007/978-3-032-02605-7

This Springer imprint is published by the registered company Springer Nature Switzerland AG
The registered company address is: Gewerbestrasse 11, 6330 Cham, Switzerland

Preface

Food security issues that arise from an increasing global population can potentially be resolved through the sustainable use of plant genetic resources and its conservation. The diversity found in plant genetic resources is important factors of improvement. It is well known that genetic diversity is necessary for existence of plant species in their natural habitats. It depends on various environmental parameters which regulate the allele number on the genome loci. Therefore, in order to develop effective management plans for the plant species conservation, it is crucial to comprehend the genetic makeup of those species from their natural habitat. Assessing the species genetic diversity is crucial for both gene editing and conservation.

Technological advancements in the field of biotechnology, including molecular markers, in vitro culture, DNA banking, and cryopreservation have significantly improved the strategies for the efficient conservation of endangered and rare and plant genetic resources. It is possible for scientists to search for novel genes and their products with this method of conserving biological information. This book discusses future biotechnological strategies for plant species conservation and further modification to improve yield and quality characteristics. Academics studying, teaching, or conducting research in the fields of biotechnology, plant science, and environmental science will find the book particularly interesting.

Gaya, Bihar, India Nitish Kumar

Acknowledgments

First and foremost, I would like to praise and thank God, the almighty, who has granted countless blessing, knowledge, and opportunity to accomplish the book project work. Thanks to all the authors of the various chapters for their contributions. It had been a bit of a long process from the initial outlines to developing the full chapters and then revising them in the light of reviewer's comments. I sincerely acknowledge the author's willingness to go through this process. I also acknowledge the work and knowledge of the members of our review panels, many of which had to be done at short notice. Thanks to all the people at Springer Nature, especially Dr. Christian Witschel and Mr. Yogesh Padmanaban, and Angham Mohamed with whom I corresponded for their advice and facilitation in the production of this book.

Gaya, Bihar, India Nitish Kumar

Contents

Conservation of Natural Plant Resources for Sustainable Development: An Overview

Ankan Das, Dibyendu Seth, Sourish Pramanik, Rajashree Mitra Thakur, Sandip Debnath, and Mehdi Rahimi

Abstract

The conservation of natural plant resources is pivotal for maintaining ecological integrity, ensuring food security, and fostering sustainable development. However, contemporary challenges such as climate change, habitat degradation, and anthropogenic pressures necessitate the adoption of advanced and integrative approaches. This chapter delves into state-of-the-art methodologies in plant resource conservation, emphasizing their implications for sustainable management. Cutting-edge techniques such as synthetic biology, high-throughput genotyping, and phenotyping, as well as cryobiotechnology, have revolutionized both ex situ and in situ conservation strategies. The integration of omic technologies, including genomics, metabolomics, and epigenomics, offers unprecedented insights into genetic variability, adaptive mechanisms, and stress tolerance in plant species. Additionally, the application of precision tools such as drones, Internet of Things (IoT) devices, and big data analytics enhances real-time ecosystem monitoring and targeted habitat restoration. The chapter also explores the significance of community-based conservation models and international policy frameworks, underscoring their synergy with scientific advancements. By synthesizing recent advancements and addressing existing gaps, this chapter provides a comprehensive framework for conserving plant biodiversity to align with global sustainability agendas.

Keywords

Synthetic biology · Cryobiotechnology · Conservation genomics · Precision agriculture · Plant biodiversity

1 Introduction

In addition to offering a variety of other environmental and cultural functions that are valuable to humans, plants are vital sources of food, condiments, medicine, fuel, shelter, feed, cattle fodder, and pollinator (Willis 2017). Tragically, invasive species, pollution, overharvesting, habitat destruction and degradation, and climate change pose threats to many useful plants. Additionally, these plants are underrepresented in ex situ conservation collections. International agreements like the Convention on Biological Diversity, the Sustainable Development Goals, and the International Treaty on Plant Genetic Resources for Food and Agriculture have established determined goals for biodiversity conservation in accordance with the enormous value of useful plants and the growing threats to their perseverance. These targets must be determined by measurable indicators to promote the growth and execution of tactics focused on protecting their genetic diversity. A variety of indicators are being put out, are being developed, or have been implemented to track progress towards these goals on both a national and international level. The number of accessions maintained in gene banks, the percentage of species and habitat protection in the wild, IUCN

A. Das · R. M. Thakur
Palli Siksha Bhavana (Institute of Agriculture), Visva-Bharati, Sriniketan, West Bengal, India

D. Seth
Department of Genetics and Plant Breeding, Chaudhary Charan Singh Haryana Agriculture University (CCSHAU), Hisar, Haryana, India

S. Pramanik
Department of Plant Pathology, Punjab Agricultural University, Ludhiana, India

S. Debnath (✉)
Department of Genetics and Plant Breeding, Palli Siksha Bhavana (Institute of Agriculture), Visva-Bharati, Sriniketan, West Bengal, India
e-mail: sandip.debnath@visva-bharati.ac.in

M. Rahimi
Department of Biotechnology, Institute of Science and High Technology and Environmental Sciences, Graduate University of Advanced Technology, Kerman, Iran

N. Kumar (ed.), *Biotechnology in Plant Conservation*, Sustainable Landscape Planning and Natural Resources Management,
https://doi.org/10.1007/978-3-032-02605-7_1

Red Listing indices, and the extent of national adoption of the Global Plan of Action for Genetic Resources for Food and Agriculture (CBD 2018) are some examples of indicators. More parameters must include all taxa of importance to the objectives and, crucially, to offer a more direct measure of trends with relation to the maintenance of genetic diversity, even though these are practical and complimentary.

The genetic foundation for agricultural development is genetic diversity (Bhandari et al. 2017). Plant breeders can create or enhance crop varieties with desired traits owing to diverse plant genetic resources (PGRs). Farmers' preferences, such as quality, high-yielding varieties, and resistance to diseases and insect pests, must be considered while creating new cultivars. In the past, people used the population's inherent genetic variety to choose favourable genotypes. As the environment changes throughout time, so does the desire for the creation of new crop kinds. Breeders can create climate-resilient cultivars with all the desired characteristics, such as tolerance to a range of biotic and abiotic stressors. Modern crop varieties are improved and developed by the utilization of genetic diversity, which can take the form of breeding stocks, wild species, mutant lines, etc. (Bhandari et al. 2017). Novel genes that are resistant to biotic and abiotic challenges must be preserved for use in breeding programmes in the future to create climate-resilient cultivars. Furthermore, the accessible germplasm should retain the plant genotypes with genes for quality attributes and visual appeal. Plant breeders can choose superior genotypes through genetic diversity both within and across plant species. These genotypes can subsequently be utilized to produce genetic stock for hybridization projects or the introduction of a crop variety (Bhandari et al. 2017). PGRs can adapt to a range of environmental circumstances according to their genetic diversity (Wernberg et al. 2018). Moreover, increasing variation in genes might lessen the detrimental effects of inbreeding in populations. Increased genetic diversity in PGRs promotes integrity, community organization, ecosystem functions, and resistance to unfavourable environmental changes (Hoban et al. 2021). Furthermore, it facilitates the use of genetically varied parents by plant breeders in a breeding programme to increase the yield of horticultural and agricultural crop types. Three degrees of genetic variety are often discernible: species-to-species diversity, population-to-population diversity, and individual-to-population diversity. In plant species, evolutionary flexibility, robustness, and adaptation are derived from genetic variety. Breeders and biotechnologists employed a variety of methods to characterize the germplasm to determine the genetic diversity prior to the discovery of diverse parents in plant breeding operations (Begna et al. 2021). Numerous approaches, including phenotypic or morphological features, biochemical or allozyme techniques, and molecular techniques, are employed to characterize the genetic diversity of PGRs and identify superior genotypes.

One of the biggest challenges is maintaining the genotypes/varieties of beneficial plants. Over the past 50 years, there has been a considerable shift in the notion of conservation and its relationship to the elusive goal of sustainable development. Two opposing trends have also emerged throughout this period. The way that most countries deal with these issues has been influenced by the rise of environmentalism, the establishment of institutions and organizations devoted to conservation, the drafting and implementation of several treaties, and other instruments. However, the magnitude and intensity of human–environment interactions have led to extensive habitat loss, degradation, and disintegration, which have exacerbated genetic diversity and species extinction. Thus, while the traditional problems of dealing with species loss, habitat loss, diversity conservation and administration, and species recovery still exist, there is a growing sense of urgency that is generating conversations about how to set priorities, questioning conventional wisdom (including the role of protected areas), and searching for new tools and techniques for evaluation as well as preservation and management choices. Plant genetic resources for food and agriculture are also the foundation of global food security. These comprise the range of genetic material present in traditional varieties, crop wild relatives, modern cultivars, and other wild species. Farmers and plant breeders can use genotypes to choose and develop new, better crops that are resistant to aggressive pests and diseases and adaptable to changing environmental circumstances. This chapter briefly talks about general approaches and cutting-edge methods for sustainable management and conservation of plant genetic resources.

2 Cutting-Edge Approaches in Plant Resource Conservation

Through the use of diverse in vitro culture techniques, plant biotechnology provides new prospects for the conservation of genetic resources. PGR storage and conveyance have benefited considerably from a variety of biotechnological methods, including cell and tissue culture and other micropropagation techniques. In order to produce PGRs in large quantities quickly for subsequent preservation and aseptic transfer, cell and tissue culture procedures are employed. In addition, new approaches and choices for more precise genetic resource conservation have been made possible by next-generation sequencing (NGS), cell fusion methods, recombinant technologies, proteomic structural biology, protein engineering, and genome editing tools. These methods aid in the preservation of medicinal, forest, decorative, uncommon, and endangered species as well as other asexually propagating plant materials. Biotechnological technologies like in vitro culture, cryopreservation, and molecular biology can be employed when PGR biological material (such seeds or organs) cannot be reproduced and preserved using conventional techniques. Biotechnological solutions can occasionally address sexual barrier issues in uncommon and endangered plant species.

The protection of rare and endangered species is now complementary in light of developments in molecular biology. One economical way to preserve PGRs is through DNA conservation of genetic resources. Many species in biodiversity are in danger of going extinct and are challenging to protect. One of the strongest options for preserving the genetic variety of these resources—which may include unique genes or alleles that might contribute to future food security—may be DNA preservation. Genomic fragments that include individual genes or whole genotypes are preserved in gene libraries or libraries of DNA samples within gene banks. DNA, RNA, and cDNA are three different formats in which genetic information may be stored. The main source of significant germplasm for upcoming scientific studies conducted globally is these libraries. The effectiveness of plant breeders in creating and enhancing crop varieties has greatly grown since the introduction of contemporary biotechnology methods. PGRs can be used more effectively thanks to high-throughput sequencing (HTS) and high-throughput phenotype (HTP) following NGS technologies. In crop development projects, DNA-based strategies are more popular and dependable than other current methods. Plant developmental phases and environmental changes have little effect on molecular markers, in contrast to other markers. The molecular description of genetic diversity across and within plant populations frequently makes use of molecular markers, including random amplified polymorphic DNA (RAPD), restriction fragment length polymorphisms (RFLPs), simple sequence repeats (SSRs), amplified fragment length polymorphisms (AFLPs), single nucleotide polymorphisms (SNPs), inter-simple sequence repeats (ISSRs), diversity array technology (DArT), etc. SSR markers are one of the many molecular markers that are often employed to describe genotypes. When assessing the genetic diversity of rare and endangered species, DNA-based markers are more effective. The ability to analyse genetic variety from a small amount of plant material is one of the primary benefits of molecular markers. Molecular markers have been widely used to study genetic diversity; rice, wheat, mungbean, soybean, common bean, pigeon pea, chilli, potato, carrot (Baranski et al. 2012), tomato, oil palm (Bakoumé et al. 2015), cotton (Huang et al. 2017), mulberry, barnyard, legume crops, and other vegetable and horticultural crops are among the crops for which various core collections of PGR accessions have been developed. The use of SNP markers in crop development initiatives is now more preferred owing to the advancements in high-throughput sequencing methods. By creating a biparental population utilizing PGR populations, several quantitative trait loci (QTLs) have been identified (Bhat et al. 2022). In addition to the molecular characterization of PGR germplasm for genetic diversity studies, molecular markers are widely used in plant breeding techniques, including evaluation of breeding materials for parental involvement selection, determining the level of genetic diversity, analysing heterosis, identifying genomic regions under selection, and evaluating cultivar purity and cultivar identity; marker-assisted gene pyramiding; marker-assisted backcross breeding (MABB); marker-assisted recurrent selection (MARS) (Bankole et al. 2017); and genomic selection for complex traits (Bhat et al. 2016).

The establishment of breeding techniques to enhance crop agronomic efficiency and accrue advantageous alleles for the desired variety has been at the centre of plant domestication since the start of the green revolution. Useful genetic variety for crop development is produced by the ongoing push to increase selection and widen a crop's genetic basis. In order to investigate significant economic features and support molecular breeding in agriculture, it is important to create genetic diversity in plants at the allelic level. Conventional breeding has up to now depended on the inherent genetic variety seen in both wild and farmed material. Furthermore, in order to introduce the desired characteristics into an elite cultivar, extensive backcrossing techniques are needed. However, the viability of this strategy is limited by the reduced supply of putative alleles in nature. To get around this, deliberate variety introduction for crop improvement has been employed. Applying various random mutagenesis techniques, including as chemical mutagenesis and γ-radiation, can introduce new advantageous alleles into cultivated plants; nevertheless, this is labour-intensive and time-consuming, particularly when it comes to screening for favourable mutants. Since the development of high-throughput methods, plant biotechnology has advanced beyond mutagenesis procedures. Currently, genome editing tools are being tested for precise and focused alteration of particular genomic regions. One such cutting-edge method of genome editing that has the ability to produce long-lasting and desired alterations in an organism is the CRISPR/Cas9 system (clustered regularly interspaced short palindromic repeats/CRISPR-associated protein 9). These genetic alterations in plants are more accurate, effective, and quick than those brought about by conventional breeding and spontaneous mutation. Furthermore, successive generations of plants are genetically identical to their traditionally produced counterparts due to the ease with which DNA sequences employed throughout the editing process can be separated and removed. By creating precise and predictable mutations (indels, insertions, deletions) in widely used plant kinds, CRISPR can expedite population improvement initiatives. New advances in CRISPR technology provide a practical and possibly effective means of adjusting our crops to problems including shifting plant–pathogen dynamics, food shortages, and climate change. A genome editing technique, when paired with functional genomics, has opened the door for biotechnology by providing the opportunity for extensive reverse genetics research, particularly in plants with enormous genome sizes. Furthermore, the primary objective of contemporary agricultural breeding is to provide the required allelic variety in crop plants. For crops where there is little variation across plant kinds, this is especially advantageous. Additionally, the genome-wide SNP discovery and genotyping method known as genotyping-by-sequencing (GBS) are based on NGS and have time- and cost-effective features in order to minimize genomic complexity and better sequence lower copy-genic areas of the plant genome; it may sequence several samples at once utilizing NGS libraries created with methylation-sensitive restriction enzymes. The most used method for identifying SNPs is GBS as it may be used on any species, whether or not it has a reference genome. Thus, by utilizing these modern technologies, identification and characterization of genotypes useful for future breeding pipelines can be further selected for conservation. These are techniques that only minimize the human effort of selection of particular effective cultivars; however, identifying the existence and presence of their usefulness is still man-driven efforts and should not be overlooked.

3 Advanced Ex Situ Conservation Strategies

The developments in ex situ conservation techniques show how biology, technology, and international cooperation may coexist. By preserving biodiversity, these strategies operate as a safeguard against the negative effects of both human activity and environmental change. Addressing upcoming conservation concerns requires ongoing innovation and interdisciplinary interaction. The preservation of species outside of their natural habitats, including in controlled settings like seed banks and botanical gardens, is known as ex situ approaches. The value of in vitro biotechnological techniques for the preservation and sustainable use of plant genetic resources has been highlighted by recent research. These methods enable the controlled preservation of a variety of plant materials, including landraces and endangered species (Tarraf and De Carlo 2024). The ability to preserve genetic variation over long periods of time has improved with the use of cryopreservation and micropropagation technology. These developments might spur related breakthroughs in conservation biology. It has been revolutionary to incorporate cutting-edge genetic techniques into conservation tactics. To increase the genetic diversity of ex situ populations—which is essential for reintroduction efforts and guaranteeing adaptability in changing environments—high-throughput sequencing and genome editing are being used (Guarino et al. 2025).

For severely endangered species, ex situ conservation has proven essential in reducing the risk of extinction. It has been investigated how to use nanotechnology, especially in the form of sophisticated coatings, to shield ex situ specimens against environmental deterioration. These methods are being modified to preserve the integrity of biological materials that are being stored. Cryogenic preservation seed banks continue to be a vital component of plant conservation initiatives. The viability and genetic integrity of stored seeds have been greatly enhanced by advancements in this field, such as novel cryoprotectants and cooling methods (Feng et al. 2019). Given the present rate of global warming, it is essential to develop ex situ conservation techniques that are climate resilient. Ex situ populations are protected against upcoming environmental changes through the use of artificial intelligence (AI)-driven predictive models (Ullah et al. 2024). Global networks like the Millennium Seed Bank Partnership have facilitated cooperation and made it possible for conservation organizations all around the world to share genetic resources and expertise. The effectiveness of ex situ tactics as a whole depends on these frameworks.

3.1 Cryobiotechnology in Germplasm Conservation

A key area of conservation biology is cryobiotechnology, which uses ultralow temperature methods to preserve plant germplasm. Science has progressed thanks to methods like vitrification and encapsulation-dehydration (Fig. 1), which increase the viability and genetic fidelity of preserved materials. To overcome the difficulties brought on by species-specific requirements and scalability, more innovation and research are needed. One essential method for conserving germplasm is cryopreservation. Genetic materials can be preserved for a long time by keeping plant and animal tissues at extremely low temperatures. For example, droplet vitrification techniques have been successfully used to cryopreserve rare ornamental plants like orchids, guaranteeing the preservation of true-to-type genetic material.

The process of vitrification turns a liquid into a glass-like consistency without creating ice crystals, which can harm cells. Advanced Plant Vitrification Solutions (PVS) have been developed to enhance post-thaw cell viability, especially in commercially and medicinally significant species. High recovery rates of stored tissues are guaranteed by this technique, which is frequently used in germplasm banks. Plant material is embedded in alginate beads and then gradually dehydrated and frozen as part of the encapsulation-dehydration process. For species with delicate germplasm, like the rare *Viola stagnina* and Thai orchid species, this method has been especially successful. Preconditioning procedures like cold acclimation affect these approaches' effectiveness rates (Żabicki et al. 2021). Cryopreservation-created synthetic seeds are showing promise as a germplasm conservation method. The potential for scalable conservation

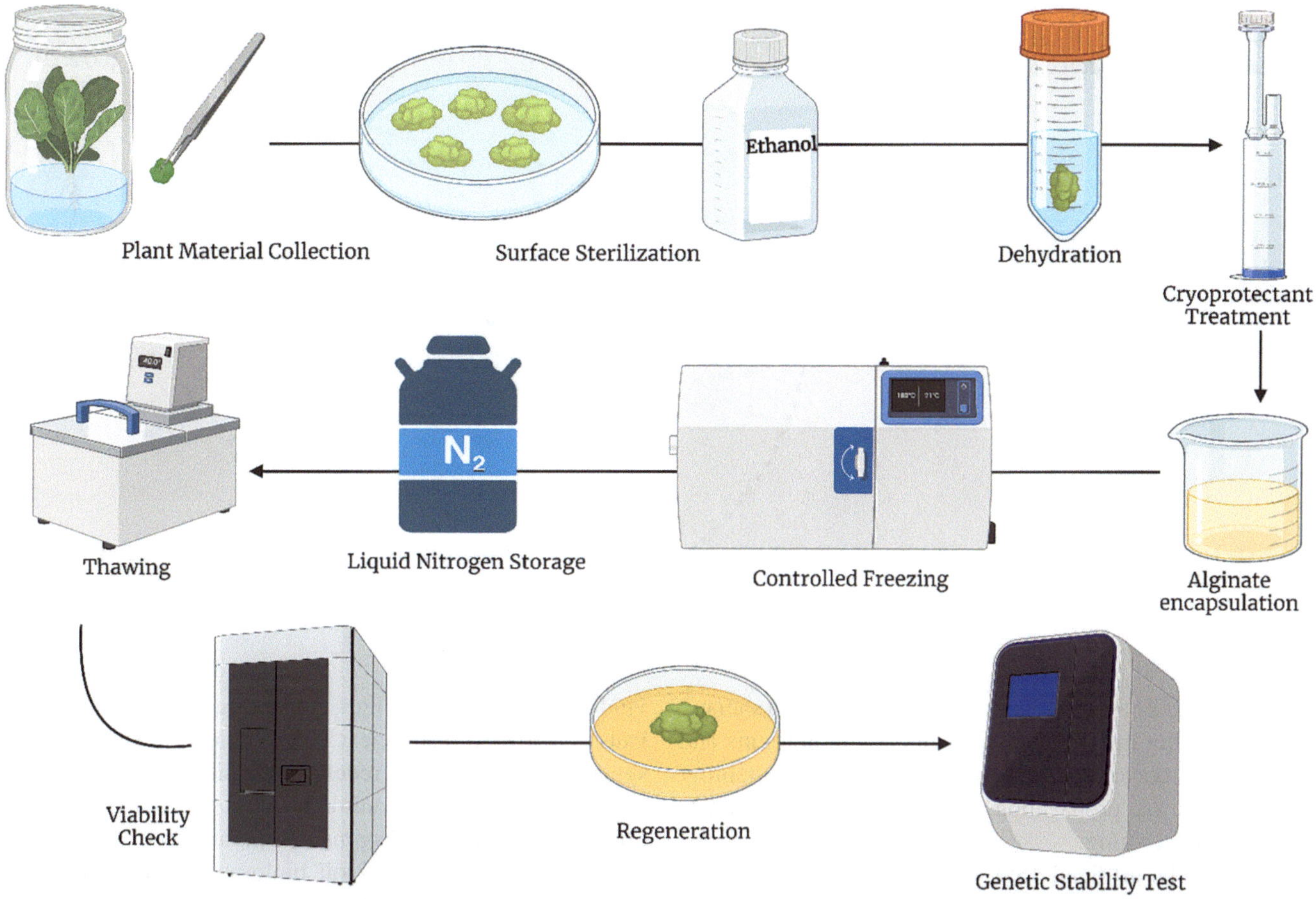

Fig. 1 Schematic workflow of cryobiotechnology for plant genetic resource conservation. *Source* This figure was created with BioRender.com.

measures has been demonstrated by the long-term storage of synthetic seeds for important species like agave and aloe, thanks to encapsulation-vitrification and encapsulation-dehydration procedures.

This illustration (Fig. 1) outlines key steps in cryopreservation, from plant material collection and cryoprotectant treatment to ultra-low temperature storage in liquid nitrogen and post-thaw regeneration. Critical tools such as programmable freezers, cryovials, and tissue culture media are highlighted, demonstrating the integration of biotechnological methods to preserve genetic diversity, combat genetic erosion, and ensure the long-term viability of endangered or economically vital plant species.

Even though cryobiotechnology has a lot of potential, there are still issues, such as the need to standardize procedures for different species. For example, because of their distinct climatic adaptations, native species from Australia need specific vitrification procedures. Applications of cryobiotechnology have been discovered in the preservation of therapeutic plants, particularly in the creation of plant resources that can prevent diabetes. Recent developments better comprehend the genetic reactions of preserved tissues to freezing and thawing by combining cryobiology and omic technology. The success rates of cryopreservation for tropical and subtropical crops have increased as a result of such integrated approaches.

3.2 Artificial Seed Technology

Artificial seed technology is a transformative advancement in biotechnology, enabling the efficient propagation, conservation, and genetic exchange of plant species. By encapsulating plant material, particularly somatic embryos, in protective coatings, this method has proven invaluable for conserving endangered species, supporting large-scale propagation, and maintaining genetic fidelity. Notable applications include the conservation of medicinal plants like *Mentha arvensis* and *Murraya koenigii*, where synthetic seeds ensure scalability and uniformity for commercial and restoration purposes. Encapsulation materials, particularly alginate-based polymers as shown in Fig. 2, have been optimized with nutrients and biodegradable components to enhance germination and pathogen resistance, further broadening the technology's impact.

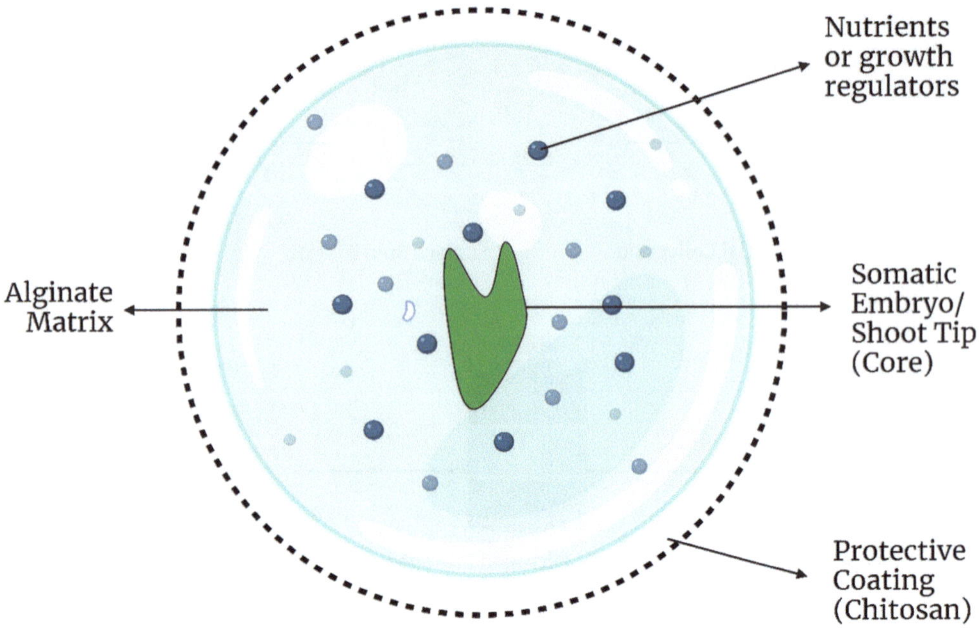

Fig. 2 Schematic diagram of an artificial seed. *Source* This figure was created with BioRender.com.

The hydrogel-based alginate matrix encapsulates a somatic embryo or shoot tip, often supplemented with nutrients and growth regulators to support germination and genetic stability during storage and propagation. The outer chitosan layer acts as a protective coating.

Additionally, artificial seed technology integrates seamlessly with cryopreservation techniques, offering extended storage durations and maintaining genetic integrity, especially for species prone to genetic drift. Its role in germplasm conservation is underscored by successful preservation of forest species like *Onobrychis viciifolia*, highlighting its potential for biodiversity preservation. Despite these advancements, challenges such as standardizing encapsulation protocols and improving post-thaw recovery persist. However, ongoing innovations in polymer science and molecular biology promise to enhance the scalability and applicability of this revolutionary conservation tool.

4 In Situ Conservation and Habitat Restoration Technologies

In situ conservation and habitat restoration have emerged as critical paradigms for safeguarding imperilled plant species that defy ex situ preservation, particularly those reliant on intricate environmental cues to sustain their ecological and biochemical functions. Consider, for instance, a rare herb endemic to a fragile ecosystem, whose survival hinges on a precise interplay of soil microbiota, microclimate conditions, and spectral light gradients—factors that artificial cultivation cannot replicate. For such species, the absence of native habitat stimuli can cripple their ability to synthesize vital secondary metabolites, rendering traditional conservation methods insufficient. This chapter explores three synergistic technologies redefining in situ strategies: remote sensing to map and monitor spectral signatures ensuring optimal light and environmental parameters, microbiome interactions to rejuvenate soil biology and nutrient cycles, and nanobiosensors to track real-time climatic interactions and stressors threatening the plant population. Together, these pillars form a dynamic, data-driven framework that not only preserves the physical habitat but also actively sustains the ecological processes essential for the plant's survival as shown in Fig. 3. By integrating macroscopic environmental monitoring with microscopic soil and climate interventions, this approach transcends mere protection—it fosters a resilient, self-regulating ecosystem where the plant thrives in its evolutionary niche, securing both its future and the biodiversity it anchors.

4.1 Drone and UAV-Based Ecosystem Monitoring

The assessment and evaluation of ecological health, especially with regard to natural disturbances and human influences, are made possible by efficient mapping and monitoring, which are essential elements of natural resource management. Comprehensive views of a variety of ecosystems, including forest canopies, aquatic vegetation in inland and coastal wetlands, grasslands, rangelands, and desert settings, are made possible by aerial imaging and are essential for conservation initiatives. These aerial photos can be transformed into quantitative data by integrating remote

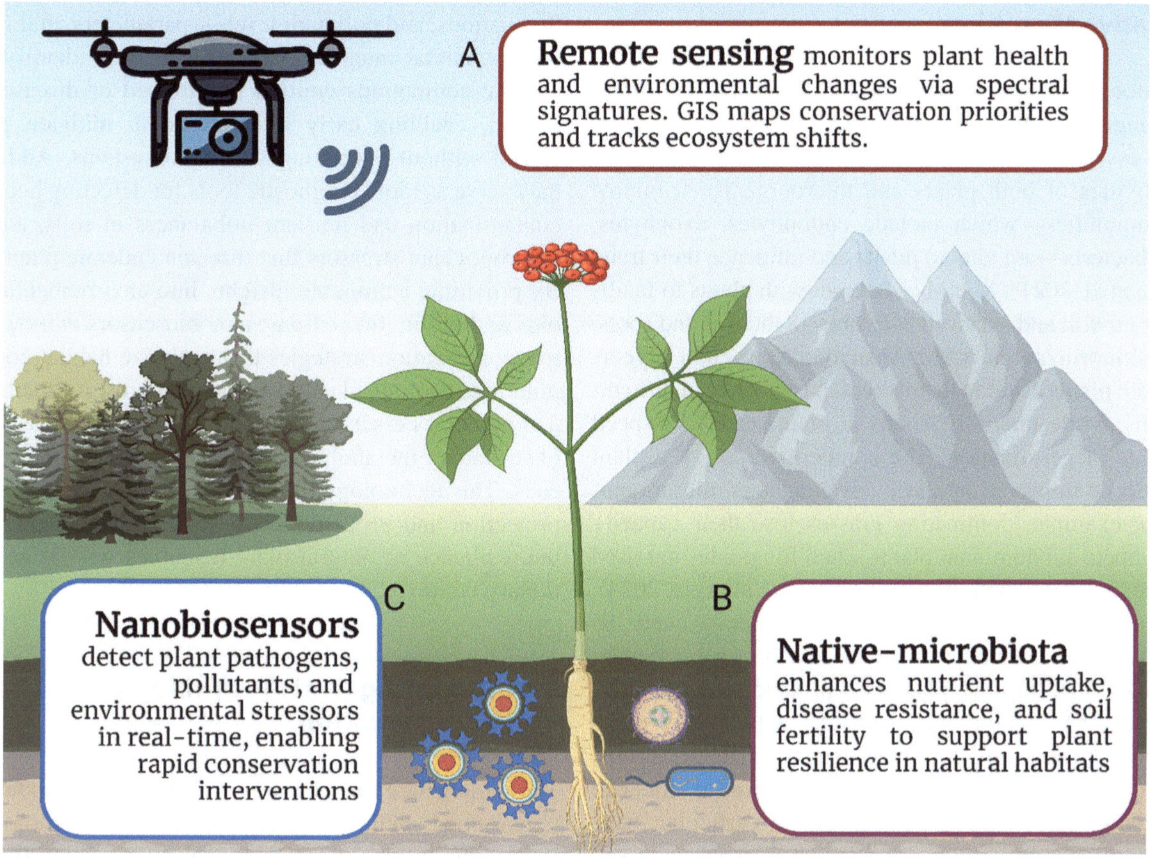

Fig. 3 Integrating remote sensing, microbiome engineering, and nanobiosensors for in situ conservation of natural plant resources. **a** Satellite/drone-derived spectral signatures map vegetation health and environmental stressors in a protected grassland ecosystem. **b** Soil cross-section depicting microbiome interactions, including symbiotic root-fungi networks (mycorrhizae) and soil inhabiting bacteria enhancing nutrient cycling. **c** Nanobiosensors embedded in soil/foliage transmit real-time microclimate and pathogen data to a field device, enabling adaptive conservation interventions. *Source* This figure was created with BioRender.com.

sensing, photogrammetry, and spatial information systems. This makes them appropriate for decision-making, predictive modelling, and geospatial analysis.

Grassland vegetation monitoring using satellite imagery from MODIS, Landsat, and SPOT has been the subject of numerous research conducted in China and abroad. The restricted geographical and spectral resolutions of these satellite systems, however, provide difficulties in precisely differentiating the species composition of grassland vegetation (Ali et al. 2016). In contrast, although they have a smaller monitoring range, stent-mounted hyperspectral imagers offer more precise species identification. Alternatively, spectral differences between vegetation species can be detected by using UAVs (unmanned aerial vehicles) fitted with high-resolution sensors to take centimetre-level photos over wide regions. By improving the precision of species identification in smaller vegetation units, this capacity makes it possible to monitor the composition of grassland vegetation in detail.

There is a lot of promise for biodiversity monitoring and conservation with UAV remote sensing technologies. A more comprehensive picture of regional biodiversity can be generated by integrating ground-based observations with index information extracted from UAV-collected hyperspectral pictures. This will enable more efficient monitoring and protection initiatives. For example, in a large-scale steppe restoration project in Wood River, Nebraska, UAVs increase the precision of biodiversity indicator computations by supplying higher-resolution data, providing deeper insights for grassland conservation decision-making.Grassland vegetation species are evaluated using a variety of indicators, such as species richness (the number of distinct species), uniformity (e.g., Pielou's J index), and diversity (e.g., Shannon index). Expanding the use of UAV remote sensing to monitor a wider variety of plant species indicators is crucial as research into this technology advances. This will improve the efficacy of monitoring grassland biodiversity.

4.2 Native Microbiota

In order for plants to develop, grow, compete, and withstand stress, fungi and other microorganisms, especially prokaryotes, are essential (Gupta et al. 2016). By synchronizing the gene activities of both plants and microorganisms, microbial communities—which include endophytes, exophytes, and soil bacteria—are vital to plants and influence their traits (Vandana et al. 2021). Microbes interact with plants to facilitate their growth and survival. Microbes in the soil and roots, such as actinorhizas and VAM (Vesicular-Arbuscular Mycorrhiza), help plants develop themselves, particularly in difficult conditions, while microorganisms in plant seeds can speed up the rate of germination. The competitiveness of a plant might also be impacted by these advantageous microorganisms. For example, dominating grasses lose their capacity to outcompete subdominant plants when fungicides are used to reduce fungal symbionts (Ettinger and LaForgia 2024). By stimulating nitrogen fixation, nutrient uptake, and the synthesis of plant hormones, rhizobacteria that promote plant development can improve plant growth. Additionally, interactions between microorganisms affect plant viability, especially under stressful circumstances. It has been demonstrated that endophytic fungi help plants become more resilient to stress; endophytes from saline habitats, in particular, have been found to improve plant growth when exposed to salt stress. Microbes will play a critical role in helping plants adapt to the increasing stresses brought on by climate change, including heat, drought, and salinity (Ullah et al. 2021). Plant diseases can also drastically lower plant survival and growth. Native plant populations have been decimated by the spread of exotic pathogens like chestnut blight, whereas invasive plant species frequently flourish because they have eluded their native infections. In ecosystems that are changing, microorganisms provide intriguing ways to improve plant health and conserve resources by promoting plant resilience.

4.3 Nanobiosensors

Nanobiosensors represent a transformative advancement in in situ conservation of natural plant resources, enabling precise, real-time surveillance of ecological and environmental variables critical to sustaining wild plant populations. Comprising a biological probe, transducer, and data recorder, these sensors detect target analytes within natural habitats and convert biological interactions into quantifiable signals. Their nanoscale design ensures exceptional sensitivity and specificity, allowing detection of subtle biochemical changes in soil, plant tissues, and microclimates. When integrated with satellite-based remote sensing and geospatial mapping, nanobiosensors facilitate comprehensive monitoring of soil nutrient dynamics, microclimate fluctuations, and pollutant levels—parameters vital for maintaining habitat integrity. For instance, they identify volatile organic compounds emitted by stressed or diseased plants in situ, enabling early intervention to mitigate pathogen spread without disrupting native ecosystems. Additionally, they serve as rapid diagnostic tools for detecting heavy metal contamination and nutrient imbalances in soils, addressing anthropogenic stressors that threaten endemic plant species. By providing actionable insights into environmental thresholds and biotic interactions, nanobiosensors empower adaptive conservation strategies that optimize habitat conditions, minimize ecological degradation, and preserve the biochemical processes essential for plant survival—such as synthesis of secondary metabolites dependent on native environmental cues. This technology thus bridges the gap between passive protection and proactive ecosystem management, ensuring the resilience of natural plant resources within their evolutionary contexts.

5 Integrating Omic Technologies in Conservation

The emergence of genomic sciences has fundamentally transformed methodologies in plant conservation biology, leveraging cutting-edge DNA analysis and high-throughput sequencing to advance our comprehension of genetic architecture, adaptive potential, and evolutionary dynamics. These innovations provide granular insights into population genetics, species resilience, and ecological interactions, forming a critical foundation for biodiversity preservation (Table 1). Contemporary genomic platforms utilize diverse molecular markers—including RAPD, RFLP, AFLP, SSCP, microsatellites, and SNPs—to map genetic variation, identify adaptive loci, and assess inbreeding risks (Husien et al. 2024). Techniques such as genome-wide association studies (GWAS), quantitative trait locus (QTL) mapping, and marker-assisted selection (MAS) have elucidated mechanisms underpinning disease resistance, environmental adaptation, and hybridization effects. DNA barcoding, particularly through mitochondrial gene cytochrome c oxidase I, has refined species identification, enabling precise biodiversity inventories and the resolution of cryptic species complexes. Comparative genomic analyses further illuminate hybridization and introgression processes, guiding strategies to maintain genetic integrity in conservation units. Advances in next-generation sequencing have accelerated forest conservation initiatives, facilitating rapid assessments of genetic erosion and microevolutionary processes like mutation and recombination (Hohenlohe et al. 2021). Collectively, these tools enhance the monitoring of genetic diversity and inform adaptive management frameworks to counteract anthropogenic and climatic pressures.

Table 1 Application of omic-based techniques in natural plant resource conservation

Crop plant	Omic technique	Key findings	Reference
Rhazya stricta	RNA-seq analysis	Identification of genes involved in salt stress response	(Tiwari and Rajwanshi 2022)
Palm cultivars	Genomic and RNA sequencing	Understanding genetic diversity and variation within cultivars	
Solanum lycopersicum	GC-MS analysis	Identification of biochemical markers for phenotypic characterization	
Legumes	GC-TOF-MS and LC-FT-MS	Metabolomic profiling for evaluating plant phenotypes	

Transcriptomic approaches have similarly emerged as indispensable for conservation, decoding gene expression dynamics to unravel stress-response mechanisms and adaptive traits. High-throughput RNA sequencing enables systematic profiling of transcriptional activity across tissues and developmental phases, revealing molecular pathways activated under stressors such as drought or salinity. Investigations into species like *Rhazya stricta* have identified pentatricopeptide repeat (PPR) proteins as critical mediators of salt tolerance (Hajrah et al. 2017), illustrating how transcriptomic data direct conservation prioritization of resilient genotypes. In fragmented populations of *Cypripedium forrestii*, RNA-seq analyses have delineated genetic differentiation and population structure, informing habitat restoration and genetic rescue protocols. Transcriptomics also underpins the study of secondary metabolite biosynthesis in medicinal plants, aiding the preservation of pharmacologically valuable species through targeted breeding. Parallels with animal studies, where transcriptional plasticity drives environmental adaptation, underscore the cross-taxa relevance of these approaches in evolutionary conservation.

Proteogenomic and metabolomic frameworks further enrich conservation science by integrating multi-omic data to decode plant resilience and ecological interactions. Proteogenomics combines poly-A RNA sequencing with shotgun proteomics to construct species-specific protein databases, enabling precise identification of stress-responsive proteins. Applications in model species like *Solanum lycopersicum* have optimized protein-coding contig analyses, streamlining the discovery of adaptive molecular signatures. Concurrently, metabolomic profiling via GC-MS and NMR spectroscopy maps biochemical networks, linking metabolite diversity to environmental adaptability (Bueno and Lopes 2020). For instance, eco-metabolomic studies identify stress biomarkers—such as osmoprotectants or defence compounds—that inform breeding programmes aimed at enhancing crop resilience. These integrative approaches elucidate plant responses to biotic and abiotic challenges, offering mechanistic insights into survival strategies under climate variability.

By synthesizing genomic, transcriptomic, proteomic, and metabolomic data, conservation biologists gain a holistic understanding of plant diversity, from genetic variation to phenotypic plasticity. This multi-omic paradigm not only identifies critical pathways for resilience but also guides the development of dynamic conservation strategies, ensuring the sustainable stewardship of plant resources amid escalating global environmental change. Such interdisciplinary methodologies represent a paradigm shift in conservation practice, bridging molecular innovation with ecological management to safeguard biodiversity for future generations.

6 Socioeconomic and Policy Dimensions

The preservation of plant genetic resources is deeply intertwined with socioeconomic considerations and policy frameworks. Early conservation strategies, emerging in the mid-twentieth century, prioritized ex situ methods, reflecting the priorities of plant breeding programmes in industrialized nations (Fig. 4). Notable examples include the creation of germplasm repositories such as the Vavilov Institute in St. Petersburg (1920) and the Commonwealth Potato Collection in Cambridge (pre-1939), alongside initiatives funded by entities like the Rockefeller Foundation (Engels and Ebert 2021). Nikolai Vavilov's pioneering research in 1926 underscored the critical value of genetic diversity and the risks posed by genetic erosion, forming a theoretical basis for subsequent global conservation endeavours. By the mid-twentieth century, the establishment of international agricultural research institutions, including the International Rice Research Institute and the International Maize and Wheat Improvement Center, highlighted the role of genetic resources in addressing food security through crop enhancement.

The 1960s marked a pivotal era for international collaboration, beginning with the FAO Technical Meeting on Plant Exploration and Introduction (1961), which catalysed the formation of the FAO Panel of Experts in 1965. This body sought to standardize protocols for conservation and utilization (Engels and Ebert 2021). The 1967 FAO/IBP Technical Conference advocated for a coordinated global ex situ conservation network, although in situ approaches remained marginal in discourse. Subsequent conferences, such as the 1971 FAO Technical Conference, called for actionable frameworks, including systematic criteria for conserving genetic

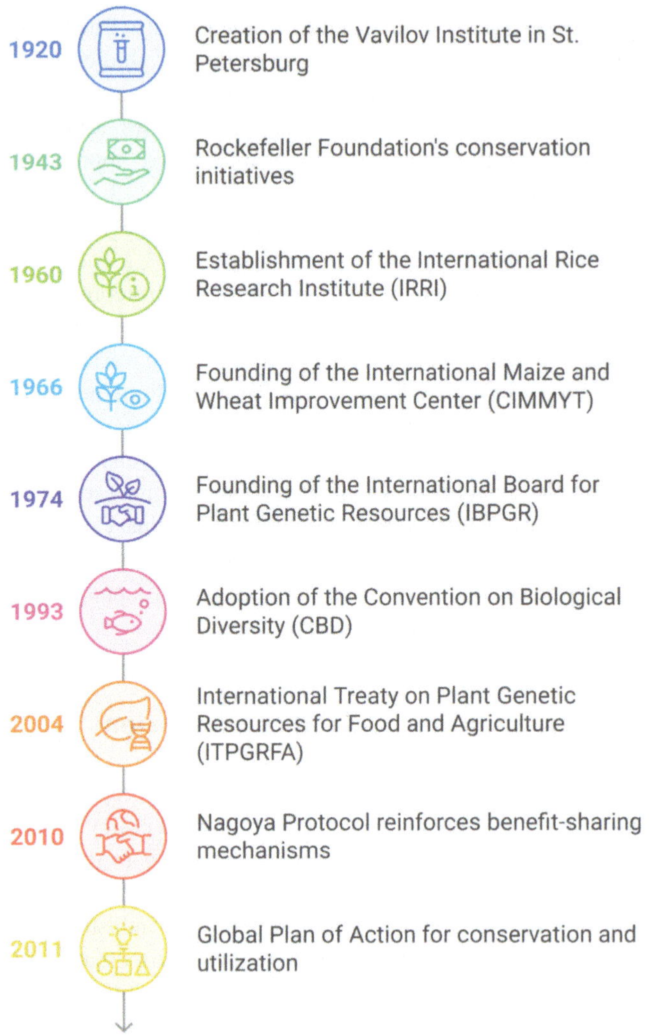

Fig. 4 Evolution of global plant genetic resource conservation system. *Source* It is noted that AI-assisted tools were used in the figure's creation. The authors directed the concept and design, and utilized the Napkin AI image generator to produce preliminary imagery for specific elements, which were then curated, edited,and composited to create the final figure

material and classifying ex situ collections (Engels and Ebert 2021).

A landmark development occurred in 1974 with the founding of the International Board for Plant Genetic Resources (IBPGR), which centralized global efforts in germplasm collection, exploration, and gene bank establishment. The 1981 FAO/UNEP/IBPGR Technical Conference expanded the scope to address technical aspects of gene bank operations, integrating emerging methodologies like in vitro conservation. The 1983 International Undertaking on Plant Genetic Resources aimed to foster unrestricted access and multilateral cooperation, although its implementation faced evolving complexities (Engels and Ebert 2021).

The adoption of the Convention on Biological Diversity (CBD) in 1993 redefined global conservation paradigms, prioritizing national sovereignty over genetic resources and emphasizing in situ conservation and on-farm management. This shift recognized Farmers' Rights, acknowledging the contributions of indigenous and local communities. Subsequent agreements, such as the Nagoya Protocol, reinforced equitable benefit-sharing mechanisms, although concerns arose regarding administrative burdens. The International Treaty on Plant Genetic Resources for Food and Agriculture (ITPGRFA) sought equilibrium between access for research and the rights of origin countries through its Multilateral System. Concurrently, the Global Plan of Action outlined strategies for conservation, sustainable utilization, and capacity development, supported by entities like the Global Crop Diversity Trust, which ensures the viability of global gene banks (Engels and Ebert 2021).

Despite progress, persistent challenges include tensions between intellectual property regimes and open access to genetic materials, uneven implementation of ITPGRFA provisions, and the integration of socioeconomic needs of communities reliant on plant biodiversity. Evolving regulatory frameworks, while promoting equity, have introduced procedural complexities that may hinder research and innovation. Advances in genomic technologies not only offer transformative potential for conservation but also pose ethical dilemmas, such as risks associated with genetically modified organisms. Addressing these issues demands holistic strategies that harmonize ecological, social, and economic priorities, underpinned by robust international cooperation, capacity building in resource-limited regions, and inclusive participation of local stakeholders in governance processes. Sustaining progress necessitates adaptive policies that balance innovation with equity, ensuring the preservation of plant genetic heritage for future generations.

7 Emerging Frontiers and Associated Challenges

Recent advances in technology have significantly enhanced conservation efforts by introducing innovative tools for biodiversity monitoring and protection. Among these, blockchain technology, environmental DNA (eDNA), and artificial intelligence (AI) stand out for their transformative potential. Blockchain's decentralization and immutability enable traceability and transparency in conservation funding, as well as the creation of digital identities for endangered species, although challenges such as high energy costs and data integration persist (Westerlaken 2024). eDNA offers a noninvasive method for biodiversity assessment, allowing the detection of rare or cryptic species and monitoring the recovery of plant communities' post-restoration. However, its effectiveness is affected by contamination risks and the inability to distinguish closely related species.

AI and machine learning revolutionize conservation by automating data analysis and enabling predictive habitat modelling under future climate scenarios. Tools powered by AI provide real-time species detection from cameras, drones, and acoustic sensors, although their success relies on the availability of high-quality datasets and robust infrastructure (Fonseca et al. 2024). Despite these technologies' immense promise, their implementation faces barriers including technical, ethical, and infrastructural challenges. Collaborative efforts between conservationists, technologists, and policymakers are essential to fully harness these innovations for sustainable biodiversity protection and resource management.

8 Role of Indian Knowledge System (IKS) in Plant Resource Conservation

The Indian Knowledge System (IKS) plays a vital and practical role in conserving plant genetic resources, especially in a country as ecologically and culturally diverse as India. Across regions, traditional communities have preserved landraces of rice, millets, pulses, and other crops through informal yet resilient systems like community seed banks, home gardens, and seasonal crop calendars. These seeds, often adapted to local conditions, hold traits now crucial for climate-resilient agriculture. The Protection of Plant Varieties and Farmers' Rights (PPVFR) Act 2001 enables farming communities to register and protect these traditional varieties. Sacred groves, still thriving in states like Kerala, Maharashtra, and Meghalaya, serve as in situ biodiversity sanctuaries—conserved not by law, but by cultural belief and ritual. Similarly, organizations like Navdanya (https://www.navdanya.org/) have helped strengthen such farmer-led conservation efforts across India. These models align directly with global development goals such as SDG 2 (Zero Hunger), SDG 13 (Climate Action), and SDG 15 (Life on Land).

9 Towards a Global Bio-Cultural Network

While modern science continues to offer powerful tools for conserving biodiversity, there is a growing need to recognize that ecological sustainability is deeply linked with cultural continuity. Across the world, communities have developed their own ways of living with and protecting nature—through rituals, taboos, seasonal calendars, oral traditions, and sustainable land use. These practices, although often informal, serve as effective systems of ecological governance. India's sacred groves, traditional seed saving, and indigenous farming wisdom are strong examples of such bio-cultural systems, but similar traditions exist worldwide.

In Iran, sacred natural sites around shrines and trees are still preserved by communities through belief systems that act as powerful conservation tools. The ancient *qanats*—underground water channels designed and maintained by generations of local knowledge—demonstrate how traditional ecological engineering can sustain biodiversity in arid zones. Nomadic tribes like the Bakhtiari and Qashqai have practiced *transhumance*, a seasonal migration pattern that ensures pasture regeneration and biodiversity conservation across landscapes. These examples show that the relationship between culture and biodiversity is universal, although expressed in different ways.

To protect and promote these living knowledge systems, the time has come to propose a *Global Bio-cultural Network*. This platform could serve as a shared international space to document, connect, and support community-led conservation practices rooted in cultural identity. Much like how UNESCO safeguards world heritage sites, this network would recognize not just places, but living traditions—sacred groves in India, *qanats* in Iran, rice terraces in the Philippines, or agroforestry rituals in parts of Africa. It could involve collaborative mapping, intercultural knowledge exchange, policy advocacy, and participatory research.

Such a network would not only enrich global biodiversity strategies but also bring long-overdue recognition to the cultural and spiritual dimensions of conservation. By linking diverse yet interconnected practices across borders, a *Global Bio-cultural Network* would reflect a simple but powerful truth: protecting biodiversity is also about protecting the cultures that protect it.

10 Conclusion

Conservation of plant genetic resources for sustainable development calls for a meaningful blend of modern science and time-tested wisdom. Technologies like synthetic biology, high-throughput genotyping, precision agriculture, drone-based monitoring, and omic tools have greatly expanded our ability to conserve biodiversity. Alongside these, ex situ methods such as cryopreservation and artificial seeds offer important safeguards for plant genetic resources. However, the real strength lies in integrating these innovations with historical experiences, traditional ecological knowledge, and community-based practices—especially those rooted in the Indian Knowledge System. The IKS offers rich insights into sustainable land use, seed preservation, and local biodiversity management that have evolved over generations. When such indigenous practices are respected and combined with scientific advancements, the outcome is more inclusive, ethical, and culturally rooted. Equally important are policies that are just and participatory, giving voice to local communities who are often the true custodians of biodiversity. Ultimately, conservation is not just a technical task—it is a social and ethical responsibility. By aligning modern tools with the wisdom of traditional systems like IKS, and ensuring active community engagement, we can foster a resilient future where ecological sustainability and human well-being go hand in hand.

References

Ali I, Cawkwell F, Dwyer E, Barrett B, Green S (2016) Satellite remote sensing of grasslands: from observation to management. J Plant Ecol 9(6):649–671

Bakoumé C, Wickneswari R, Siju S, Rajanaidu N, Kushairi A, Billotte N (2015) Genetic diversity of the world's largest oil palm (Elaeis guineensis Jacq.) field genebank accessions using microsatellite markers. Genet Resour Crop Evol 62:349–360

Bankole F, Menkir A, Olaoye G, Crossa J, Hearne S, Unachukwu N, Gedil M (2017) Genetic gains in yield and yield related traits under drought stress and favorable environments in a maize population improved using marker assisted recurrent selection. Front Plant Sci 8:808

Baranski R, Maksylewicz-Kaul A, Nothnagel T, Cavagnaro PF, Simon PW, Grzebelus D (2012) Genetic diversity of carrot (Daucus carota L.) cultivars revealed by analysis of SSR loci. Genet Resour Crop Evol 59:163–170

Begna T (2021) Conventional breeding methods widely used to improve self-pollinated crops. Int J Res 7(1):1-16

Bhandari HR, Bhanu AN, Srivastava K, Singh MN, Shreya HA (2017) Assessment of genetic diversity in crop plants-an overview. Adv Plants Agric Res 7(3):279–286

Bhat JA, Ali S, Salgotra RK, Mir ZA, Dutta S, Jadon V, et al. (2016) Genomic selection in the era of next generation sequencing for complex traits in plant breeding. Front Genet 7:221

Bhat R, Singh AK, Mushtaq M, Salgotra RK, Sharma M, Bhat BA, et al. (2022) Identification of QTLs for yield and associated traits in F 2 population of Rice. Phyton 91(11):2439

Bueno PC, Lopes NP (2020) Metabolomics to characterize adaptive and signaling responses in legume crops under abiotic stresses. ACS Omega 5(4):1752–1763

CBD (Convention on Biological Diversity) (2018) Frequently asked questions on access andbenefit-sharing (ABS). Montreal, Canada. https://www.cbd.int/abs/doc/abs-factsheet-faq-en.pdf

Engels JM, Ebert AW (2021) A critical review of the current global ex situ conservation system for plant agrobiodiversity. I. History of the development of the global system in the context of the political/legal framework and its major conservation components. Plants 10(8):1557

Ettinger CL, LaForgia ML (2024) Invasive plant species interact with drought to shift key functions and families in the native rhizosphere. Plant Soil 494(1):567–588

Feng J, Hu W, Fanai ML, Zhu S, Wang J, Cai J, Lu F (2019) Mechanical process prior to cryopreservation of lipoaspirates maintains extracellular matrix integrity and cell viability: evaluation of the retention and regenerative potential of cryopreserved fat-derived product after fat grafting. Stem Cell Res Ther 10:1–11

Fonseca L, Oliveira E, Pereira T, Sá JC (2024) Leveraging ChatGPT for sustainability: a framework for SMEs to align with UN sustainable development goals and tackle sustainable development challenges. Manag Mark 19(3)

Guarino C, Cipriani G, Jorrin Novo JV, Carputo D, Di Vaio C (2025) Agrobiodiversity at different scales for improving conservation strategies. Front Plant Sci 15:1457713

Gupta A, Gupta R, Singh RL (2016) Microbes and environment. In: Principles and applications of environmental biotechnology for a sustainable future. Springer, Singapore, pp 43–84

Hajrah NH, Obaid AY, Atef A, Ramadan AM, Arasappan D, Nelson CA, et al. (2017) Transcriptomic analysis of salt stress responsive genes in Rhazya stricta. PLoS One 12(5):e0177589

Hoban S, Campbell CD, da Silva JM, Ekblom R, Funk WC, Garner BA, et al. (2021) Genetic diversity is considered important but interpreted narrowly in country reports to the convention on biological diversity: current actions and indicators are insufficient. Biol Conserv 261:109233

Hohenlohe PA, Funk WC, Rajora OP (2021) Population genomics for wildlife conservation and management. Mol Ecol 30(1):62–82

Huang C, Nie X, Shen C, You C, Li W, Zhao W, et al. (2017) Population structure and genetic basis of the agronomic traits of upland cotton in China revealed by a genome-wide association study using high-density SNP s. Plant Biotechnol J 15(11):1374–1386

Husien HM, Saleh AA, Hassanine NN, Rashad AM, Sharaby MA, Mohamed AZ, et al. (2024) The evolution and role of molecular tools in measuring diversity and genomic selection in livestock populations (traditional and up-to-date insights): a comprehensive exploration. Vet Sci 11(12):627

Tarraf W, De Carlo A (2024) In vitro biotechnology for conservation and sustainable use of plant genetic resources. Plants 13(14):1897

Tiwari S, Rajwanshi R (2022) Overview of omics-assisted techniques for biodiversity conservation. In: Molecular genetics and genomics tools in biodiversity conservation. Springer Nature, Singapore, pp 63–78

Ullah A, Bano A, Khan N (2021) Climate change and salinity effects on crops and chemical communication between plants and plant growth-promoting microorganisms under stress. Front Sustain Food Syst 5:618092

Ullah F, Saqib S, Xiong YC (2024) Integrating artificial intelligence in biodiversity conservation: bridging classical and modern approaches. Biodivers Conserv:1–21

Vandana UK, Rajkumari J, Singha LP, Satish L, Alavilli H, Sudheer PD, et al. (2021) The endophytic microbiome as a hotspot of synergistic interactions, with prospects of plant growth promotion. Biology 10(2):101

Wernberg T, Coleman MA, Bennett S, Thomsen MS, Tuya F, Kelaher BP (2018) Genetic diversity and kelp forest vulnerability to climatic stress. Sci Rep 8(1):1851

Westerlaken M (2024) Digital twins and the digital logics of biodiversity. Soc Stud Sci:03063127241236809

Willis KJ (ed) (2017) State of the world's plants 2017. Royal Botanic Gardens, Kew Report

Żabicki P, Mikuła A, Sliwinska E, Migdałek G, Nobis A, Żabicka J, Kuta E (2021) Cryopreservation and post-thaw genetic integrity of Viola stagnina kit., an endangered species of wet habitats–a useful tool in ex situ conservation. Sci Hortic 284:110056

Role of In Vitro Technology in Plant Conservation: Relevance to Management of Natural Plant Resources and Diversity

Tania Sagar, Sandeep Kaur, Shajaat Hussain, Nipunta, Nisha Kapoor, and Ritu Mahajan

Abstract

Human existence has been seriously threatened by the ongoing loss of biodiversity. One-third of plant species are endangered globally due to various causes. Since biodiversity protection is a worldwide issue, a number of tactics have been implemented throughout recognizing and preserving plant species. Protection of biodiversity can be accomplished through both in situ and ex situ approaches. Methods for preserving rare as well as endangered genetic resources in plants, conventional knowledge regarding germplasm, and their value management have been significantly enhanced and expanded by recent biotechnology advancements, such as molecular markers, cryopreservation, and in vitro culture. Therefore, the present chapter discusses how in vitro technologies for conserving plant resources are used as an effective method in bio-cultural conservation and further used to enhance the conservation procedures for preserving plant biodiversity.

Keywords

Conservation · In vitro · Biodiversity · Cryopreservation · Genetic transformation

1 Introduction

As living resources in the ecosystems, plants are essential to maintaining ecological equilibrium. The quality and number of resources found within a geographic limit are based on the variety found in plants. Even though many plant species are imperiled, endangered, or extinct, conserving plant germplasm has gotten less attention throughout time. Due to their rising vulnerability, plants are essential to the resilience of ecosystems and the operations of human society. The rate at which biodiversity is disappearing from the planet is unparalleled. The International Union for the Conservation of Nature and Natural Resources (IUCN) has listed several plants under extinct category. Their primary goal is to identify and classify species that are more likely to become extinct worldwide (those classified as vulnerable, endangered, and critically endangered species). Biodiversity loss severely impacts social, economic, organizational, political, scientific, and communication contexts, which has become a complex problem worldwide. Several human-caused factors are to be blamed for the current rate and extent of plant variety loss (Barnhill-Dilling and Delborne 2021). Extirpation and species extinction in addition to habitat degradation are examples of biodiversity loss, which are generally decreasing variety from genes to ecosystems. The ongoing decline in biodiversity is well known, and even more alarmingly, as instead of slowing down, it is getting worse. Decreases in biodiversity lead to fewer ecosystems, less resilient nature, and fewer species interactions and ecosystem services. Therefore, the human incursions and climate change may be the most serious threats to biodiversity on land, given the evidence that anthropogenic pressure and climate change have significantly exacerbated the extinction of wild plant species. Given the likelihood of future environmental changes, it is imperative that conservation activities be prioritized effectively in order to ensure the preservation of biodiversity. Anthropogenic stressors, massive fires, the introduction of new species from other areas, and the invasion of dominant species and cultivars have a significant effect on plant biodiversity and resulted in the increase in number of vulnerable species (Kaya 2024).

T. Sagar · S. Kaur · S. Hussain · Nipunta · N. Kapoor · R. Mahajan (✉)
Plant Molecular Biology Lab, School of Biotechnology, University of Jammu, Jammu, India
e-mail: ritufeb@gmail.com

N. Kumar (ed.), *Biotechnology in Plant Conservation*, Sustainable Landscape Planning and Natural Resources Management,
https://doi.org/10.1007/978-3-032-02605-7_2

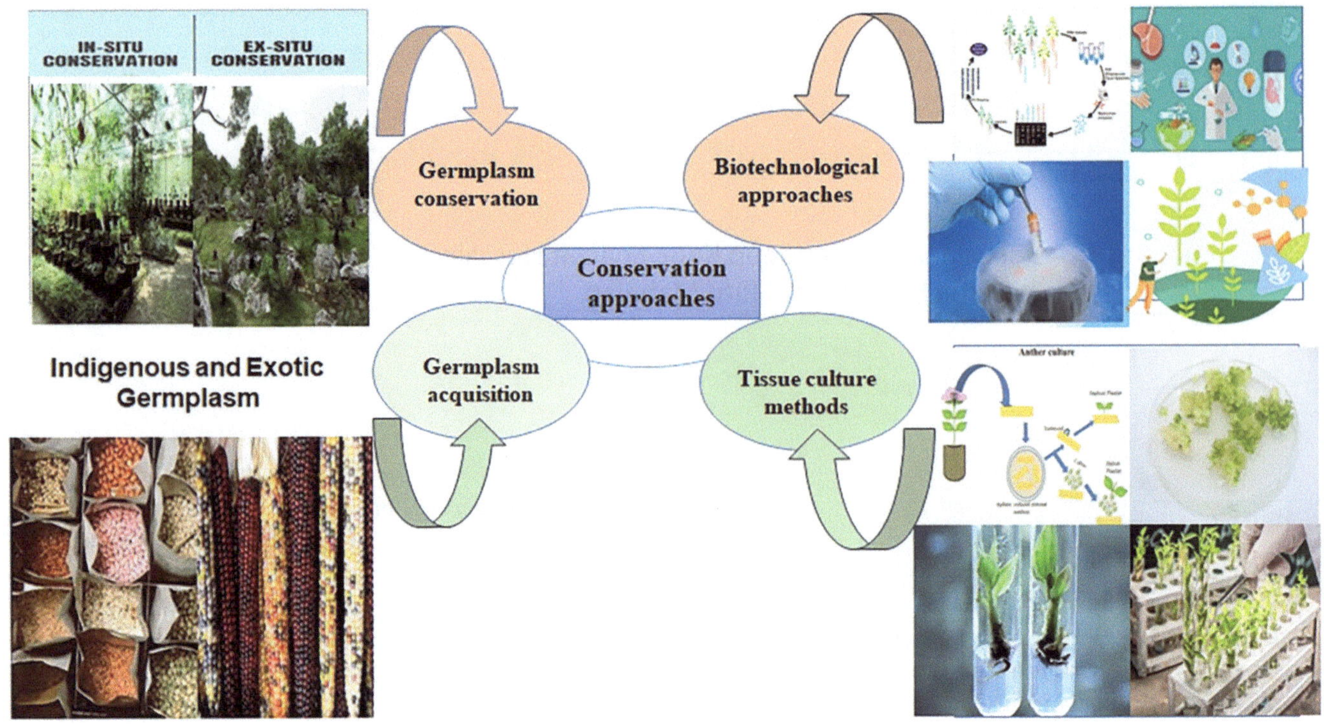

Fig. 1 Approaches for germplasm conservation

Biotechnology plays a positive role in plant conservation strategies, those results in quicker germplasm propagation and preservation of variety of species that are not successfully served by seed-based methods. However, due to ambiguous perception, implementation, regulation, and finance systems, there are differences in societal acceptance and success (Pence et al. 2020). Plant biodiversity conservation schemes use conventional techniques, such as ex situ (outside the natural ecosystem) and in situ (natural habitat). The best environmental conditions determine the plant's survivability in both approaches. Therefore, biotic (insects, herbivorous animals, etc.) and abiotic (temperature, drought, salinity, etc.) factors rarely affect plant species that are protected by these techniques. The Convention on Biological Diversity (CBD) took action to protect both native and agriculture biodiversity since biodiversity conservation is a worldwide priority. In terms of producing plants or plant propagules quickly and preserving them over time, traditional plant propagation techniques have some drawbacks. For ex situ conservation, biotechnological approaches including plant tissue culture, plant cell culture, anther culture, embryo culture, etc. are appropriate and practical (Fig. 1). On the other hand, the use of plant biotechnology tools has led to the improvement in the production of exceptional quality seeds. Several low temperature storage and cryopreservation procedures, somatic embryogenesis, cell culture technologies, and in vitro methods include embryo rescue techniques and micropropagation based on axillary meristems or shoot apical used for preserving plant biodiversity (Mahajan et al. 2016). Since the global population is rising at an alarming rate, so will be the need for genetic resources for providing food securit and bioactive metabolites and in health industry. Numerous significant genetic resources, including better breeding material, landraces, wild close relatives, wild species, genetic stock, and modern variants (Salgotra and Chauhan 2023), have already gone extinct as a result of genetic erosion and environmental changes.

2 Diversity and In Vitro Plant Conservation

Plants are essential to all ecosystems and to life. Despite their significance, plant biodiversity is under danger worldwide, and the number of endangered species is rising sharply each year. As a result of human overexploitation, natural populations or even entire species are typically lost due to habitat destruction and modification. Recently, the loss of genetic variety has also been exacerbated by pollution and climate change. Since many of these species are indigenous, they are rare, and only a limited number of wild populations are frequently resistant. The degree to which individuals of a variation or population within a species differ genetically is known as genetic diversity. Genetic drift, gene flow, inheritance, and mutations all result in the recombination of genetic material (DNA). This process results in variations in physical traits, physiological traits, protein structure or isoenzymes, DNA sequence, and epigenetic profiles. The genetic material found in the reproducing individuals of the population determines the variation found in plant and animal populations (Mahajan et al. 2018). For conservation, regions of species richness, dynamic functional ecosystems, and living communities that are important to the economy, medicine, and environment are the main priorities. Therefore, sustainable use of plant genetic resources (PGRs) is imperative to address food security concerns as well as other difficulties brought on by population expansion. If appropriate and strict PGR conservation methods and policies are not put in place, these resources will eventually disappear. The genetic resources of rare and endangered plant species, as well as resistant seed and vegetatively propagated species, and technological products like elite genotypes and genetically altered material are all examples of plant biodiversity that may be effectively conserved by in vitro procedures. Using in vitro methods, it is possible to effectively gather explants from vegetatively propagated and resistant seed species in field natural environments. Plant tissue culture has been employed in micropropagation, conservation, and the overproduction of some pharmacological compounds from medicinal plants. Assessment of diversity followed by targeted genome editing and other novel biotechnology-based breeding methods can be used to create customized medicinal plants with unique secondary metabolite profiles. While conventional biotechnological methods have been used in modern plant breeding, for a very long time, people have utilized medicinal herbs to produce food, medicine, and other purposes. Additionally, plant tissue culture provides a number of methods for ex vitro conservation of rich phyto-diversity, regeneration of genetically modified superior clones, quick clonal proliferation, germplasm conservation, and the manufacture of phyto-constituents (Kaur et al. 2018).

3 Cryopreservation

The best method for long-term preservation of seeds that are susceptible to desiccation is cryopreservation, which involves placing the seeds in either the vapor phase of nitrogen (− 140 °C) or liquid nitrogen (− 196 °C). Such high temperatures effectively stop plant seeds' cellular division and metabolic activity, putting them in a condition of "pseudo-death" that allows for long-term preservation. Concurrently, cryopreservation acts as a safeguard against further culture and the buildup of mutations in natural environments (Zeng et al. 2024). Field collections are long-term protected against the loss of plant germplasm via cryopreserved materials. A systematic and intentional cryo-banking of plant genetic resources can be made possible by employing appropriate cryopreservation techniques that offer high regrowth rates to prevent genetic variation in tissue culture cells throughout extended subcultures. Because shoot tips have a high level of genetic stability, they are excellent cryopreservation candidates (for clonally grown plants) and their structured tissues are made up of meristematic cells with dense cytoplasm and tiny vacuoles (Normah et al. 2019). The primary benefit of this strategy is the reduction of in vitro culture costs, space needs, contamination, and the danger of somaclonal variation by retaining wild species (for breeding purposes) which are currently endangered and storing them in Liquid nitrogen (LN) would aid in maintaining the genetic diversity. Various species, including distinct clonal lineages within the same species, have varying capacity for surviving cryopreservation conditions. To get the highest recovery rates following cryo-storage, protocols must be tailored specifically for the species or clone and kind of explant to be cryopreserved. Due to their worth and extensive understanding of their growth circumstances, crops and commercially significant plant species are the subject of several cryopreservation research. To increase the effectiveness of the cryopreservation process, a number of pre- and post-cryopreservation conditions and factors are to be taken into account. These include the size of the tissue explants, the formulation of the growth medium, the use of preconditioning procedures, and the standardization of the culture conditions of the material generated in vitro such as age, morphogenetic state, subculture duration, light and nutritional needs, and others. Several cryopreservation techniques include the vitrification-based techniques as well as traditional, older techniques based on freeze-induced cell dehydration. Encapsulation dehydration, vitrification, encapsulation-vitrification, desiccation, pre-growth, pre-growth-desiccation, and droplet freezing are few examples of cryopreservation techniques. Numerous plant species' cells, tissues, and organs have been successfully treated and preserved using these techniques. For organisms that are reproduced vegetatively, cryopreservation is a well-established technique while it is far less developed for resistant seed species due to their structural complexity, heterogeneity, and extreme susceptibility to desiccation.

4 In Vitro Methods

4.1 Micropropagation

Micropropagation of plants is an effective approach for mass propagation of medicinal and ornamental plants, crops, and other economically significant plants. It refers to an in vitro clonal propagation of the plants on a nutrient medium under aseptic conditions by utilizing various tissue culture techniques. It entails specific stages like shoot induction and multiplication, rooting, and acclimatization. The plant tissue employed for in vitro propagation is called an explant, which is surface sterilized using numerous sterilizing agents (0.1% mercuric chloride, 5% sodium hypochlorite, and bavistin), before being inoculated on a selected medium supplemented with sucrose, growth regulators (auxin, cytokinins), and vitamins. Plants exhibit exceptional developmental plasticity. Therefore, the cultured explant cells express totipotency and morphogenetic response and they follow a specific developmental route of dedifferentiation, redifferentiation, and organization of meristematic centers in response to growth regulators/phytohormones. Axillary buds and shoot tips are the most frequently employed explants at the initiation stage because they already have existing buds. Conversely, regeneration of some explants, such as stem, leaf, or root tissue, entails adventitious bud development through dedifferentiation and redifferentiation and necessitates a thorough evaluation of the genetic stability of the culture to avoid somaclonal variation.

In addition to clonal propagation that guarantees the germplasm conservation, micropropagation offers several benefits such as mass propagation of plant species recalcitrant to traditional propagation approaches, seasonal independence, development of pathogen-free plants, limited space, and labor requirement and provides plant material appropriate for genetic transformations and mass production of plant-derived bioactive metabolites. Hadi et al. (2023) developed a micropropagation protocol for the germplasm conservation of *Aconitum violaceum*, a medicinal plant widely utilized in ancient Amchi and Tibetan's medicinal system. It is depleting due to unsustainable harvesting of its rhizome and habitat destruction. They successfully established an in vitro roots and shoots of nodal explant on basal Murashige and Skoog (MS) medium enriched with phytohormones, namely kinetin and L6-benzylaminopurine.

4.2 Callus Culture

It is defined as the process of formation of a callus, a mass of undifferentiated and disorganized cells, from specific plant tissues in response to abiotic or biotic stimuli. The whole plant can be regenerated from callus either by somatic embryogenesis or organogenesis. In vitro callus induction begins by subjecting explants to particular growth regulators and hormones, stimulating cell dedifferentiation and differentiation. The phytohormones that play a significant role in callus formation are auxin and cytokinin (Chandimali et al. 2023). The callus development requires a precarious equilibrium between concentrations of these growth regulators, whereas a sublime ratio can cause shoot and root regeneration. The callus has pervasive use in both industrial applications and elementary scientific research. In vitro clonal propagation and phytochemicals synthesis through callus culture are alternative approaches for producing phytochemicals on a large scale and disease-free plants; thus, callus culture has a significant effect on the conservation of potent medicinal plants as well as their ameliorative potential. Abdelgawad et al. (2024) established a callus culture to elicit its phytochemical profile and observed that elicitation of the callus culture with 200 mg/L polyethylene glycol along with ultraviolet exposure for 60 min resulted in the elevated flavonoid and phenolic contents (approximately 5 times higher) than the plant extract.

4.3 Somatic Embryogenesis

It is one of the potent plant biotechnological approaches for clonal regeneration, germplasm conservation, and genetic improvement. It is an in vitro regeneration technique in which somatic cells are exposed to ideal in vitro experimental parameters and underwent morphological and biochemical changes that prompted the synthesis of somatic embryos (SEs) that thereafter mature into whole plants.

Although SEs are derived from vegetative cells, they trace a similar pattern of primordial embryonic development as their respective zygotic embryogenesis. It is theorized that SEs are bipolar in structure, having a shoot/root axis and a closed vascular system. SE is generally affected by myriad factors, like explant age and type, phytohormones, and genotype, and is classified into two types: direct somatic embryogenesis (DSE) and indirect somatic embryogenesis (ISE). In DSE, dedifferentiation stage is not involved and the explant directly commences the initiation of SE development with nominal genetic reprogramming, thereby circumventing the callus formation, contrary to ISE that starts with the callus formation. SE serves as an exceptional in vitro experimental system to understand the physiological, molecular, and biochemical aspects of plant embryo development. Several transcription factors/gene regulators are involved in the regulation of somatic embryogenesis such as baby boom (BBM), somatic embryogenesis receptor-like kinase 1 gene (SERK1), wuschel homeobox 2 (WOX2), abscisic acid insensitive3 (ABI3), and leafy cotyledon2 (LEC2). Wang et al. (2023) used the tissue culture approach to optimize a competent regeneration procedure for induction of SEs derived from stem explants in *Punica granatum* and thus laid the foundation for prospective bioengineering improvement and genetic transformation of *P. granatum* with beneficial agrarian traits. They demonstrated that the development of callus and SEs was observed in MS media enriched with 1.0 mg/L NAA and 0.5 mg/L 6-BA. Earlier, Wang et al.

(2022) demonstrated that auxin stimulates the acquisition of somatic cell totipotency through evoking huge changes in chromatin accessibility and transcriptome and the key LEC2–WOX2/3 axis acts as a molecular bridge between early embryonic development pathway and totipotency-related transcription factor genes.

4.4 Somaclonal Variations

A high level of genetic homogeneity across the regenerated plants is required for clonal multiplication and the continuing existence of elite genotypes selected for their superior traits. However, due to gene mutations or modifications in epigenetic markers, plant tissue culture may produce genetic variability or somaclonal variants. The pace of somaclonal changes is persuaded by both quantity band duration of subcultures, particularly cell suspension and callus cultures. In somaclonal variations, genetic variation is included in every conventional crop breeding program and crop improvement cycle that lasts for 10–15 years. Tissue culture-induced somaclonal variation presents a chance to identify natural variability for possible use in crop improvement. Many reports of novel variations among somaclones indicate that passage through tissue culture can change the incidence and circulation of genetic reshuffling events. India produced the sugarcane somaclonal varieties (Co94012 and VSI434), which had favorable traits such a high yield, a higher sucrose content, and a tolerable resistance to red rot. Another sugarcane cultivar introduced in India that uses somaclonal variety is Somaclone VSI434. For some plant species, particularly ornamental plant crops, a widespread range of somaclones have been released in which industry offers many opportunities because ornamental species are so diverse. The genetic variety of ornamental plant species in various nations is influenced by factors such as climate, altitude, and local culture. In contrast to screening a comparable quantity of plants in the field, that takes extra time and area, the in vitro selection technique allows millions of cells or protoplasts to be screened against a range of biotic and abiotic stressors in a single petri dish (Gosal and Wani 2018). The formation of new phenotypic traits, such as variations in plant size, color, and texture, is facilitated by the genetic inconsistency that takes place through in vitro technology resulting in the outline of new decorative plants. New somaclones from various cultures have introduced traits that are beneficial for breeding, like high productivity, tolerance to abiotic stressors, and resistance to infections. The source and genetic composition of somaclonal variants, together with detection methods, are included in the instrument for agricultural advantages. In vitro plant cultivation involves DNA-based studies for analyzing the genetic stability and detecting somaclonal changes depending on the culture and evaluation area. Several molecular markers have confirmed the mechanism underlying somaclonal variation, either through DNA methylation or modifications to the base pair sequence of DNA. Identification of the trait's genetic basis is not required in the following events of transformation, isolation, and cloning. If somaclones are grown in cell culture, there is no chance of achieving chimeric expression. Crops with insignificant genetic bases or limited genetic systems (such as apomicts and vegetative reproducers) have been benefited mostly from somaclonal variation. Many commercial firms routinely use in vitro-generated diversity in their breeding processes for generating ornamental plants. Since somaclonal variation is random and non-reproducible, so it is more challenging to employ. Furthermore, even though somaclonal variation produced the new varieties, improved variants were not often chosen because mostly the variations were unstable after crossing or selfing. The majority of explants that underwent micropropagation showed signs of somaclonal variations that occurred when replicas of genetically indistinguishable plants showed distinct phenotypes upon regeneration. It becomes more noticeable when cells are cultured for prolonged periods and explants or micropropagated plants experience multiple subcultures.

4.5 Anther/Pollen Culture

Anther culture is an alternative method for producing entirely homogenous lines. Upgraded crop varieties with desired qualities like superiority, productivity, and resilience to environmental stressors are required for agricultural activities. The use of haploids has become a crucial crop enhancement tactic and is a useful tool for screening of desired features or introducing a mutation in their genetic. Conversely, six to seven generations of self-crossing are necessary for the traditional breeding strategy. For agricultural development, haploids have been positively produced in vitro culture crop varieties, including rice, wheat, barley, maize, tomatoes, potatoes, grapes, and sunflower. Tian et al. (2015) reported that the F1 offspring of the crossbreed "Nanjing 46" × "Zaxima" which utilized as reference in order to comprehend the connection among callus induction and pollen fertility from rice anthers during low-temperature treatment (LTT). The group also investigated the pollen fertility, anther, callus induction, and the histochemical and microstructural analysis of rising rice plant anthers treated with LTT. Additionally, pretreatment, culture media, phytohormones, the physiological phase of plant, developmental phase of pollen, genotypes, and culture conditions have all been examined in relation to successful anther culture. True-breeding (doubled haploid) plants, which are typically obtained after seven to eight generations using conventional approaches, are guaranteed by these methods in less than a year (Gosal and Wani 2018). Haploids are valuable resources for mutant breeding and the screening of desirable features. The most popular and successful haploid generation techniques are isolated microspore culture and in vitro anther culture. Islam et al. (2023) reported haploid plant regenerated from in vitro pollen microspores ultimately increases plant homozygosity. Researchers and entrepreneurs are employing tissue culture techniques to introduce, generate, relocate, and protect the current plant gene pool. Two enhanced anther culture media, namely, Ali-1(A1) and Ali-2 (A2), modified from N6 medium, were used to study the effectiveness of callus establishment and plant rejuvenation in *Indica-Japonica*, *Japonica*, and *Indica* crossbreeds. The results of this study showed that the effectiveness of green plantlet regeneration and the frequency of callus induction were significantly influenced by genotype and media. Thus, doubled haploid rice plant varieties can be produced using the enhanced anther culture medium. This can be helpful in various breeding projects which will allow the rapid production of crop varieties for farmers with limited resources. Also, the haploid transformation based on microspore or anther culture effectively saves time, space, and money.

4.6 Embryo Culture and Embryo Rescue

A multicellular structure formation which has potential to produce into a new plant is called an embryo. The practical method for producing interspecific and intergeneric hybrids between cross-parents is embryo culture. Desired genes from indigenous species have been successfully incorporated into cultivated grain and vegetable crop varieties (Gosal and Wani 2018). Since 1940s, embryo culture has adapted more to identifying the nutritional and physical requirements of embryonic development, discounting seed dormancy and delaying brief reproductive cycles, assessing the viability of seeds, supplying material for plant tissue culture, and preserving immature mixed seeds from incompatible hybrids. Plant breeders have been using embryo cultivation since twentieth century. Barley embryos were used for embryo cultivation in the first 10 years of the previous century (Dwivedi et al. 2021). Typically, two types of seed cultivation exist: one is carried out with developed embryos and primarily aids in disabling seed dormancy to decrease the germination period known as mature embryo culture. Therefore, in demand to produce viable seedlings, they will be expanding in a simple viable shape with the inclusion of vitality origin which is possible with the mature embryos found in nature. Marasek-Ciolakowska et al. (2021) extensively reviewed different crosses between interspecific hybrids from *Tulipa* species to use embryo and ovule cultures. The most ancient and most effective type of tissue culture techniques is embryo rescue, which helps to develop plant embryos and create interspecific cultivar. Consequently, the development of numerous interspecific and intergeneric crop hybrids through current plant breeding depends on severely on embryo rescue. In addition to making successful crosses easier, embryo rescue has the probable chances to produce haploids, doubled haploids, and ploidy levels that can be altered for chromosome engineering using replacement lines. Moreover, embryo rescue can be utilized to propagate rare plants and reduce the reproductive cycle. The phases of embryonic development have also been extensively studied using particularly in embryo-lethal mutations. The applications of embryo rescue involve the shortening of breeding cycles, overcoming embryo abortion, self and cross incompatibility, developing interspecific and intraspecific hybrids, production of triploids and haploids, seedless in fruit crops, resistance against biotic and abiotic stresses, production of synthetic seeds, seed germination in parastitic plants, determining seed viability, breaking seed dormancy, conversation, and reciprocity of germplasm. The most popular embryo rescue technique involves taking immature embryos and culturing them directly onto culture media. The reproductive cycle has also been shortened in unusual plants that have been propagated by using seed germination physiology. The embryo nourishment has been comprehended, and the morphogenesis of embryos from the zygote to the cotyledon stage has been studied using embryo rescue techniques. Krishna et al. (2022) cultivated young embryos using a variety of embryo rescue approaches, such as placenta and ovary cultures. Many scientists prefer to use embryo rescue and in vitro culture strategy to stunned embryo abortion in a variety of horticultural crops, especially peach, avocado, and plum. By preventing the post-zygotic barriers from being activated, the widely used biotechnological technique regenerates the entire plant from an underdeveloped embryo.

5 Plant Genetic Transformation Techniques

Genetic transformation is a vital approach for increasing the quality, yield, and tolerance of plants to biotic and abiotic stress. New genes can be progressively inserted into the nuclear gene pool of diverse species of plants using an array of approved approaches for genetic transformation. These approaches are divided into two categories, i.e., direct and indirect. Direct method employs outside factors to introduce desired genes of interest into plant cells, like electroporation, silicon carbide, microinjection, particle bombardment, etc., while indirect approach uses vector, including *Agrobacterium*-mediated introduction of genes into desired cells. The primary hindrance for successful genetic transformation in plants is how biomolecules penetrate the plant cells via tough multi-layered cell wall and then regeneration of genetically modified plants either through somatic embryogenesis or de novo organogenesis from an in vitro-propagated explant (Xu et al. 2022).

5.1 Indirect Genetic Transformation Method

Agrobacterium-mediated genetic transformation in plants is mainly referred to as indirect genetic transformation approach. Plasmids that induce hairy roots (Ri) or tumors (Ti) are found in *Agrobacterium* species (*rhizogenes* and *tumefaciens*). Using plasmid modifications, T-DNA segment on the Ri/Ti plasmid can be incorporated into the plant genome and the desired gene can be combined with the T-DNA segment into the genome of plant. When the damaged portion of the plant releases phenols or acidic sugars, VirA detects them and phosphorylates VirG to activate it. VirG further stimulates the expression of *Vir* gene in *A. tumefaciens*. Afterwards the T-DNA segment of the Ri/Ti plasmids is cleaved at the right and left border repeat sequences by the joint action of induced VirD2 and VirD1. VirD2 is covalently bonded to the T-DNA 5'end during cleavage. On the contrary, VirD2/T-DNA departs the bacteria through the Type IV secretion system or T4SS. Also, the T-chain in the plant cells may be non-covalently coated by VirE2, producing the VirD2/VirE2/T-DNA-T intricate and facilitating the integration of T-DNA into the genome of the plant (Ramkumar et al. 2020). In contrast to *A. tumefaciens*, which carries the Ti plasmid that causes tumors on the crown area of host plants, the hairy roots developed by the plasmid (Ri) harbored by *Agrobacterium rhizogenes* revealed non-geotropism, multidirectional rapid growth, and multilateral roots on the medium without any phytohormone. However, Cao et al. (2023) employed *A. rhizogenes* to culture explants that resulted in transformed roots which were able to accomplish heritable alteration of the plant species successfully, mainly woody plants herbaceous plants and some tuberous roots like sweet potato. This technique makes it possible to genetically transform some species that were previously challenging to do so under unsterilized conditions.

Majority of dicots and few monocot plants can be transformed by *Agrobacterium*-mediated genetic transformation approach, which provides the benefit of simple operation, genetic stability, and high efficiency. Though, the effective use of this approach in monocots is affected by various factors. Unlike dicots, genetic transformation mediated by *Agrobacterium* in monocots is limited since most monocot plants lack an apparent division cell, which makes them less susceptible to natural infection caused by *Agrobacterium*. Breakthrough in monocot transformation using *Agrobacterium* is due to key elements or characteristics needed for efficient infection as well as gene transfer, including the use of explants with a significant number of actively dividing cells or premature embryos. Additionally, it has been observed that for *Agrobacterium* to effectively transform cereals, highly toxic strains of the *Agrobacterium* needs to be used along with appropriate vectors. An essential aspect for complete transformation along with regeneration of the transgenic plants is an effective selection of altered cells from non-altered cells. In contrast, transformation in early dicots was dependent on resilience to aminoglycosides (G418, neomycin, and kanamycin), which was mainly unsuccessful in the majority of cereals. Transformants for barley, wheat, and maize were selected using herbicide-resistant markers. The transformation effectiveness by *Agrobacterium* in monocots was enhanced by optimization of vector, use of hyper-virulent strains of *Agrobacterium* and application of appropriate selectable markers. Yu et al. (2024) observed overexpression of *TaLAX1* gene that significantly enhances the *Agrobacterium tumefaciens*-mediated genetic transformation and revealed the molecular mechanisms that underpin wheat regeneration. In contrast, Zhao et al. (2024) produced transgenic sprouts or calli through *Agrobacterium rhizogenes*-mediated transformation approach in watermelon. However, for crops with poor regenerative ability, delivery efficacy is an essential aspect for the future.

5.2 Direct Genetic Transformation Methods

5.2.1 Particle Bombardment

It is a physical technique for direct delivering the foreign DNA into the genome of the plant, also referred to as "gene-gun." Plant transformation based on this approach is not only constraint to the receptor material but also calli, immature embryos, and organs can be utilized as targets for gene-gun-mediated approach. This method develops a DNA-coated micro-carrier by coating the desired gene on the tungsten surface or gold powder and accelerated the micro-carriers by using high-pressure helium gas or an electric discharge into the gas acceleration conduit. These particles acquire enough momentum to penetrate the recipient cells at high velocity, whereas the desired gene encased on outside stays in the cell and is subsequently incorporated into the chromosome of plant to produce the transformed plant. It has gradually emerged as one of the most prevalent transformation approaches after indirect genetic transformation and has been efficiently employed to diverse species of plants such as poplar, tomato, and soybean. Qiu et al. (2022) established a particle-bombardment-mediated momentary expression strategy of *GRF* interacting factor 1 intricate which enhanced the regenerative capability of wheat and accomplished reliable genome editing. Parameters including pressure, distance, and number of bombardments must be adjusted when applying this approach on diverse crops as these variables can impact the transformation efficiency at final. But it has certain drawbacks in comparison to *Agrobacterium*-mediated transformation, including greater expenses, low transformation rate, and unprotected foreign DNA.

5.2.2 Silicon–Carbide–Whisker (SCW)-Mediated Genetic Transformation

SCWs are needle-like micro-whiskers that vary in length from 10 to 80 μm and possess a diameter of approximately 0.5 μm. The sharp cutting edges of the whiskers penetrate the plant cell wall and ultimately the nucleus. This genetic transformation is accomplished by putting DNA and embryogenic calli or suspended cells in SCW containing centrifuge tube, as SCW has the ability to disrupt the cell membrane by creating the needle-like pores that allow exogenous DNA to enter the desired cells. The transformation mediated by SCW is quick and simple and does not need any specialized instrument. However, the operation procedure must be done with great care because harm to cells caused during operation lowers their capability to regenerate, ensuing in a poor conversion efficiency. Arshad and Asad (2019) described unique characteristics of calli, when utilized as explants in SCW-mediated transformation in cotton with diverse genes by employing embryogenic calli as constant supplier of explants.

5.2.3 Pollen Tube Pathway and Microinjection-Mediated Transformation

In higher plants, pollination occurs when pollen generates the pollen tube on the stigma surface and expands it along the style to the ovule, where the pollen nucleus fertilizes the ovule. In this transformation method, the recipient plant's stigma is removed immediately as pollination occurs and solution of foreign DNA is employed dropwise to the severed style of the recipient plant. Pollen tube growth carries the exogenous DNA to the ovary of the recipient plant, where it combines with the undivided fertilized egg. This integration of the foreign DNA into the recipient genome occurs during the embryogenic phase and exists in the altered seed. In contrast to particle bombardment as well as *Agrobacterium*-mediated transformation, this approach is relatively easy and does not entail cell culture, plant regeneration, or protoplast manipulation processes. Additionally, the disadvantages of genetic variation like methylation and mutation, genotype limitation, and poor regenerative ability are frequently avoided with this approach. Zhou et al. (2023) confirmed that the transformation strategy is efficient for transgenic peanut plant production. However, this approach has not been extensively employed, since exogenous gene transformation by this method has only been fruitful in few monocots and dicots due to natural flowering cycle constraints.

Microinjection-mediated transformation is a technique in which the DNA is injected into the nucleus or cytoplasm of a single plant or protoplast by utilizing a glass micro-capillary pipette. It involves fixing the target cell under a microscope by using two micro-manipulators, one is a micro-capillary tube that contains small quantity of DNA solution and the other is a holding pipette that holds the cell. When the transfer of gene is finished, the transformed cells are cultured and developed into transgenics. This technique has been extensively employed in animal cells, as plants have thick cell wall, so it is challenging for the syringe to enter the cell wall and inject the foreign DNA. However, plant cell wall hydrolysis by hydrolase enzyme results in the death of protoplasts, which is the primary barrier to genetic modification.

5.2.4 Electroporation-Mediated Genetic Transformation

Electroporation technique uses high field pulses to temporarily puncture the cell membrane of target cell and increases the flexibility of plasma membrane of the host cell. DNA and cell or protoplast is cultured together and then transient pores are generated in the target cell membrane by high-field and short pulses. However, on the bilayer of lipids, the water pores developed by electric stimulation can be differentiated into two phases. Firstly, an aqueous pore is formed when water molecules penetrate the lipid bilayer. Secondly, to introduce the DNA into target cell, the head regions (polar) of nearby lipids are shifted towards the molecules of water, generating hydrophilic pores that permit the transmembrane passage of additional impermeable molecules. These small pores can be repaired under an appropriate electrical pulse and reverting the cells to their initial condition. Low cost, stable transformation rate, and rapid application are the advantages of electroporation transformation technique over particle bombardment as well as *Agrobacterium*-mediated transformation. Moreover, electroporation mainly yields single-copy plasmid segments as opposed to particle bombardment, which frequently introduces huge plasmid concatemers. The primary drawback of electroporation is that it operates only with few receptor species and is challenging to transform plants with thick cell wall.

5.2.5 Nanoparticle-Mediated Transformation

Nanoparticles retain ability as biomolecular transport material due to their wide-spectrum applicability and ability to penetrate and precisely target the plant cells to control tissue distribution and gene expression. Initially, this technique required external forces to introduce proteins or plasmids mediated by nanoparticles into plant cells. Consequently, a transient system was created using a mesoporous silica nanoparticle to introduce exogenous DNA into *Arabidopsis* roots without the requirement for mechanical force. This approach can effectively transfer DNA for intermittent elevated RNA and protein expression in tobacco, wheat, arugula, and cotton leaves or protoplasts and significantly advances the understanding the role of genes in these plants. Initially stable transformation with this method was accomplished by transferring magnetic nanoparticles containing exogenous DNA to the pollen, using a magnetic flux and scanning for viable genetically modified seeds. To overcome these difficulties, genotype limitations, and tissue culture in maize genetic transformation, a novel genotype-independent maize transformation mechanism was established by employing nano-magnetic beads under ideal circumstances for pollen sustainability and pore opening (Wang et al. 2020). However, the nanoparticle-mediated transformation is typically modest, although if the nanoparticles are able to enter the plant cells, transformation success rate is constrained. Also, it is difficult to stabilize nanoparticles in plants as they can aggregate, precipitate, or decompose and this unstable environment makes genetic transformation less stable. To prevail over the obstacles and limitations in this genetic transformation method, ongoing research and technological advancements are required. One such advancement is polyvinylimine-transformed single-wall nanotubes of carbon, which have been demonstrated to electrostatically assimilate nucleic acids and safeguard them from nuclease breakdown. Additionally, it has been observed that pollen magnetic transfection significantly increases the effectiveness of transformation, eradicates species reliance, and accomplishes multigene co-transformation and high-throughput screening. They generally speed up molecular breeding processes, and such techniques can eventually be used to produce delivery systems mediated using nanoparticles in the future.

5.2.6 CRISPR/Cas and ZFN-Mediated Targeted DNA Transfer

Traditional gene transfer techniques often incorporate DNA at arbitrary locations throughout the genome, leading to damage or suppression of some important functional genes, thus altering the agronomic characteristics of plants. It is a great choice for introducing DNA into appropriate plant genomic locations. To achieve high efficiency and broad array of targeted genomic regions, researchers are looking into a number of specific DNA insertion techniques in plants such as Clustered Regularly Interspaced Short Palindromic Repeats (CRISPR)/Cas9 and zinc finger nucleases (ZFNs).

Cas nuclease along with guiding RNA molecules constitutes the CRISPR/Cas network, which directs Cas to produce double-strand break (DSB) on the genome target with a specific nucleotide sequence. Homologous recombination or end joining of foreign donor DNA can modify a genome by replacing alleles or inserting a targeted transgene. CRISPR/Cas is an incredibly programmable approach, as the guide RNA variable section can be readily modified to change the recognition specificity. Targeted DNA insertion is one of several applications for this method. To commence gene targeting in plant species, *Agrobacterium* or particle bombardment introduced the donor DNA and the CRISPR/Cas gene as a recombinant T-DNA locus. Following the recognition of plants exhibiting the anticipated directional insertion, genetic separation was used to remove the primary T-DNA from the genome. Also, CRISPR/Cas9 and chemically altered donor DNA can improve the efficacy of inserting sequences into the genome of plants by an order of degree.

ZFN is a recombinant nuclease that possesses both generalized DNA cleavage domain and protein DNA-binding motif. It was initially proved that ZFNs may be used to target DNA introduction in plants and induce homologous recombination. Bonawitz et al. observed that how ZFNs were used to target a 16.2 Kb segment of DNA containing four transgenes into the genome of soybean. The efficient application of ZFNs facilitates the advancement of genetic transformation approach.

6 In Vitro Conservation of Plants: Limitations and Opportunities

One of the measures taken to address the loss of biodiversity is the conservation. In the fields of conservation-policy science and legislation), plants are typically viewed as either resources or as naturally occurring elements with inherent value. The study and practice of characterizing, elucidating, valuing, preserving, and sustaining biological diversity—an intrinsically multidisciplinary and intentional endeavor of plants—are commonly referred to as conservation. Preservation and transfer of plant germplasm in a genetically stable and disease-free state through tissue culture are the primary goals of in vitro conservation. For effective in vitro conservation, plants or plant parts to be regenerated into whole plants use the genetic resources for growth and development. Since the 1980s, when tissue culture techniques became widely suitable for the multiplication of clonal crops, several nations have made investments in tissue culture labs. Viable conservation strategies like micropropagation are employed to reintroduce endangered plants into their natural habitats.

Many of the labs combine objectives for rapid multiplication, gene bank conservation, and pathogen cleansing under in vitro conditions. The efficiency of biotechnological techniques for germplasm management includes methods such as cryopreservation, in vitro propagation, DNA banks, and genetic transformation which enable the small, medium, and long-lasting preservation of elite plants, pathogen-free material, and genetic diversity. Therefore, it is advisable to cultivate these species in order to curb overexploitation practices which pose a severe danger to the useable sustainable resources. To satisfy the world's future food security needs, genetic resources are a crucial component. They serve as a storehouse of beneficial genes and alleles that greatly aid in the genetic enhancement of crop plants and supply the basis for initiatives aimed at improving them. Food security and agro-biodiversity are two major factors that prerequisite the use of a wider variety of genetic diversity and it depends upon the protection of plant genetic resources. Through the selection and breeding of novel, more productive crops that are resilient to biological and environmental challenges, genetic variety offers opportunities for development. Modern technology mainly in vitro culture processes and molecular biology offers some crucial instruments for better plant genetic source and conservation. In vitro cultures provide a workable substitute for plant genetic conservation, when there is nonavailability of seed banks. Either sluggish growth techniques (plantlets on medium) or cryopreservation (long-standing storage in liquid nitrogen) is used. In order to clearly outline the fair and equitable distribution of the rewards resulting from biotechnology on a worldwide basis, the Nagoya Protocol was enacted in year 2010 (Lachenmeier and Montagnon 2024). Benefit sharing is still inadequate even when in vitro technologies need smaller quantity of plant material during the harvesting phase, particularly in nations for which the Nagoya Protocol has not been approved. Analysis of plant tissues can be used to locate the origin and impose conservation measures. This demonstrates how biotechnologies can complement conservation initiatives. It implies that conservationists work in dynamic ecological and sociocultural environments where technology may have conflicting functions. In vitro technology can have certain drawbacks and restrictions. Plants in the wild endure their combination with dominating microbiomes and abiotic factors, which include a variety of microbes along with the ability to cope with predators. It could be challenging towards project adaptation for achievement from outcomes attained in wet labs into inhabited environments. Therefore, lengthy, multi-season, or multi-year schedules may be necessary for conservation efforts. Few reports show caution on the structure of ex vivo micropropagation culture media and the accumulation of growth regulators which could change the biological, morphological, and genomic integrity of certain plants. So, individual species must be examined separately to determine the ideal conditions required for its growth and survival. Furthermore, in ex vivo culture, the explants get separated from its natural ecological environment. As a consequence, the intricacy of plant interactions confined inside social networks or ecosystems still remains in reduced form.

7 Conclusion

Conservation of plant diversity is crucial to the working of human society, and success for conserving the plant germplasm is generally determined by various integrative approaches. The use of biotechnological approaches is an effective technique of exploring the plant genetic diversity for its storage and conservation. Further successful production as well as multiplication of genetically reliable plants from plant cultures is essential for successful conservation under in vitro conditions. Also, in vitro propagation and genetic transformation are imperative to enhance the secondary metabolites production and develop the valuable content/bioactive molecules present in the plants. Further efforts are required to encourage various strategies for in vitro conservation of available plant germplasm and expand the utilization of various approaches/protocols to other plant species which are in verge of extension. Furthermore, there is need to assure the legitimate application of available plant genetic resources for the coming generations.

Acknowledgments The authors thankfully acknowledge various funding agencies (Govt. of India) such as UGC, JK-DST, and NMPB for research grants in the form of major research projects and School of Biotechnology, University of Jammu for its support.

References

Abdelgawad FAM, El-Hawary SS, El-Kader EMA, Alshehri SA, Rabeh MA, Essa AF, El-Mosallamy AE, El Gedaily RA (2024) Eliciting callus cultures for the production of cytotoxic polyphenolics from *Maesaindica* Roxb. Sweet Plants 13(14):1979

Arshad M, Asad S (2019) Embryogenic calli explants and silicon carbide whisker-mediated transformation of cotton (*Gossypium hirsutum* L.). Methods Mol Biol 1902:75–91

Barnhill-Dilling SK, Delborne JA (2021) Whose intentions? What consequences? Interrogating "intended consequences" for conservation with environmental biotechnology. Conserv Sci Pract 3(4):e406

Cao X, Xie H, Song M, Lu J, Ma P, Huang B, Wang M, Tian Y, Chen F, Peng J, Lang Z (2023) Cut–dip–budding delivery system enables genetic modifications in plants without tissue culture. Innovation 4:100345

Chandimali N, Park EH, Bak SG, Lim HJ, Won YS, Lee SJ (2023) Seaweed callus culture: a comprehensive review of current circumstances and future perspectives. Algal Res 103376

Dwivedi SL, Spillane C, Lopez F, Ayele BT, Ortiz R (2021) First the seed: genomic advances in seed science for improved crop productivity and food security. Crop Sci 61:1501–1526

Gosal SS, Wani SH (2018) Cell and tissue culture approaches in relation to crop improvement. In: Gosal SS, Wani SH (eds) Biotechnologies of crop improvement. Springer Cham, pp 1–55

Hadi A, Rafiq S, Singh S, Ali S, Nawchoo IA, Wagay NA (2023) Seed viability, seed germination, and micropropagation of *Aconitum violaceum* Jacq.exStapf—a rare and threatened medicinal plant of Ladakh Himalaya. In Vitro Cell Dev Biol Plant 59(3):315–328

Islam MT, Arif MR, Hasan MT, Robin AHK (2023) Anther culture in crop plants: progress and perspectives. Plant Breed Biotech 11(2):69–96

Kaur R, Aslam L, Kapoor N, Mahajan R (2018) Phytochemical analysis and antioxidant activity of wild pomegranate collected from Patnitop, Jammu and Kashmir. Biosci Biotech Res Asia 15(2):335–341

Kaya E (2024) Importance of plant biodiversity and long-term conservation of plant genetic resources via biotechnological strategies. J Biosci Med 12(11):584–591

Krishna R, Khandagale K, Benke AP, Soumia PS, Manjunathagowda DC, Ansari WA, Mokat DN, Gawande SJ, Ade AB, Singh M (2022) Embryo rescue: a potential tool for improvement of economically important crops. In: Advances in plant tissue culture, pp 259–282. doi: https://doi.org/10.1016/b978-0-323-90795-8.00009-6

Lachenmeier DW, Montagnon C (2024) Convention on biological diversity (CBD) and the Nagoya protocol: implications and compliance strategies for the global coffee community. Foods 13(2):254

Mahajan R, Kapoor N, Singh I (2016) Somatic embryogenesis and callus proliferation in *Picrorhiza kurroa* Royle ex. Benth. J Exp Biol Agric Sci 4(2):201–209

Mahajan R, Javed A, Kapoor N (2018) Characterization of genetic diversity of wild pomegranate collected from Himachal Pradesh, India. Ann Plant Sci 7:2042–2046

Marasek-Ciolakowska A, Sochacki D, Marciniak P (2021) Breeding aspects of selected ornamental bulbous crops. Agronomy 11:1709

Normah MN, Sulong N, Reed BM (2019) Cryopreservation of shoot tips of recalcitrant and tropical species: advances and strategies. Cryobiology 87:1–14

Pence VC, Ballesteros D, Walters C, Reed BM, Philpott M, Dixon KW, Vanhove AC (2020) Cryobiotechnologies: tools for expanding long-term ex situ conservation to all plant species. Biol Conserv 250:108736

Qiu F, Xing S, Xue C, Liu J, Chen K, Chai T, Gao C (2022) Transient expression of a *TaGRF4-TaGIF1* complex stimulates wheat regeneration and improves genome editing. Sci China Life Sci 65(4):731–738

Ramkumar TR, Lenka SK, Arya SS, Bansal KC (2020) A short history and perspectives on plant genetic transformation. In: Rustgi S, Luo H (eds) Biolistic DNA delivery in plants. Humana, New York, pp 39–68

Salgotra RK, Chauhan BS (2023) Genetic diversity, conservation, and utilization of plant genetic resources. Genes 14(1):174

Tian QQ, Lu CM, Li X, Fang XW (2015) Low temperature treatments of rice (*Oryza sativa* L.) anthers changes polysaccharide and protein composition of the anther walls and increases pollen fertility and callus induction. Plant Cell Tissue Organ Cult 120:89–98

Wang FX, Shang GD, Wang JW (2022) Towards a hierarchical gene regulatory network underlying somatic embryogenesis. Trends Plant Sci 27(12):1209–1217

Wang J, Xia X, Qin G, Tang J, Wang J, Zhu W, Qian M, Li J, Cui G, Yang Y, Qian J (2023) Somatic embryogenesis and plant regeneration from stem explants of pomegranate. Hortic 9(9):1038

Wang N, Arling M, Hoerster G, Ryan L, Wu E, Lowe K, Gordon-Kamm W, Jones TJ, Chilcoat ND, Anand A (2020) An efficient gene excision system in maize. Front Plant Sci 11:1298

Xu N, Kang M, Zobrist JD, Wang K, Fei SZ (2022) Genetic transformation of recalcitrant upland switchgrass using morphogenic genes. Front Plant Sci 12:781565

Yu Y, Yu H, Peng J, Yao WJ, Wang YP, Zhang FL, Wang SR, Zhao Y, Zhao XY, Zhang XS, Su YH (2024) Enhancing wheat regeneration and genetic transformation through overexpression of *TaLAX1*. Plant Commun 5(5)

Zeng L, Sun Z, Fu L, Gu Y, Li R, He M, Wei J (2024) Cryopreservation of medicinal plant seeds: strategies for genetic diversity conservation and sustainability. Plants 13(18):2577

Zhao Y, Zhu H, Lu X, Anees M, He N, Yang D, Chen Z, Hong Z, Zhang J, Liu W (2024) Streamlined *agrobacterium* rhizogenes-mediated hairy root transformation for efficient CRISPR/Cas9-based gene editing evaluation in diverse *Citrullus* varieties. Hortic Plant J 11(2):816–826

Zhou M, Luo J, Xiao D, Wang A, He L, Zhan J (2023) An efficient method for the production of transgenic peanut plants by pollen tube transformation mediated by *agrobacterium tumefaciens*. Plant Cell Tissue Organ Cult 152(1):207–214

Conservation of Endemic Plants: Conventional and Biotechnological Approaches

D. Thirisha Rani, M. Yagna priya, S. Priyadharshini, and Parthiban Brindha Devi

Abstract

Endemic plants are species that are native to a specific geographical region and are found nowhere else. These plants are crucial for maintaining biodiversity and ecosystem balance; some of them have special characteristics like *Rafflesia arnoldii* (Indonesia), which has the largest individual flower, and *Cycas beddomei* (India), an endangered cycad species endemic to India, important for its prehistoric lineage and medicinal values like rosy periwinkle (*Catharanthus roseus*), which has a source of alkaloids used in cancer treatment, particularly in drugs like vincristine and vinblastine, which are vital in treating leukemia and Hodgkin's disease. Neelakurinji (*Strobilanthes kunthiana*), endemic to India, is traditionally used in Ayurvedic medicine for its supposed anti-inflammatory and rejuvenating properties. Conservation of endemic plants is essential to preserve the unique ecosystems they support and to prevent the loss of irreplaceable plant species. Conventional conservation approaches include in situ methods like protecting natural habitats, creating national parks, and establishing reserves to safeguard plants in their natural environments. Ex situ methods, such as seed banks, botanical gardens, and nurseries, focus on conserving plants outside their natural habitats for research and restoration purposes. In contrast, biotechnical approaches offer innovative solutions for conservation, utilizing tools like tissue culture, cryopreservation, and genetic engineering. These techniques allow for the propagation and preservation of endangered species, helping to overcome challenges like habitat fragmentation and limited population sizes. Combining both conventional and biotechnical methods is key to ensuring the long-term survival of endemic plants, as each approach complements the other by addressing different aspects of conservation challenges.

Keywords

Endemic plants · Biodiversity · Conservation · In situ conservation · Ex situ conservation · Medicinal plants · Tissue culture · Cryopreservation · Genetic engineering

1 Introduction

The conservation of biodiversity has become a critical global priority as ecosystems face unprecedented threats from human activities and environmental changes. Among the various components of biodiversity, "endemic plants" hold a particularly significant position. These plants, which are restricted to specific geographical regions, play vital roles in maintaining ecological balance and sustaining local ecosystems. The need to conserve these plants has grown urgent due to increasing threats such as habitat destruction, climate change, and invasive species (Hata et al. 2024). This paper explores the significance of endemic plants, the challenges they face, and the approaches—both conventional and biotechnological—used to conserve them. A comprehensive understanding of these strategies is essential to develop sustainable conservation frameworks that can address the unique requirements of endemic species (Hata et al. 2024).

The conservation of endemic plants is critical not only for preserving biodiversity but also for maintaining ecosystem resilience and supporting human well-being. Understanding their ecological roles, the threats they face, and the objectives of conservation provides a foundation for developing effective strategies. In the following sections, we will explore conventional and biotechnological approaches to

D. Thirisha Rani · M. Yagna priya · S. Priyadharshini · P. B. Devi (✉)
Department of Bio-Engineering, School of Engineering, Vels Institute of Science, Technology and Advanced Studies (VISTAS), Chennai, India
e-mail: pbrindhadevi@gmail.com

N. Kumar (ed.), *Biotechnology in Plant Conservation*, Sustainable Landscape Planning and Natural Resources Management,
https://doi.org/10.1007/978-3-032-02605-7_3

ensure the survival of these invaluable species (Martin and Coulon 1995).

The global distribution of endemic plants highlights their remarkable diversity and evolutionary significance. These plants are not only icons of their respective regions but also crucial to maintaining ecosystem functionality and providing valuable services. However, their limited ranges make them vulnerable to threats, underscoring the importance of conservation. By understanding the ecological roles and contributions of endemic plants, we can better appreciate their irreplaceable value and the urgency of protecting them for future generations (Martin and Coulon 1995).

1.1 Importance of Endemic Plants in Biodiversity

Endemic plants are the backbone of many ecosystems, often contributing to unique ecological niches and supporting specialized interactions with other species. Their importance can be summarized as follows:

Ecological Roles: Endemic plants frequently act as keystone species, providing food, shelter, and resources for endemic and native fauna. For example, the "*Welwitschia mirabilis*," endemic to the Namib Desert, sustains a variety of desert organisms (Martin and Coulon 1995).

Genetic Diversity: These plants often possess unique genetic traits that can contribute to overall biodiversity. Their genetic resources can be valuable for research, particularly in agriculture, medicine, and climate adaptation (Martin and Coulon 1995; Kumar 2024).

Cultural Significance: Many endemic plants hold cultural, spiritual, and medicinal importance for indigenous communities. For instance, the sacred lotus (*Nelumbo nucifera*) in parts of Asia represents both spiritual and ecological significance (Kumar 2024).

Climate Regulation: By maintaining ecosystem integrity, endemic plants contribute to climate regulation through carbon sequestration, water cycle management, and soil preservation. Without these plants, ecosystems can suffer cascading effects, leading to biodiversity loss at multiple levels (Kumar 2024).

1.2 Definition and Characteristics of Endemic Plants

Endemic plants are species that are confined to a specific geographical region and are not naturally found elsewhere. Their distribution can range from vast areas, like continents, to highly restricted locales, such as a single island or valley.

Characteristics:

Restricted Habitat Range: Endemic plants are often adapted to specific environmental conditions, such as the dragon's blood tree (*Dracaena cinnabari*) on Socotra Island.

Ecological Specialization: These species are highly specialized to their native ecosystems, often making them sensitive to environmental changes (Kumar 2024).

Vulnerability: Due to their limited range, endemic plants are more susceptible to threats like habitat destruction and climate change (Kumar 2024).

Evolutionary Significance: Many endemic plants represent unique evolutionary lineages, providing insights into speciation and adaptation (Cogoni et al. 2021)

1.3 Threats to Endemic Plants

The survival of endemic plants is jeopardized by a combination of natural and anthropogenic factors.

Habitat Destruction: Urbanization, deforestation, and agriculture often lead to the destruction or fragmentation of habitats. For instance, deforestation in the Amazon Basin threatens endemic understory plants adapted to rainforest conditions.

Climate Change: Rising temperatures, altered precipitation patterns, and extreme weather events disproportionately affect endemic species because they cannot easily migrate or adapt. Alpine plants, such as those in the Himalayas, are particularly vulnerable.

Invasive Species: Non-native species can outcompete endemic plants for resources. For example, invasive grasses in Hawaii have displaced several endemic flowering plants.

Overexploitation: Many endemic plants are harvested for their economic value. For instance, the overharvesting of Sandalwood (*Santalum* spp.), endemic to certain regions, has led to population decline.

Pollution: Soil and water pollution from industrial activities often degrade habitats, directly affecting the survival of endemic flora.

Limited Reproductive Capacity: Due to their specialized nature, many endemic plants have limited reproductive mechanisms, making them slow to recover from population losses.

Objectives of Conservation: The conservation of endemic plants aims to safeguard not only their survival but also the integrity of the ecosystems they support.

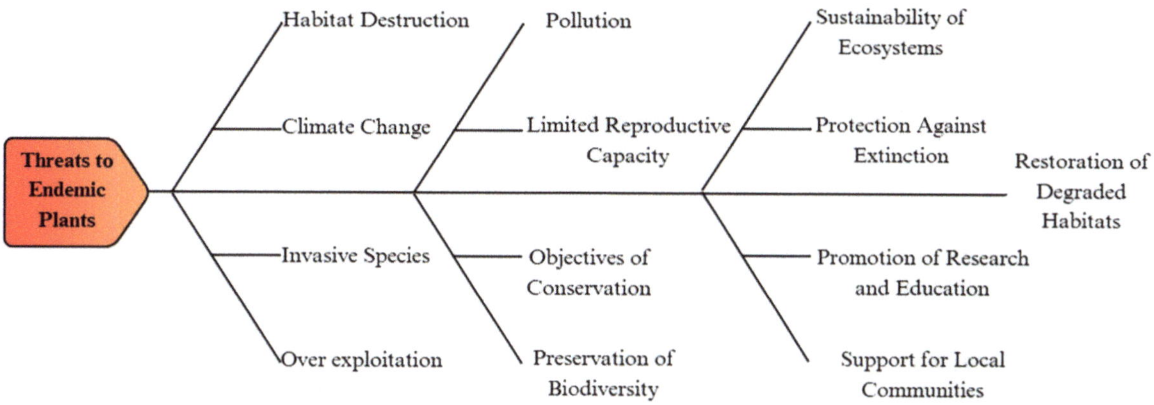

Fig. 1 Threats to endemic plants

Preservation of Biodiversity: Conserving endemic plants ensures the survival of unique genetic resources, which can have ecological, medicinal, and agricultural applications.

Sustainability of Ecosystems: Endemic plants often stabilize ecosystems. For example, mangroves endemic to coastal regions protect shorelines and support marine biodiversity.

Protection Against Extinction: The primary goal is to prevent extinction by addressing threats and implementing protective measures.

Restoration of Degraded Habitats: Reintroducing endemic species into their native habitats can aid ecosystem recovery.

Promotion of Research and Education: Conservation programs foster scientific research and public awareness about the importance of endemic plants.

Support for Local Communities: Conservation efforts can provide economic and cultural benefits to indigenous and local communities, promoting sustainable livelihoods (Fernández-Concha et al. 2021; Garzón-Machado et al. 2012). A representation of threats to endemic plants is shown in Fig. 1.

2 Overview of Endemic Plants

Endemic plants are vital components of global biodiversity, confined to specific geographical areas and often existing under unique ecological conditions. Their restricted distribution makes them significant from both evolutionary and ecological perspectives. This section explores the global distribution of endemic plants, highlights iconic examples from different regions, and underscores their crucial contributions to ecosystem services. Endemic plants are species that are native to a specific geographic area and are not naturally found anywhere else in the world. These plants have evolved and adapted to their unique environments over long periods of time, often in response to specific climatic, geological, or ecological conditions. The concept of endemism is crucial in understanding biodiversity and conservation efforts, as endemic species contribute significantly to the ecological uniqueness of their habitats. The term "endemic plants" refers to the varieties of plants which tend to be specific to a certain geographical stamp. It cannot be found in the conquest of any other area on the planet. Such plants usually arise after if they were present in one or more isolated environments for a considerably longer time period. Culturally and typologically, the concept of plants in endemic areas is highly related to the geo and climatic history of the region as well as their isolation from other regions. Such plants remain to be very important as they have a direct impact on the ecosystem that they are a part of. More often than not, they form the required micro niches and help in the ecosystem eviction. Due to their distinct features, such plants are primarily the target of anthropogenic conservation as they tend to be fragile and endangered due to external factors like changes of the environment, destruction of the habitat, and non-spore species of plants being rolled in. Anthropogenic activities have resulted in the disintegration of a large number of endemic species. The likes of urbanization, deforestation, agricultural expansion, and climate change have contributed to the obliteration of the natural habitats where such plants used to grow. Not only this, the introduction of invasive varieties has also increased the competition for resources which has invariably threatened their existence (Cheminal et al. 2022).

2.1 Global Distribution of Endemic Plants

The distribution of endemic plants is influenced by factors such as geographical isolation, climatic conditions, and historical biogeography. Regions with high levels of endemic plant species are often referred to as biodiversity hotspots, areas that are both rich in species diversity and under significant threat from human activities.

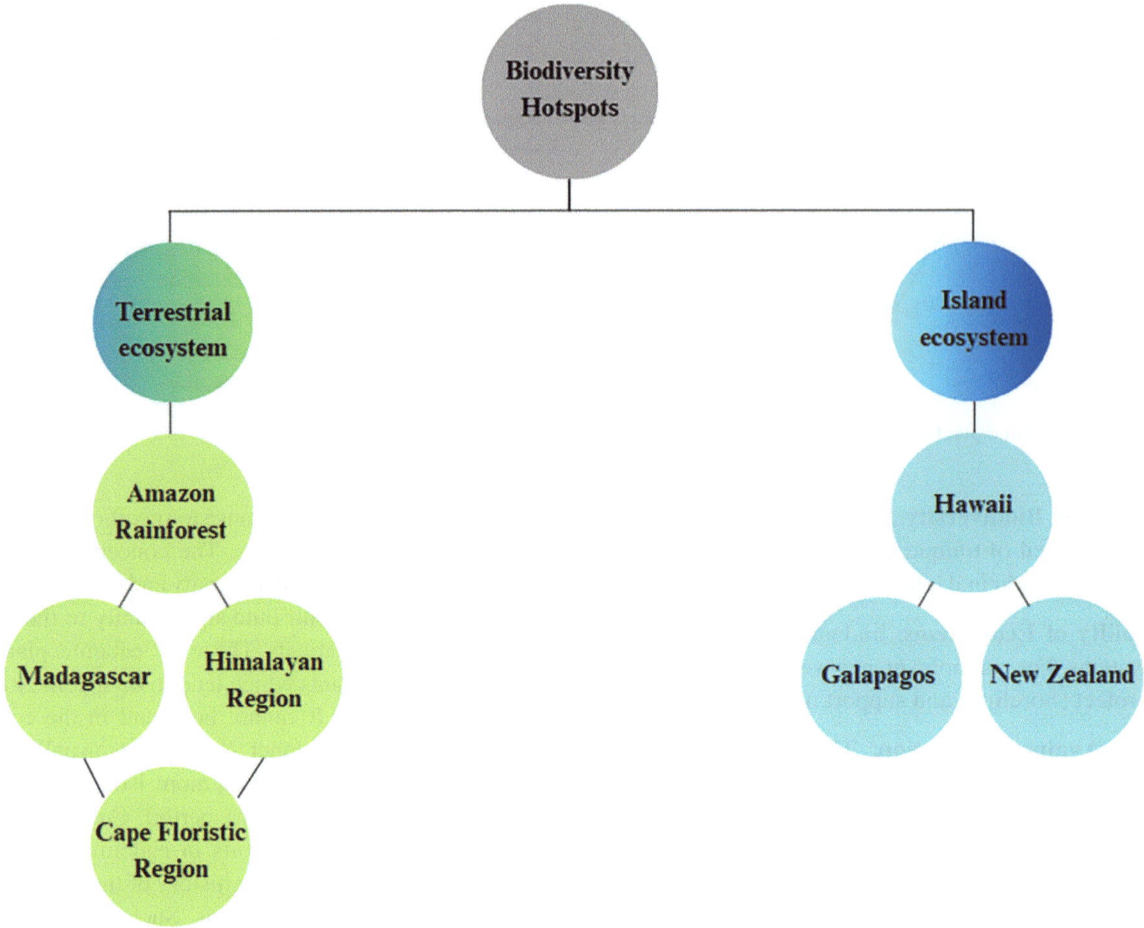

Fig. 2 Global distribution of endemic plants

Biodiversity Hotspots:

Amazon Rainforest: The Amazon Basin harbors thousands of endemic plant species, including unique trees like ***Bertholletia excelsa*** (Brazil nut tree).

Madagascar: Approximately 90% of Madagascar's plant species, such as the **traveler's palm (*Ravenala madagascariensis*)**, are endemic, shaped by millions of years of isolation.

Himalayan Region: The Himalayas are home to high-altitude endemic flora like ***Rhododendron arboreum*** and ***Saussurea obvallata*** (Brahma Kamal).

Cape Floristic Region: South Africa's fynbos biome includes numerous endemic species, such as the **king protea (*Protea cynaroides*)**, the national flower of South Africa.

Islands as Endemic Plant Hotspots: Islands are natural laboratories of evolution, fostering unique plant species due to isolation. Examples include **Hawaii**: Home to plants like the **Hawaiian silversword (*Argyroxiphium sandwicense*)**; **Galapagos Islands**: The islands are home to endemic species like ***Scalesia* spp.**, often referred to as the "Darwin's finches" of the plant world due to their adaptive radiation; and **New Zealand**: Features endemic conifers like the **kauri tree (*Agathis australis*)**. A representation of the global distribution of endemic plants is shown in Fig. 2.

2.2 Examples of Iconic Endemic Plants from Different Regions

Highlighting specific examples from various parts of the world illustrates the incredible diversity of endemic plants.

Americas: **Sierra Nevada Blue Oak (*Quercus douglasii*)**: Found only in California, USA, this oak is crucial for supporting native wildlife and preventing soil erosion. ***Puya raimondii***: Endemic to the Andes, this giant bromeliad is one of the world's largest flowering plants, standing up to 10 m tall.

Africa: ***W. mirabilis***: A living fossil found in the Namib Desert, this plant can survive for over 1000 years by extracting moisture from fog. **Baobab trees (*Adansonia* spp.)**: Endemic to Madagascar, these trees store water in their massive trunks and support local ecosystems (van der Werff and Consiglio 2004).

Asia: **Nepenthes Rajah**: Found in Borneo, this carnivorous plant has adapted to nutrient-poor soils by trapping insects in its pitcher-like structure. **Lotus (*N. nucifera*)**: Although culturally significant across Asia, certain varieties are endemic to specific regions like India and China (van der Werff and Consiglio 2004).

Oceania: **Kakabeak (*Clianthus maximus*)**: Endemic to New Zealand, this striking plant with red, beak-shaped flowers is critically endangered due to habitat loss. **Tasmanian Blue Gum (*Eucalyptus globulus*)**: Native to Tasmania, this tree is valued for its ecological and medicinal properties (van der Werff and Consiglio 2004).

Europe: **Edelweiss (*Leontopodium alpinum*)**: Found in the Alps, this iconic flower has adapted to survive in harsh mountain climates. **Canary Island Pine (*Pinus canariensis*)**: This endemic tree of the Canary Islands is uniquely fire-resistant, showcasing adaptation to volcanic landscapes (van der Werff and Consiglio 2004).

2.3 Endemic Plants and Ecosystem Services

Endemic plants play a pivotal role in maintaining ecosystem health and resilience. Their contributions can be categorized into provisioning, regulating, supporting, and cultural services (Cogoni et al. 2021).

Provisioning Services: **Food and Medicine**: Many endemic plants serve as sources of food and traditional medicine. For instance, the **Madagascar periwinkle (*Catharanthus roseus*)** has yielded life-saving cancer drugs. **Raw Materials**: Endemic trees like **sandalwood (*Santalum* spp.)** in India are highly valued for their fragrant wood and essential oils (Cogoni et al. 2021).

Regulating Services: **Climate Regulation**: Endemic trees and shrubs, such as the **Amazonian rubber tree (*Hevea brasiliensis*)**, sequester carbon, mitigating climate change. **Erosion Control**: Plants like the **Hawaiian naupaka (*Scaevola* spp.)** stabilize coastal soils, preventing erosion (Cogoni et al. 2021).

Supporting Services: **Soil Formation and Fertility**: Endemic plants contribute to nutrient cycling, particularly in fragile ecosystems like tropical rainforests. **Pollination Networks**: Many endemic plants have coevolved with specific pollinators, ensuring ecosystem productivity. For example, the ***Hibiscus arnottianus***, endemic to Hawaii, relies on native honeycreepers for pollination (Cogoni et al. 2021).

Cultural Services: **Spiritual and Cultural Importance**: Endemic plants are integral to cultural identities. For example, the **Sacred Bodhi Tree (*Ficus religiosa*)** is revered in Buddhism and is endemic to parts of India and Nepal.

Tourism and Education: Unique endemic plants attract ecotourists and serve as symbols of environmental education. The **King Protea (*P. cynaroides*)** is a major draw for tourists visiting South Africa's fynbos biome (Cheminal et al. 2022).

3 Factors Endangering Endemic Plant Species

Endemic plants, with their restricted distribution and specialized ecological roles, are highly susceptible to a range of threats. These threats often act synergistically, compounding their vulnerability and increasing the risk of extinction. This section explores the primary factors threatening endemic plants, including habitat destruction, climate change, overexploitation, and the impact of invasive species (Zhang et al. 2018).

3.1 Habitat Destruction and Fragmentation

Habitat destruction is the most significant threat to endemic plants, driven by human activities such as agriculture, urbanization, and infrastructure development. Forests, wetlands, grasslands, and other ecosystems are converted into farmland or industrial areas at alarming rates. For example:

Deforestation in the Amazon: Expansion of soy farming and cattle ranching has led to massive forest loss, threatening endemic plants like *B. excelsa* (Brazil nut tree). **Coastal Development**: Construction in coastal areas has severely impacted mangrove forests and coral reefs, endangering endemic species like the mangrove palm (*Nypa fruticans*) (Zhang et al. 2018).

3.1.1 Fragmentation and its Effects

Habitat fragmentation occurs when large ecosystems are broken into smaller, isolated patches due to roads, dams, or urban sprawl. This isolation disrupts the ecological networks on which endemic plants depend (Zhang et al. 2018; Walker and Kendrick 1998):

Pollinator Disruption: Fragmentation reduces the populations of specialized pollinators, leading to decreased reproductive success in plants like the Joshua Tree (*Yucca brevifolia*) in the Mojave Desert (Walker and Kendrick 1998).

Edge Effects: Small, fragmented habitats are more exposed to environmental stressors like wind, temperature fluctuations, and invasive species (Walker and Kendrick 1998).

Case Studies: **Cape Floristic Region**: Urbanization and agriculture have destroyed over 30% of the natural fynbos habitat, threatening endemic plants like the silver tree (*Leucadendron argenteum*).

Western Ghats, India: Coffee and tea plantations have fragmented forest habitats, endangering endemic species like *Syzygium travancoricum*, a critically endangered tree (Walker and Kendrick 1998).

3.2 Climate Change and its Impacts

Rising Temperatures: Global warming is altering ecosystems at an unprecedented pace. Endemic plants, often adapted to narrow climatic conditions, struggle to survive even slight changes in temperature (Mahato 2014).

Alpine and Arctic Plants: Species like *S. obvallata* (Brahma Kamal) in the Himalayas are forced to move to higher altitudes, where suitable habitats eventually disappear (Mahato 2014).

Altered Precipitation Patterns

Changes in rainfall affect water availability, leading to drought stress or flooding. For example, quiver tree (*Aloe dichotoma*) in Southern Africa is experiencing population decline due to decreasing rainfall and desertification (Trenberth 2011).

Extreme Weather Events

Storms, hurricanes, and wildfires are becoming more frequent and intense due to climate change, disproportionately affecting plants with limited ranges:

Hurricane Impacts: The endemic plants of the Caribbean, such as *Buxus vahlii*, are often uprooted or destroyed by hurricanes (Alarcón et al. 2023).

Australian Bushfires: The 2019–2020 fires in Australia destroyed large swathes of habitat, threatening endemic species like the Wollemi pine (Alarcón et al. 2023).

Shifts in Pollination and Phenology

Climate change disrupts the timing of flowering and fruiting, creating mismatches with pollinators. For example, *Rhododendron* species in Nepal now flower earlier, which may lead to a lack of synchrony with their pollinators (Alarcón et al. 2023).

3.3 Overexploitation and Illegal Trade

Unsustainable Harvesting: Endemic plants are often harvested for timber, food, medicine, or ornamental purposes, leading to population declines:

Sandalwood (*Santalum* spp.): Endemic to parts of India and Southeast Asia, these trees are overharvested for their valuable heartwood and essential oils (Flores-Palacios and Valencia-Díaz 2007).

Orchids: Many endemic orchid species, such as *Paphiopedilum* spp., are illegally collected for their beauty, pushing them toward extinction (Flores-Palacios and Valencia-Díaz 2007).

3.3.1 Illegal Trade

The illegal plant trade, driven by global demand for exotic species, has devastating effects on endemic flora. Plants like *Diospyros celebica* (black ebony) from Indonesia are targeted for their high-value wood (Flores-Palacios and Valencia-Díaz 2007). **Cactus Smuggling**: In Mexico, endemic cacti such as *Ariocarpus fissuratus* are poached for the ornamental trade (Flores-Palacios and Valencia-Díaz 2007).

3.3.2 Medicinal and Cultural Overharvesting

Traditional uses of plants can lead to overharvesting. For example, yew trees (*Taxus wallichiana*) in the Himalayas are harvested for taxol, a compound used in cancer treatment, despite conservation regulations (Liu et al. 2019).

Case Studies: **Baobab Trees in Madagascar**: Overharvesting of fruits and seeds for international markets is putting these iconic trees at risk. ***Nepenthes* Species in Southeast Asia**: Carnivorous plants are being illegally traded as collector's items (Liu et al. 2019).

3.4 Invasive Species and Competition

Non-native plants, animals, and pathogens often outcompete endemic plants for resources or introduce diseases. **Guava (*Psidium cattleianum*) in Hawaii**: This invasive species has overtaken native forests, threatening endemic plants like *H. arnottianus*.

Predation and Herbivory: Invasive herbivores and predators pose direct threats to endemic plants. For example, goats in the Galápagos Islands: Introduced goats have devastated vegetation, including endemic plants like *Scalesia* spp. (Thomson 2005).

3.4.1 Pathogens and Diseases

Non-native pathogens can devastate endemic plant populations:

Phytophthora cinnamomi: This soil-borne pathogen has caused widespread mortality in Australian endemic species like the Banksia. **Habitat Alteration**: Invasive species often modify habitats, making them unsuitable for endemic plants. **Cogongrass (*Imperata cylindrica*)**: This aggressive grass has invaded tropical ecosystems suppressing the growth of native and endemic species (Thomson 2005).

Case Studies.

Island Ecosystems: The introduction of invasive species has been catastrophic for island endemics, such as the near extinction of the St. Helena Olive Tree (*Nesiota elliptica*) due to invasive plants and habitat loss. **South Africa**: The invasive Australian Acacia species has outcompeted native fynbos plants, altering water availability and nutrient cycling. Endemic plants face a complex web of threats, ranging from habitat destruction and climate change to overexploitation and invasive species. Each of these threats alone can be devastating, but their combined impact significantly amplifies the risk of extinction. Addressing these challenges requires an integrated approach that includes habitat protection, sustainable practices, climate adaptation strategies, and the control of invasive species. Only through concerted global and local efforts can we ensure the survival of these invaluable components of biodiversity (Esquivias et al. 2015).

4 Conventional Approaches to Conservation

The traditional approach to conservation is based on the protection of biodiversity through the conservation of natural habitats and plant species management in their natural or ex situ environment. These methods have been developed and perfected over decades and form the basis of global conservation efforts. This chapter looks at in situ and ex situ conservation methods, explains their importance, and describes the policies and frameworks that support them. The protection of the endemic plants and the restoration of their necessary conditions have often revolved around their natural habitats. The establishment of national parks, wildlife reserves, and nature reserves where the human nature contact is minimal or controlled so that habitat destruction is minimal is one of the most focused strategies. Such zones serve as protection from human encroachment and exploitation, thus allowing endangered flora to survive and multiply. In situ conservation is another conventional method which is self-explanatory by its name, i.e., conserving biological species in their home, pick a cool spot. Restoration of degraded ecosystems in order to improve for the endemic plants the conditions thereof is also included. Things like reforestation, soil erosion control, and watershed recovery can help the equilibrium of the ecosystem return as well as its provisions to the tips of the flora. However, another effective strategy focuses on securing the ex situ conservation. It includes seed banks, tissue culture, and other botanical gardens where flora are grown in gardens outside of their natural environments. And the endemics were set out in the botanical gardens indoors and where they can be biologically viewed without harming them.

4.1 In Situ Conservation

This means preserving species in the wild and ensuring they are able to keep interacting with their ecosystems as they evolve. Such a method is effective for endemic plants as it maintains all the complex ecological networks which they rely on (Kala 2005).

Protected Areas and Natural Reserves

Protected areas are the core of in situ conservation. They consist of national parks, wildlife sanctuaries, and biosphere reserves set up to conserve biodiversity (Kala 2005).

National Parks: Plants endemic to California are conserved within the national parks, and their ecosystems remain largely intact. Examples include *Sequoia sempervirens* (coastal redwood) (Kala 2005).

Biosphere Reserves: The Nilgiris Biosphere Reserve in India conserves endemic species like *Bentinckia condapanna* (Nilgiri palm). These reserves also serve as a haven for associated flora and fauna, thus offering a holistic model of conservation. One such example is the Galapagos Islands National Park, which protects endemic plants such as *Scalesia* spp. along with unique fauna (Kala 2005). A representation of in situ conservation methods is shown in Fig. 3.

4.1.1 Community-Based Conservation Efforts

Local communities are important partners in conservation efforts due to their traditional ecological knowledge. Community-based projects empower locals to use natural resources sustainably: **Sacred Groves**: Natural reserves in some parts of India where plants like *Myristica magnifica* are conserved because of cultural and religious beliefs. **Kenya's**

Fig. 3 In situ conservation

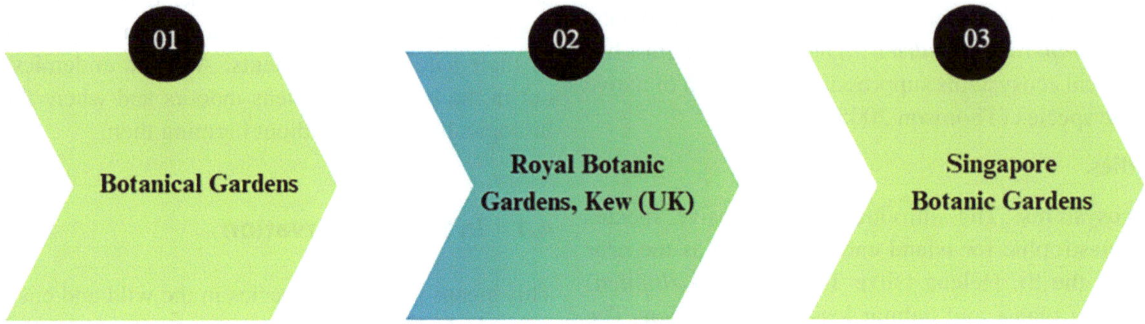

Fig. 4 Ex situ conservation

Arabuko-Sokoke Forest: A community-managed conservation of biodiversity in the protection of endemic species like *Brachystegia spiciformis* (Kastolani 2018).

4.1.2 Advantages of in Situ Conservation

The ecosystem and its dynamics are retained, facilitate the evolutionary adaptation of plants, and preserve the genetic diversity required for long-term survival. However, habitat fragmentation, encroachment, and lack of finance often hinder the success of in situ conservation (Kastolani 2018).

4.2 Ex Situ Conservation

Ex situ conservation refers to the protection of plant species outside their natural habitats. This is especially useful for critically endangered or extinct-in-the-wild species.

Botanical Gardens: Botanical gardens play a crucial role in conserving plant diversity and educating the public about the importance of endemic species. These gardens maintain living collections of plants for research, conservation, and restoration projects.

Royal Botanic Gardens, Kew (UK): Kew Gardens hosts over 8.5 million plant specimens, including rare and endemic species like the Frankincense Tree (*Boswellia sacra*).

Singapore Botanic Gardens: Known for its collection of Southeast Asian orchids, including endemic species such as *Papilionanthe Miss Joaquim*. Botanical gardens also act as repositories for genetic material, facilitating restoration efforts for degraded habitats. A representation of ex situ conservation methods is shown in Fig. 4.

4.2.1 Seed Banks and their Role

Seed banks store seeds under optimal conditions to ensure their viability for future use. These are genetic libraries that protect biodiversity from all sorts of threats, from climate change and habitat destruction. **Svalbard Global Seed Vault (Norway)**: This is popularly called the "Doomsday Vault," as millions of seed samples from every corner of the world are stored here, including species endemic to the place. **Millennium Seed Bank Partnership (UK)**: This project has preserved seeds of more than 40,000 species, including endangered and endemic plants such as *Adenium obesum* from Africa. Seed banks are also a source of resources for reforestation, habitat restoration, and agricultural research, which will ensure the survival of endemic species against environmental uncertainties (León-Lobos et al. 2012).

4.2.2 Tissue Culture and Cryopreservation

Advanced ex situ techniques include tissue culture and cryopreservation that store plant tissues, embryos, and cells at subzero temperatures: Tissue Culture: Tissue culture has successfully been used to propagate endemic plants such as the carnivorous species *Nepenthes khasiana* from India. Cryopreservation: This is a method of storing plant tissues like zygotic embryos of endangered species such as *Wollemia nobilis* or Wollemi pine. Tissue culture and cryopreservation are cutting-edge biotechnology techniques that have played a major role in the in vitro conservation of endemic plants, particularly the rare and endangered species. These methods have become a key component of the conservation strategies implemented worldwide, rendering it possible to maintain plant species in the most adverse of conditions for prolonged periods of time. Growing plant cells, tissues, or organs in a sterile, controlled environment which often employs a nutrient medium is referred to as tissue culture. This advancement allows the massive multiplication of plants using minimal tissues such as a stem or a single leaf. Conservation of such seriously endangered or endangered species of plants or species that are difficult to multiply is possible through tissue culture by mass producing plants without the need for flowers or seeds that can be hard to function. This technology is also important for maintaining genetic information as it can generate a large number of genetically identical plants which is referred to as clonal propagation thereby guaranteeing the retention of desirable traits or diversity of genes. The importance of plant tissue culture is wide; however, one aspect that stands out is its ability to produce disease-free plantlets. This is essential as many plants that are endemic to given areas may be affected by certain infectious diseases that threaten their continued survival in the wild; thus, tissue culture is useful in creating healthy plants that are free from infection. Furthermore, tissue culture can be employed as a form of genetic bank for future studies or reconstruction projects even if the plants are extinct from their natural environment. Cryopreservation is yet another important aspect of plants. This technique involves the storage of plant genetic materials in suspension at such extreme temperatures, usually heated at liquid nitrogen that they can survive for long. This procedure can be performed for seeds, embryos, and plant masses including those of numerous rare and threatened phyla. Cryopreservation essentially freezes the biochemical clock of the plant material, rendering it biologically inactive, and thus, it can be kept forever without spoiling. Thus, in cases of loss of ecosystems for members of a species or threat toward it for reasons such activities, cryopreservation of genetic material for the given seed species can be completed and later "tissue culture" can be used for regeneration of plants. One of the advantages of cryopreservation is that it provides a safe, long-term means of preserving genetic diversity without the need for maintaining living plants in an environmental medium. It also minimizes the chances of genetic drift occurring in small plant populations. Cryopreservation techniques continue to improve, and there is now the possibility of storing an increasingly diverse range of plant material, including pollen, seeds, and even meristematic tissues—the tissue responsible for plant growth. These methods of biotechnology are more relevant to species threatened by habitat destruction, climate change, or other threats that render the standard conservation approaches inadequate. Tissue culture and cryopreservation allow scientists to make backup populations of endangered plants that can be released into the wild if necessary, ensuring the survival of species that otherwise would be put at risk of extinction (Bajaj 1995; León-Lobos et al. 2012).

4.2.3 Advantages of Ex Situ Conservation

It offers a backup for the critically endangered species, assists in reintroduction and restoration programs, and aids research on plant physiology, genetics, and adaptation. Although ex situ conservation has several benefits, there are disadvantages, such as high costs and difficulties in maintaining genetic diversity outside of natural conditions (Bajaj 1995).

4.3 Policies and Legal Frameworks for Plant Conservation

Plant conservation needs robust policies and international cooperation to deal with the various challenges that threaten the endemic plants. The role of policies and legal frameworks is important in plant conservation, since they provide a framework for the effective protection and management of plant species and their habitats in a sustainable manner. At the international level, there are a few important agreements and conventions that were established with the aim of achieving the conservation of plant biodiversity, including endemic species, and guiding the implementation of national policies. The most important of these international frameworks is the CBD, adopted in 1992 and ratified by nearly every country, which emphasizes the conservation of biodiversity, the sustainable use of natural resources, and the equitable sharing of benefits arising from the use of genetic resources. It contains clear targets concerning plant conservation: protection of endemic species, restoration of degraded ecosystems, and promotion of land-use practices that will bring benefit to both people and the environment. Another major international convention is the Convention on International Trade in Endangered Species of Wild Fauna and Flora, commonly known by the acronym CITES. CITES was seeking to ensure that international trade of plant and animal species does not threaten their survival in the wild. A large proportion of endemic plant species may be vulnerable to overexploitation, mainly those plants used for medicinal, ornamental, or commercial reasons. CITES regulates trade by listing these species in one of three appendices according to the degree of protection. This ensures that illegal or unsustainable trade in such endemic plants is prohibited and helps in sustaining the wild populations. Apart from the international conventions, national policies and legal frameworks play an important role in protecting the plants of restricted distribution. The governments mostly formulate legislations with regard to biodiversity conservation, habitat protection, and prohibition on illegal plant collection or trading. These may include national laws that establish the status of protected areas, forbid the collection of certain plant species from the wild, or require licenses for the collection of endangered species for research, conservation, or commercial purposes. The effectiveness of such laws depends on their enforcement, while the existence of penalties will deter most illegal activities affecting endemic plant populations. At the national level, most countries have established a body or agency specifically entrusted with the care and conservation of plants, normally in cooperation with local authorities, NGOs, and other forms of community groups. They are involved in monitoring the populations of plant species, studying those that are in danger, and undertaking various programs to protect them. These agencies might also be responsible for educating the general public on the importance of plant conservation and encouraging local communities toward sustainable practices. In some regions, indigenous peoples and local communities have long been stewards of endemic plants, and their traditional knowledge can be an important asset in plant conservation efforts. Legal frameworks that recognize and protect the rights of indigenous communities to manage their natural resources are essential in integrating traditional conservation practices with modern scientific approaches. This collaboration creates better and more culturally appropriate conservation strategies that are likely to be supported by local people. Conservation of endemic plants will also be supported by regional agreements and initiatives besides national laws. For instance, some regions have developed conservation frameworks that target specific ecosystems or plant groups. Many of these agreements provide for joint planning for the protection of habitats and species across shared borders, realizing the transboundary nature of much ecological concern. Such collaboration is all the more regional in nature, where one considers endemic species whose existence may cross national borders to guarantee protection is coordinated and all-encompassing. Another important aspect of legal frameworks in plant conservation is the integration with broader environmental policies, such as climate change, deforestation, and land-use planning. Incorporation of plant conservation into policies related to agriculture, forestry, and urban development would help to ensure that such objectives are not relegated in the face of competing interests. This may include the encouragement of sustainable agriculture, control of urban sprawl, or more stringent control of logging and land clearing in sensitive areas. The success of these policies and legal frameworks depends on sufficient funding, political will, and the involvement of different stakeholders: governments, NGOs, local communities, and the private sector. Public awareness and education are also important in garnering support for plant conservation efforts and promoting responsible behavior, such as the sustainable harvesting of plant resources or the use of conservation-friendly practices in agriculture and forestry. It follows, therefore, that the need to conserve endemic plants depends mostly on strong policies and legislative frameworks, which give the skeleton on which such species and their habitats are protected. From international conventions down to national laws, regional agreements, and right down to local regulations, there is a very important role played in ensuring that this will be preserved for humanity in the future, balancing between human populations and environmental needs. In particular, successful implementation, stakeholder involvement, and public participation are crucial to these legal frameworks, and more is still to be done to bring them up to date with the dynamic challenges facing plant conservation (Chaudhary et al. 2024; Ren et al. 2019).

4.3.1 International Treaties and Agreements

There are several international agreements, which are more concerned with biodiversity conservation and the flora, including the endemic species:

Convention on Biological Diversity (CBD) encourages the nations to formulate strategies to conserve their biodiversity, with a primary focus on endangered and endemic species.

CITES (Convention on International Trade in Endangered Species) controls international trade in orchids, cacti, and many other plants so that there is no overexploitation.

Global Strategy for Plant Conservation (GSPC) aims to halt the loss of plant diversity by 2030 through targets such as conserving 75% of threatened plant species in ex situ collections.

National Policies: Countries implement localized policies to protect endemic species and their habitats:

India's Biological Diversity Act (2002) protects species like *Eriocaulon ratnagiricum*, endemic to Maharashtra.

Endangered Species Act (USA) provides listings of plants like *Hibiscus brackenridgei*, the state flower of Hawaii, under federal law.

4.3.2 Protected Area Networks

Policies usually demand that protected area networks are to be designed to help preserve endemic species. Such protected area networks ensure the survival of endemic plant species like *Armeria maritima* as in European coastal areas with the network known as Natura 2000 in European Union (Hoveka et al. 2020).

4.3.3 Challenges and Limitations

Lack of Enforcement: Policies fail mainly because of poor monitoring and enforcement. Development projects often conflict with conservation objectives, thus threatening protected areas (Hoveka et al. 2020).

4.4 In Situ and Ex Situ Conservation Integration

The integration of in situ and ex situ methods increases the effectiveness of conservation efforts. For example:

Reintroduction Programs: Seeds or plantlets grown in botanical gardens are reintroduced to their natural habitats to revitalize populations. The reintroduction of the toromiro tree (*Sophora toromiro*) to Easter Island is one of the most successful stories.

Restoration of Degraded Ecosystems: Seed banks offer genetic material for reforestation of degraded areas, such as the use of endemic tree species in Madagascar's reforestation projects. The in situ and ex situ strategies for conventional conservation are an integral part of the protection of endemic plants. Protected areas, botanical gardens, seed banks, and supportive legal frameworks play a critical role in conserving plant diversity. Nonetheless, challenges such as habitat loss, inadequate funding, and enforcement gaps must be surmounted to ensure the effectiveness of these efforts. By implementing these methods and fostering global cooperation, we can ensure the long-term survival of endemic plants and their invaluable contributions to ecosystems and human well-being (Feng et al. 2023).

5 Biotechnological Strategies to Conservation

Biotechnology has been an outstanding tool in conserving plant biodiversity especially for endemic species under pressure from habitat destruction, climate change, and genetic erosion. Biotechnological approaches save plant genetic resources besides offering the possibility of improving the adaptability of these resources to harsh conditions and making restoration easy. This section presents the core biotechnological methods with their applications in conservation (Mandal 1999).

5.1 Plant Tissue Culture Techniques

The tissue culture is the basis of biotechnological conservation: It allows for the production of plants from cells, tissues, or organs in controlled environments. The most important feature of plant tissue culture techniques in the study and propagation of plants is very significant for those species which are considered rare or even endangered and cannot be traditionally propagated easily. The process of culturing is, in general, accomplished under sterile conditions on nutrient media with the growth of plant cells, tissues, or organs. Tissue culture allows, under optimized conditions of light, temperature, and humidity, the development and growth of plant material in vitro. One of the most widely used techniques in tissue culture is micropropagation, which involves taking a small piece of plant tissue and culturing it in vitro. The explant can be taken from any part of the plant, including leaves, stems, roots, or even seeds. The tissue is placed on a nutrient agar medium that provides the necessary nutrients and hormones to stimulate cell division and growth. Given appropriate conditions, the explant goes into a mass of callus cells that then could be induced to shoot and root and finally develop a whole plant. Micropropagation allows for the rapid multiplication of plants in a relatively short period. It is especially applicable to plants that grow very slowly or have low natural seed viability, which applies to many endemic species. This technique ensures that large numbers of genetically identical plants can be produced from a single parent, which is especially valuable for conservation purposes when dealing with limited or endangered populations. Another important technique within plant tissue culture involves the process of somatic embryogenesis, a process by which embryos are formed from somatic cells; basically, any plant cells other than reproductive cells will do, such as those in the roots, leaves, or stems. Generally, the process starts with callus induction from somatic cells, which are further stimulated to develop into embryos. Later on, such embryos can be germinated into new plants. Somatic embryogenesis can be applied to plants of species that are difficult to reproduce through sexual means and also forms a very important step in the genetic preservation of endemic plants, since using this method will provide several plants that possess certain characteristics that one may desire to have. One of the important advantages of tissue culture is that it offers disease-free plants. Many plants in their natural state are afflicted with diseases originating from fungi, bacteria, and viruses. It can clean plant material of a great deal of these pathogens via tissue culture so that plants grown from culture will be healthy and disease-free. This is especially so in the case of endemic plants, which usually face heightened risks from disease and pests in their natural settings. By producing clean plant material, tissue culture maintains health and viability in rare plant populations. Tissue culture also allows the preservation of genetic diversity. For an endangered plant species, the genetic material from different individuals can be cultured and stored. This is done mostly by producing meristems, undifferentiated cells at the tips of the roots and shoots, capable of dividing to develop into new tissues. Meristems can be cultured in vitro, offering a way to preserve the plant's genetic material without the need for fully developed plants. This can serve as a genetic bank that ensures the preservation of a plant's genetic resources for future restoration efforts or research. Tissue culture can also be applied to the production of secondary metabolites, which are chemical compounds produced by plants in response to stress, disease, or environmental conditions. These may be medicinally, agriculturally, or industrially useful metabolites. By using tissue culture, large amounts of such compounds can be produced under controlled conditions without necessarily having to harvest the plant from the wild. This has implications for the propagation of endemic plants of traditional medicinal use, which could be overexploited to threaten their survival in natural environments. The technique also permits the study of plant biology and genetics. For instance, investigations into the aspect of plant development, gene expression, and environmental response can be studied in vitro, providing insights into the biology of these endemic species. Such knowledge informs various conservation strategies and helps science understand the needs of the plants in their natural setting. In addition, tissue culture provides a controlled setting whereby scientists can experiment with genetic modifications or breeding techniques to improve plant resilience, disease resistance, or other traits important for both conservation and agriculture. One challenge facing plant tissue culture, however, is the need for specialist equipment and expertise. Some common processes require sterile environments to often be done in a laminar flow hood. Nutrient levels must be finely controlled, and the resultant plants carefully monitored for evidence of growth. Techniques may be relatively simple for many species, but endemic plants most often require tailored protocol since their growth habits or ecological needs are very unusual. This makes plant tissue culture research and development expensive and time-consuming; however, the results may be priceless, especially toward the survival of rare and endangered plants. In conclusion, plant tissue culture techniques provide a powerful tool for the conservation, propagation, and study of endemic plants. Tissue culture has become indispensable in plant conservation, allowing for the rapid multiplication of plants, the preservation of genetic diversity, and the production of disease-free material. As these techniques continue to evolve, they offer new possibilities for saving endangered plant species and their long-term survival (Pence 1999; Jain et al. 2012)

5.1.1 Applications in Conservation

Propagation of Rare Species: Tissue culture techniques are used to multiply plants that are difficult to propagate through conventional methods, such as the critically endangered *N. khasiana*, a carnivorous plant endemic to India (Pence 1999).

Pathogen-Free Plants: Tissue culture plants are very often free from pathogens. Plants obtained through tissue culture are released into their natural habitats (Pence 1999).

5.1.2 Case Studies

Orchid Conservations: Tissue cultures have been used in successfully multiplying rare orchids that are highly prized in the international trade of ornamental plant, such as the *Paphiopedilum fairrieanum.* **Desert Plants**: Plants obtained through tissue culture, in particular, the *A. fissuratus*, a living rock cactus, has been successfully obtained (Khan et al. 2013).

5.1.3 Benefits

It facilitates the year-round propagation irrespective of the season. It conserves genetic material for long-term usage. It offers uniform and large-scale production of plantlets.

5.2 Micropropagation

Micropropagation is an advanced tissue culture technique involving the rapid multiplication of plants from small plant parts, including meristems or nodes.

5.2.1 Techniques in Micropropagation

Shoot Tip Culture: It is used for the production of clonal plants, which retain the genetic identity of the parent plant.

Auxiliary Bud Multiplication: Efficiently generates large numbers of plants from a single bud.

Applications

Clonal Propagation of Endemics: Plants like *Rauvolfia serpentina*, known for its medicinal properties, are conserved through micropropagation (Chatterjee and Ghosh 2020).

Reintroduction Efforts: Micropropagated plantlets of the Wollemi pine (*W. nobilis*) have been reintroduced to the wild (Chatterjee and Ghosh 2020).

5.3 Somatic Embryogenesis

Somatic embryogenesis involves the generation of embryos from somatic cells, bypassing the need for seeds (Mihai et al. 2023).

Importance in Conservation

Mass Propagation: Allows large-scale production of plants with high genetic uniformity.

Cryopreservation: Embryos can be cryopreserved for long-term storage (Mihai et al. 2023).

Case Studies

Oil Palm Conservation: Endemic palm species like *Elaeis guineensis* var. *guineensis* have been propagated through somatic embryogenesis (Wang et al. 2014).

Tropical Forest Plants: Species such as *Shorea robusta*, a core endemic tree in Southeast Asia, benefit from this method in reforestation activities (Wang et al. 2014).

Problems

Needs highly specialized skills and specialized laboratory facilities. Risk of somaclonal variation that produces genetic variation in the plants (Wang et al. 2014).

5.4 Cryopreservation of Plant Genetic Material

Cryopreservation is a method of conserving plant tissues, seeds, or embryos at ultra-low temperatures (− 196 °C) in liquid nitrogen, thereby ensuring their life span is limitless (Bajaj 1995).

Methods

Encapsulation-Dehydration: It is used in seed or shoot tip storage.

Vitrification: Plant tissues are rapidly frozen, and ice crystals are prevented to form (Bajaj 1995).

Applications

Preservation of Critically Threatened Plants: cryopreservation has been applied to the species such as *Zizania texana* or Texas wild rice.

Orchid Seeds and Pollen: Endemic orchids such as *Dendrobium nobile* are cryogenically preserved (Bajaj 1995).

Benefits

It provides long-term conservation of genetic diversity, requires minimal space compared to field gene banks, and provides a backup against natural disasters or habitat destruction (Rao 2004).

Limitations

Initial setup costs are high, and it requires periodic monitoring to ensure viability (Rao 2004).

5.5 Genetic Engineering for Conservation

Genetic engineering refers to the alteration of plant genetic material in order to make them more adaptable, resilient, or productive (Khan et al. 2013).

Enhanced Stress Tolerance: The genes from stress-tolerant plants are transferred to endemic species for better survival in unfavorable conditions (Khan et al. 2013).

Resistance to Disease: Genetic engineering enables the endemic plants to resist diseases. For instance, introducing resistance genes into species that are susceptible to *Phytophthora* (Khan et al. 2013).

Example

Golden Rice: Not native, but the genetic engineering of rice to produce Vitamin A is a model for improving the nutritional content of other plants and may benefit endemic species used in traditional diets (Bossert and Potthast 2024).

Concerns and Ethical Issues

- Genetic pollution of wild populations.
- Ethics of changing native genetic integrity.

The rapid development of genetic engineering technologies has created multiple opportunities for treating genetic diseases and improving human health. However, genetic engineering technology poses ethical, societal, and security challenges. This brief explores these risks, focusing on those related to genetically modified organisms (GMOs) and the revival of ideologies that consider some races to be "more suitable" than others. The brief also discusses security concerns, including the potential for biological warfare and bioterrorism. It underlines the necessity for comprehensive global governance to ensure the responsible and ethical use of genetic engineering technologies to mitigate risks and maximize benefits (Bossert and Potthast 2024).

5.6 Developing Stress-Tolerant Plants

Endemic plants are highly sensitive to environmental changes. Therefore, the development of stress-tolerant varieties is a priority (Rodriguez et al. 2008). The techniques include the following:

Marker-Assisted Selection (MAS): Genetic markers linked to stress-tolerant traits are identified. Marker-assisted selection in animals is done to detect genes for genetic disorders, disease resistance, and improved product quality. The use of marker-assisted selection to improve longevity, feather pecking, stress resistance, and desired behavior characteristic of animal varieties (Rodriguez et al. 2008).

Transgenic Plants: Genes from stress-resistant species are transferred to endemic plants (Rodriguez et al. 2008).

Examples: **Drought-Tolerant Maize**: The techniques developed to produce drought-tolerant crops may be applied to endemic species such as *Prosopis cineraria*, an arid land species (Rodriguez et al. 2008).

Challenges

- Risk of loss of genetic identity in highly transformed plants.
- Resistance to uptake by local communities on cultural grounds.

These stressors include extreme levels of light (high and low), radiation (UV-B and UV-A), temperature [high and low (chilling, freezing)], water (drought, flooding, and submergence), chemical factors (heavy metals and pH), salinity due to excessive Na+, deficient or in excess of essential nutrients, and gaseous pollutants (Rodriguez et al. 2008).

5.7 DNA Barcoding and Molecular Marker Techniques

DNA barcoding and molecular marker techniques enable the identification and characterization of plant species at a specific level (Srivastava and Manjunath 2020). The technique uses standardized DNA regions for identification purposes, even with minute tissue samples, assists in the discovery of cryptic species, and verifies the identity of endemic plants.

Molecular Markers

Markers such as SSR or simple sequence repeats and AFLP (amplified fragment length polymorphism) are applied for genetic studies (Srivastava and Manjunath 2020):

Genetic Diversity Analysis: It identifies genetic variation within and between populations of endemic species (Srivastava and Manjunath 2020).

Conservation Prioritization: It points out populations having the greatest genetic diversity for specific conservation activities (Srivastava and Manjunath 2020).

Applications

Rare Plant Identification: DNA barcoding has identified previously unknown populations of *Primula palinuri*, an endemic Italian plant (Kress 2017).

Trade Regulation: Molecular markers are utilized to trace the origin of illicitly traded plants, enabling enforcement of CITES regulations (Kress 2017).

Benefits

- Nondestructive sampling methods.
- Accurate identification of even degraded samples.

Biotechnological approaches provide innovative solutions for conserving endemic plants, complementing conventional methods. Techniques like plant tissue culture, cryopreservation, genetic engineering, and DNA barcoding are revolutionizing the field of plant conservation by enabling the propagation, preservation, and genetic analysis of rare and endangered species. Despite challenges like cost and ethical considerations, these methods hold immense potential for ensuring the survival of endemic plants in the face of mounting environmental threats (Kress 2017).

6 Case Studies for Plant Conservation

Case studies provide valuable insights into the practical application of both conventional and biotechnological approaches to plant conservation. These examples highlight successful conservation programs, the integration of biotechnological techniques, and the challenges encountered in applying these methods, particularly in developing regions (Baskin et al. 1997).

6.1 Challenges in Applying Biotechnological Methods in Developing Regions

While biotechnological methods have been successful in many parts of the world, applying them in developing regions presents unique challenges, often due to economic, infrastructural, and educational barriers. Application of biotechnological methods in developing regions faces numerous formidable challenges, which often come in the way of their efficiency and widespread use. Such challenges have normally been embedded in economic, infrastructural, and sociocultural factors that make the integration of advanced technologies into conservation efforts more complex. Lack of appropriate infrastructure and resources is one of the major challenges. Most developing countries lack the appropriate facilities, equipment, and technology for such sophisticated biotechnological approaches like tissue culture, genetic modification, and cryopreservation. In fact, the laboratories that can support such techniques would require laminar flow hoods, autoclaves, and refrigeration units for cryopreservation, which might not be accessible in most developing regions. Besides, keeping optimum conditions for the mentioned techniques—like uninterruptible electricity supply and controlled temperatures—is hardly feasible; this may result in inconsistent results or complete failure of conservation projects. In its absence, these basic resources make the use of biotechnological methods unrealistic or too hard to operate. In addition, most developing countries lack adequate numbers of skilled personnel possessing the technical expertise needed to efficiently apply the biotechnological techniques. Biotechnology as a field is so specialized, requiring expertise in applied microbiology, plant physiology, and genetic engineering. These specific trainings and education are few; therefore, qualified personnel who are fit to conduct and manage this biotechnological research and conservation work are very scarce. Even when available, the skills may be concentrated within specific areas or institutions, posing challenges for wider dissemination and adoption of knowledge and technologies. As a result, the skill disparities could yield poor applications of these biotechnological methods in the application or reliance upon foreign skills, which will not be practical in some situations over time. In addition, economic factors are significant obstacles in the widespread practice of the biotechnological methods of plant conservation. Poor financial resources in the developing world make it increasingly difficult to implement and continue biotechnological techniques. The investment required for new technology development, establishment of laboratories, procurement of equipment and materials, and sustaining research programs over time is considerable. Other operational costs that are ongoing in nature include salaries for the staff, funding for various research works, and the acquisition of additional resources, adding to the burden of the already tight budgets. These financial burdens make plant conservation efforts hard to prioritize for many developing countries, especially when other pressing issues such as poverty, education, or healthcare are in dire need of attention and funding. This may relegate biotechnological methods to the background while cheaper, traditional approaches to conservation are pursued. Another challenge is the complex regulatory environment that surrounds biotechnological practices, especially in regions with limited legal frameworks for biotechnology. Many times, the introduction of GMOs and other advanced biotechnological innovations requires heavy regulatory oversight regarding safety concerns, ethical considerations, and environmental impact. In developing regions, this regulatory framework is usually not fully developed or may be absent, thus raising uncertainty regarding the legal status of biotechnological research and its associated risks. This can sometimes lead to overly restrictive or lenient policies that either hinder the development of new biotechnological solutions or expose ecosystems and local communities to unforeseen consequences. The uncertainty in regulation may discourage foreign investment, international collaboration, and the adoption of new biotechnologies, further slowing progress in plant conservation. Cultural and social factors can also play a role in the acceptance and adoption of biotechnological approaches. Biotechnology, especially genetically modified plants, faces considerable resistance to its use in most developing regions. The local communities could be apprehensive about the safety and other ethical issues related to genetic manipulation, or there might be a strong preference for traditional farming or indigenous knowledge, for instance. This is sometimes exacerbated by lack of public awareness or understanding of biotechnological techniques and their potential benefits in conservation. These social barriers could be overcome by a well-planned community outreach system for education and communication aimed at building trust in and support for biotechnological interventions. Besides, there may be resistance to the use of foreign technologies perceived as incompatible with indigenous practices or values. This can seriously delay the process of adopting biotechnical methods, even when such methods have clear advantages for plant conservation. Lastly, the challenge of political instability and governance issues in some developing regions can further complicate the application of biotechnological methods. In an environment characterized by political instability, corruption, and weak governance, efforts to establish or sustain biotechnological research and conservation programs are undermined. It is difficult to ensure long-term continuity of projects, obtain consistent funding, and see effective implementation of policies in such an environment. Moreover, political instability can trigger changes in priorities, moving conservation either down the policy agenda or completely off it due to more pressing political or economic imperatives. These conditions undercut the potential for success from biotechnological efforts and make such efforts susceptible to disruption and incapable of fulfilling their promise for lasting conservation outcomes. The applications of biotechnological approaches to plant conservation in developing regions are promising, but several challenges impede the utilization of this technology, which include infrastructure, technical capacity, financial, regulatory uncertainty, cultural resistance, and political instability. These demands target investment in infrastructure, education, and capacity-building, along with efforts toward public trust building and the development of enabling legal and regulatory frameworks. Removing these barriers could unleash the full power of biotechnology for the conservation of endemic plants and other valuable plant species in developing regions (Titanji et al. 2007; Ali et al. 2022).

6.1.1 Limited Access to Technology and Expertise

Many developing regions lack the necessary infrastructure, technical knowledge, and funding to implement biotechnological conservation programs effectively. For example, **African Continent**: Countries in Africa, rich in plant biodiversity, often struggle with limited access to laboratory facilities and the high costs associated with tissue culture and cryopreservation technologies. **Asia**: While countries like India and China have made strides in biotechnology, smaller nations with high biodiversity, such as Nepal and Bhutan, face challenges in accessing cutting-edge techniques (Feng et al. 2023).

6.1.2 High Costs and Resource Limitations

Biotechnological methods, such as genetic engineering, tissue culture, and cryopreservation, often require significant financial resources. Developing nations, especially those with limited budgets and other pressing needs, may prioritize other sectors over plant conservation. **Example**: In Southeast Asia, countries with endemic species such as the ***Nepenthes*** genus face difficulties in funding large-scale tissue culture programs. Without external financial support, it is challenging to sustain such initiatives (Feng et al. 2023).

6.1.3 Cultural and Ethical Barriers

In some regions, there may be cultural resistance to the adoption of biotechnological methods, particularly genetic engineering. Communities that rely on traditional knowledge may be skeptical about the introduction of genetically modified plants, even if they offer significant conservation benefits. **Example**: Indigenous communities in the Amazon Basin may prefer traditional methods of resource management and conservation, such as protecting sacred groves, over biotechnology-based interventions. Cultural and ethical barriers play a significant role in the adoption and implementation of biotechnological methods, especially in plant conservation, where traditional practices, local beliefs, and ethical concerns can influence decisions. In many developing regions, the introduction of advanced biotechnological techniques, such as genetic modification, tissue culture, and other forms of genetic engineering, may be met with skepticism or resistance due to cultural factors. These barriers are deeply rooted in the historical, social, and economic contexts of communities, and they can hinder the acceptance of potentially beneficial conservation strategies. One major cultural barrier is the preference for traditional farming methods and local knowledge systems. In many communities, agriculture and plant conservation are closely tied to cultural identity and heritage. Indigenous and local knowledge about plant species, their medicinal uses, and sustainable harvesting practices have been passed down through generations and are integral to the way of life in these communities. As a result, there may be reluctance to adopt biotechnological methods that seem foreign or incompatible with traditional practices. For example, genetically modified crops may be viewed as unnatural or a disruption to the local ecosystem, as these technologies often involve altering plants in ways that are not traditionally practiced. This can lead to resistance, even if the biotechnological methods are designed to enhance the conservation of endangered or important plant species. Ethical concerns also play a major role in shaping attitudes toward biotechnology. The modification of plant genes or the creation of genetically modified organisms (GMOs) raises questions about the ethical implications of altering the natural genetic makeup of a species. Many people, particularly in rural and indigenous communities, may perceive genetic modification as a form of "tampering" with nature, which can be seen as morally wrong or dangerous. Ethical concerns are often amplified by fears of unforeseen consequences, such as the potential spread of modified genes into wild populations, unintended environmental impacts, or negative effects on human health. These concerns are heightened by a lack of understanding or misinformation about the science behind biotechnology and its potential benefits. Furthermore, there is often a perceived lack of control over biotechnological processes, especially when these technologies are introduced by foreign entities, multinational corporations, or governments. This sense of losing autonomy over local resources can lead to distrust and opposition to biotechnology. For instance, if biotechnological interventions are seen as being driven by external commercial interests, there may be concerns about the exploitation of local plant species for profit, potentially without fair compensation for local communities. This can result in resistance to adopting such methods, particularly in areas where the use of native plants has cultural, medicinal, or economic significance.

There is also the issue of "biopiracy," where plant species with significant cultural or medicinal value are exploited without the knowledge, consent, or benefit to the indigenous people who have traditionally used them. Many communities fear that biotechnological approaches could lead to the commercialization of their plant species, with profits going to outside corporations or entities rather than the local population. This fear is compounded by historical instances where indigenous knowledge was appropriated without consent or compensation. This issue has created a broader ethical debate about the ownership of genetic resources and the fair and equitable sharing of benefits derived from these resources. The uncertainty surrounding the long-term environmental impacts of biotechnological interventions also raises ethical concerns. While biotechnology has the potential to help conserve endangered plant species, there are worries about its ability to unintentionally disrupt ecological balance. For instance, introducing genetically modified plants into the wild could have unforeseen effects on local biodiversity, potentially leading to the extinction of native species or the creation of hybrid plants that threaten the natural ecosystem. Many communities and environmental organizations advocate for the precautionary principle, which suggests that biotechnological interventions should be undertaken cautiously and only when potential risks have been thoroughly evaluated. Additionally, there are ethical concerns about the fairness and transparency of biotechnological research and its potential consequences for vulnerable populations. In regions where biotechnology is being introduced, there is often a lack of robust public consultation or participation in decision-making processes. This can lead to feelings of exclusion and resentment, particularly if communities feel that they have not been adequately informed about the potential benefits and risks of biotechnological interventions. Engaging local communities in the discussion and decision-making process is essential to ensure that their concerns are heard and that ethical considerations are addressed in a way that respects cultural values and practices. In conclusion, cultural and ethical barriers present significant challenges in the application of biotechnological methods for plant conservation, particularly in developing regions. These barriers arise from deep-rooted cultural values, historical experiences, and ethical concerns about altering natural systems. Overcoming these barriers requires a sensitive and inclusive approach that acknowledges the importance of traditional knowledge and values while also providing clear, transparent information about the potential benefits and risks of biotechnological interventions. Respect for local customs, autonomy, and a commitment to fair and equitable practices are essential in ensuring that biotechnological methods are accepted and successfully implemented in conservation efforts (Pedrollo et al. 2016; Chen et al. 2016).

6.1.4 Political and Regulatory Challenges

In some developing regions, political instability, weak governance, and lack of regulatory frameworks pose significant challenges to the effective implementation of plant conservation programs. In many cases, there is a lack of enforcement of conservation policies, resulting in overexploitation and illegal trade of endemic plants. **Example**: In parts of South America, endemic plants like ***Cactaceae* species** are often harvested for illegal trade, and there are insufficient regulations in place to protect these plants or enforce biotechnological conservation measures (Kitschelt 1986).

These case studies highlight the potential of both conventional and biotechnological conservation approaches in protecting endemic plants. Successful programs in places like the Galápagos Islands and Australia showcase the power of in situ conservation, while biotechnological methods have played an essential role in the revival of rare species like the Wollemi pine. However, challenges in applying these techniques in developing regions, such as limited access to resources, technical expertise, and political instability, remain significant hurdles. To address these issues, global partnerships, technology transfer, and local capacity building are critical to ensuring the continued success of plant conservation efforts worldwide (Kitschelt 1986).

7 Integration of Conventional and Biotechnological Approaches

The conservation of endemic plants requires a multifaceted approach that combines traditional methods with advanced biotechnological tools. While conventional approaches such as protected areas, botanical gardens, and community-based efforts have long been the cornerstone of conservation, biotechnological methods offer precision, scalability, and innovation. Integrating these two approaches creates a synergistic system that maximizes the strengths of both, ensuring more effective and sustainable conservation outcomes (Jain et al. 2012).

8 Conclusion

The conservation of the endemic plants is of utter importance in the framework of global biodiversity preservation. In the restricted regions where only these plants are native to, they are both indispensable to their particular systems and also valuable from other perspectives of ecological, medical, and cultural relevance. This study points to traditional approaches to plant conservation need to be hybridized with biotechnological techniques in order to manage this very complex challenge which faced endemic plants. Key Findings, Recommendations for Action, and the Need for Continuous Research and Innovation to Assure the Future of Such Species Endemic plants are part of the biodiversity, which provides some unique ecological functions and reflects environmental health. By protecting these plants, ecosystems will retain resilience and continue to function, providing ecosystem services. Habitat destruction, climate change, overexploitation, and invasive species have caused endemic plants to be disproportionately harmed. Coupled with a lack of effective policies and weak enforcement mechanisms, the threats have pushed many species to the brink of extinction. Traditional and biotechnological methods must be combined, supported by public–private partnerships and community involvement, to ensure comprehensive conservation strategies. Integrated approaches can help bridge gaps in resources, knowledge, and implementation.

References

Alarcón D, Santos D, Arroyo MT (2023) Population-based evidence of climate change adaptation in an endangered plant endemic to a biodiversity hotspot. Plants 12(10):2017

Ali I, Joyia FA, Mustafa G, Mirza SA, Khan MS (2022) Emerging trends to improve tropical plants: biotechnological interventions. In: Tropical plant species and technological interventions for improvement. IntechOpen

Arnaiz Fernandez-Concha AS (2021) La lengua de señas peruana como vehículo de aprendizaje del castellano escrito como segunda lengua

Bajaj YPS (1995) Cryopreservation of plant cell, tissue, and organ culture for the conservation of germplasm and biodiversity. In: Cryopreservation of plant Germplasm I. Springer, Berlin, Heidelberg, pp 3–28

Baskin JM, Snyder KM, Walck JL, Baskin CC (1997) The comparative autecology of endemic, globally-rare, and geographically-widespread, common plant species: three case studies. Southwest Nat:384–399

Bossert LN, Potthast T (2024) Genetic engineering, nature conservation, and animal ethics: why genetically modifying wild sentient animals is not a good option. Environ Ethics

Chatterjee T, Ghosh B (2020) Micropropagation of medicinal plants: a review. Int J Econ Plants 7(2):066–072

Chaudhary RP, Joshi SP, Rimal SK, Belbase N, Subedi CK (2024) Plant diversity conservation in the Nepal Himalaya: status, policies, and legislative frameworks. In: Flora and vegetation of Nepal. Springer International Publishing, Cham, pp 383–447

Cheminal A, Kokkoris IP, Zotos A, Strid A, Dimopoulos P (2022) Assessing the ecosystem services potential of endemic floras: a systematic review on the Greek endemics of Peloponnese. Sustainability 14(10):5926

Chen SL, Yu H, Luo HM, Wu Q, Li CF, Steinmetz A (2016) Conservation and sustainable use of medicinal plants: problems, progress, and prospects. Chin Med 11:1–10

Cogoni D, Fenu G, Dessì C, Deidda A, Giotta C, Piccitto M, Bacchetta G (2021) Importance of plants with extremely small populations (PSESPs) in endemic-rich areas, elements often forgotten in conservation strategies. Plants 10(8):1504

Esquivias MP, Zunzunegui M, Díaz Barradas MC, Álvarez-Cansino L (2015) Competitive effect of a native-invasive species on a threatened shrub in a Mediterranean dune system. Oecologia 177:133–146

Feng C, Zhang J, Huang H (2023) Parallel situ conservation: a new plant conservation strategy to integrate in situ and ex situ conservation of plants. Biodivers Sci 31(9):23184

Flores-Palacios A, Valencia-Díaz S (2007) Local illegal trade reveals unknown diversity and involves a high species richness of wild vascular epiphytes. Biological conservation 136(3): 372–387

Garzón-Machado V, del-Arco-Aguilar MJ, Pérez-de-Paz PL (2012) Threat or threatened species? A paradox in conservation biology. J Nat Conserve 20(4):228–230

Hata C, Endo C, Tanaka H, Hiruma M, Kumamoto M, Takenaka I, et al. (2024) Conservation units and the origin of planted individuals of an endangered endemic species Lobelia boninensis in the Ogasawara Islands. Sci Rep 14(1):27410

Hoveka LN, Van Der Bank M, Davies TJ (2020) Evaluating the performance of a protected area network in South Africa and its implications for megadiverse countries. Biol Conserv 248:108577

Jain M, Johnson TS, Krishnan P (2012) Biotechnological approaches to conserve the wealth of nature: endangered and rare medicinal plant species, a review. J Nat Remedies:93–102

Kala CP (2005) Indigenous uses, population density, and conservation of threatened medicinal plants in protected areas of the Indian Himalayas. Conserv Biol 19(2):368–378

Kastolani W (2018) How to conservate in situ and ex situ community based biodiversity park in sumedang regency of west java province?. In: IOP conference series: earth and environmental science. IOP Publishing, Apr 2018, vol 145, no 1, p 012091

Khan SM, Page SE, Ahmad H, Harper DM (2013) Sustainable utilization and conservation of plant biodiversity in montane ecosystems: the western Himalayas as a case study. Annals of botany 112(3): 479–501

Kitschelt HP (1986) Political opportunity structures and political protest: anti-nuclear movements in four democracies. Br J Polit Sci 16(1):57–85

Kress WJ (2017) Plant DNA barcodes: applications today and in the future. J Syst Evol 55(4):291–307

Kumar V (2024) Biodiversity: importance and climate change impacts. Curr Res Med Sci 3(3):14–19

León-Lobos P, Way M, Aranda PD, Lima-Junior M (2012) The role of ex situ seed banks in the conservation of plant diversity and in ecological restoration in Latin America. Plant Ecol Divers 5(2):245–258

Liu X, He P, Chen W, Gao J (2019) Improving multi-task deep neural networks via knowledge distillation for natural language understanding. https://doi.org/10.48550/arXiv.1904.09482

Mahato A (2014) Climate change and its impact on agriculture. Int J Sci Res Publ 4(4):1–6

Mandal BB (1999) Conservation biotechnology of endemic and other economically important plant species of India. In: Plant conservation biotechnology. CRC Press, pp 237–252

Martin B, Coulon JB (1995) Facteurs de production du lait et caractéristiques des fromages. I. Influence des facteurs de production sur l'aptitude à la coagulation des laits de troupeaux. Le Lait 75(1):61–80

Mihai RA, Melo Heras EJ, Pinto Valdiviezo EA, Espinoza Caiza IA, Cubi Insuaste NS, Mejía JP, et al. (2023) Somatic embryogenesis of representative medicinal trees in South America—current status. Forests 14(10):2066

Mounce R, Smith P, Brockington S (2017) Ex situ conservation of plant diversity in the world's botanic gardens. Nat Plants 3(10):795–802

Nijman V, Shepherd CR, Sanders KL (2012) Over-exploitation and illegal trade of reptiles in Indonesia. Herpetol J 22(2):83–89

Pedrollo CT, Kinupp VF, Shepard Jr, G, Heinrich M (2016) Medicinal plants at Rio Jauaperi, Brazilian Amazon: Ethnobotanical survey and environmental conservation. Journal of Ethnopharmacology 186: 111–124

Pence VC (1999) The application of biotechnology for the conservation of endangered plants. In: Plant conservation biotechnology. CRC Press, pp 253–276

Rao NK (2004) Plant genetic resources: advancing conservation and use through biotechnology. Afr J Biotechnol 3(2):136–145

Ren H, Qin H, Ouyang Z, Wen X, Jin X, Liu H, Zhao L (2019) Progress of implementation on the global strategy for plant conservation in (2011–2020) China. Biol Conserv 230:169–178

Rodriguez RJ, Henson J, Van Volkenburgh E, Hoy M, Wright L, Beckwith F, et al. (2008) Stress tolerance in plants via habitat-adapted symbiosis. ISME J 2(4):404–416

Srivastava D, Manjunath K (2020) DNA barcoding of endemic and endangered orchids of India: a molecular method of species identification. Pharmacogn Mag 16(70)

Thomson D (2005) Measuring the effects of invasive species on the demography of a rare endemic plant. Biol Invasions 7(4):615–624
Titanji VP, Ngwa AA, Ngemenya M (2007) Applications of biotechnology techniques to the study of medicinal plants. Afr J Med Med Sci 36(suppl):23–29
Trenberth KE (2011) Changes in precipitation with climate change. Clim Res 47(1–2):123–138
van der Werff H, Consiglio T (2004) Distribution and conservation significance of endemic species of flowering plants in Peru. Biodivers Conserv 13:1699–1713
Walker DI, Kendrick GA (1998) Threats to macroalgal diversity: marine habitat destruction and fragmentation, pollution and introduced species
Wang Y, Chen F, Wang Y, Li X, Liang H (2014) Efficient somatic embryogenesis and plant regeneration from immature embryos of Tapiscia sinensis Oliv., an endemic and endangered species in China. HortScience 49(12):1558–1562
Zhang L, Hong G, Li Z, Gao X, Wu Y, Wang X, et al. (2018) Assessment of the ecosystem service function of Sandy lands at different times following aerial seeding of an endemic species. Sustainability 10(4):902

Plant Genetic Resources: Their Conservation and Utilization Through Various Strategies Involving Modern Genomic Tools and Traditional and Cultural Methodologies

Vinod Janardan Dhole, J. Souframanien, and P. Dhanasekar

Abstract

Plant genetic resources (PGRs) including landraces, wild relatives, obsolete varieties, and modern cultivars are of significant value for food and agriculture. Their conservation is non-negotiable for effectively tackling critical challenges like pests, diseases, and climate change, which are essential for achieving and sustaining global food security. Successful PGR conservation demands a robust strategy that integrates in situ (natural habitats) and ex situ (seed banks, gene banks, cryopreservation, botanical gardens) methods. Community engagement is vital to fully leverage indigenous knowledge and implement participatory breeding programmes that support conservation of locally adapted crops. To optimize the utility of PGR, national and international policies should be refined to foster access and benefit sharing while incentivizing conservation efforts. The adoption of modern technologies, including digital databases and genomic tools, is absolutely needed to improve the monitoring and maximize the use of genetic resources. Conclusively, integrating traditional knowledge with modern tools is key to preserving valuable PGR, sustaining genetic diversity, mitigating biodiversity losses, bolstering food systems, and ensuring long-term global food security and ecological balance.

Keywords

Community seed banks · Ex situ and in situ conservation · Food security · Genetic resources · Geographical indications

V. J. Dhole (✉) · J. Souframanien · P. Dhanasekar
Nuclear Agriculture and Biotechnology Division, Bhabha Atomic Research Centre, Trombay, Mumbai, India
e-mail: vjdhole@yahoo.co.in

1 Introduction

The world is experiencing a population explosion, with a projected 9 billion people by 2050, the biggest human population ever before in Earth's history. Research and development in agriculture, health, engineering, and energy sectors have all contributed significantly to the human population by increasing access to food, water, shelter, and medical facilities, thereby increasing life expectancy. Plant genetic resources (PGRs) are of invaluable significance in the history of human progress, which is often ignored and undervalued. PGR significantly contributed to food grain production through the development of high-yielding rice and wheat varieties with semi-dwarf plant type generated from PGR such as Dee-geo-woo-gen and Norin 10, respectively, which became the ultimate driving force behind the Green Revolution (Ferrero-Serrano et al. 2019). Despite increased food production, around one billion people go hungry every day, which is likely to worsen in the coming years due to climate change. Hence, crop improvement programmes are of paramount importance in agriculture, which is mainly based on genetic variability. Various PGR, like landraces and wild relatives, were effectively used to create better varieties and hybrids with improved traits of economic importance. In most of the major crops, yield levels are plateauing and must be broken by broadening the genetic base of modern-day varieties using elite PGR and modern plant breeding approaches. Genetic variation is the ultimate need of any crop improvement programme; hence, genetic resources, which are the treasure of genetic variability for valuable yield traits, including diseases, pests, drought, and salinity tolerance need to be conserved to ensure basic need like food of future generations. The plant genetic stock which is of actual or potential value for food and agriculture is referred to as plant genetic resource (ITPGRFA and FAO 2013). The importance of PGR was recognized in early stages of crop species domestication. Humans started to save the seeds or planting material of

N. Kumar (ed.), *Biotechnology in Plant Conservation*, Sustainable Landscape Planning and Natural Resources Management,
https://doi.org/10.1007/978-3-032-02605-7_4

domesticated species for future use. They started to choose better plants among the diverse population, thus giving rise to better forms, i.e., landraces, through artificial selections over the period. After the rediscovery of Mendel's principle, the scientific approach to plant breeding increased the pace of crop improvement, thereby utilizing the PGR for the development of new high-yielding varieties. Plant breeders are always on the hunt for new genetic variability, which ultimately enhances the importance and value of germplasm, their collection, documentation, and preservation. Though modern crop varieties are bred through the extensive use of germplasm, on the contrary, those improved varieties are the major threats to the germplasm conservation as landraces and obsolete varieties were displaced by them. The vanishing of landraces from the area of crop origins, due to their replacement with modern cultivars, gave rise to the new term 'genetic erosion', which was coined by Erna Bennett (Khoury et al. 2021). Later on, systematic studies and concern over the loss of genetic resources went beyond the loss of land races. Genetic erosion is mediated via evolutionary forces (mutation, selection, migration, gene flow, and genetic drift), driven by proximate forces (cultural, socioeconomic, and demographic), and climatic changes. Thus, genetic erosion is defined as the loss of crop diversity in a given area over a given amount of time, typically measured by the waning of species, variety and/or within-variety (genetic/genomic) variation (Khoury et al. 2021). The germplasm conservation is confronted with multiple obstacles, with genetic erosion being of primary concern. The genetic erosion of PGR is continuing even after the efforts of preservation of crop genetic resources throughout world. The genetic erosion takes place at various levels due to methods of germplasm conservation, whether it is ex situ or in situ. Genetic drift due to population size and sampling during collection and maintenance, migration and gene flow owing to mechanical mixture and outcrossing, germination losses during storage, losses in field due to shattering, and disease epidemics are the major factors leading to genetic erosion in the ex situ conservation of PGR. Destruction of natural habitats due to urbanization, industrialization, and construction of roads and dams posed major threats to wild species of crop plants, which are only conserved in situ in their natural habitats. Conservation of PGRs will ensure their use for the future generations. The effort of selecting, collecting, exchanging, and preserving germplasm resources was initiated at the beginning of agriculture. But, in the modern era, these activities are being coordinated nationally and internationally as the significance of PGR is recognized in the uniformity-based agriculture system for the development of high-yielding single uniform variety, which is well suited to large-scale cultivation with farm mechanization. Uniformity-based modern agriculture posed a major challenge to PGR as against diversity-based cropping systems like mixed and intercropping, which promoted conservation of diversity. Hence, traditional varieties disappeared very quickly from the cropping system. The shrinking of genetic diversity and genetic erosion was well recognized by N. I. Vavilov, who emphasized the importance of conservation of PGR. In 1974, the International Board for Plant Genetic Resources (IBPGR) was established with the mandate of promoting and assessing the collection and conservation of PGR worldwide (Table 1). The PGRs are conserved in situ, e.g. national parks, wild sanctuaries, biosphere reserves, biodiversity hot spots, wetlands and lakes, and ex situ, e.g. seed banks, gene banks, tissue culture banks, cryopreservation, botanical gardens, and pollen banks. These conservation methods are facing many challenges related to documentation, storage, management, and distribution and need strong financial and government policy support for better conservation and utilization of crop genetic resources. The strategies need to be changed with the advent of modern technologies in phenotyping, documentation, genotyping by

Table 1 Important gene banks conserving crop genetic resources in the world

Institution	Country	Accessions	Plant species
Svalbard Global Seed Vault	Norway	1,301,397	All
NPGS (National Plant Germplasm System)	USA	620,254	16,000 (200 crop species)
ICGR-CAAS (Institute of Crop Germplasm Resources, Chinese Academy of Agricultural Science)	China	450,000	180 crop species
NBPGR (National Bureau of Plant Genetic Resources)	India	468,660	1800
VIR (N. I. Vavilov Research Institute of Plant Industry)	Russia	327,800	2500
NIAS (National Institute of Agrobiological Science)	Japan	243,463	–
CIMMYT (Centro Internacional deMejoramiento de Maíz y Trigo)	Mexico	173,571	–
IPK (Leibniz-Institut für Pflanzengenetikund Kulturpflanzen for schung)	Germany	151,408	2912
ICARDA (International Centre for Agricultural Research in the Dry Areas)	Lebanon	143,000	–
ICRISAT (International Crops Research Institute for the Semi-Arid Tropics)	India	129,954	11 crops
IRRI (International Rice Research Institute)	Philippines	132,000	Rice

sequencing, and tightly linked DNA markers with economically important traits for better conservation and utilization of crop genetic resources. To make the new strategies, we should know the difficulties in procedures of conservation (exploration, collection, documentation, storage, and distribution) and the effectiveness of government policies presently implemented.

2 Categories of PGR and Major Threats in Their Conservation

2.1 Landraces

The landrace is a cultivar originated and evolved in specific eco-geographical area with better adaptations to various stresses due to their genetic heterogeneity (Casañas et al. 2017). Being genetically heterogeneous with better adaptability, landraces were considered an important plant genetic resource not only from a biodiversity point of view but also as a source for nutritional quality and sensory properties. These landraces were majorly conserved by farmers, but modern varieties and hybrids are replacing them for better production. There is a decline in landrace richness (Dyer et al. 2014), genetic variation within-landraces (Trifonova et al. 2021), and cultivated area (Gomes Viana et al. 2020). The replacement of landraces by modern cultivars and hybrids varies from crop to crop and country to country (underdeveloped, developing, and developed countries) depending upon the adoption of modern cultivars by farmers. In India, the landraces of minor crops like small millets, minor pulse crops like moth bean, rice bean, sweet pea, horse gram and major crops like rice are still in cultivation, particularly in remote and tribal areas. These land races possess genetic variation for various traits like seed coat colour, seed size taste, and aroma (Fig. 1). Whereas maize landraces are totally replaced by modern cultivars and hybrids in developed countries like the USA, landraces are still cultivated in Mexico (Brush 2004; Bellon et al. 2018). Apart from replacement by new, improved varieties, the conserved landraces are under the threat of genetic erosion during conservation and multiplication in seed banks. Maintenance and documentation of landraces become very tedious due to their heterogeneous nature and need for a greater size of population and seed sampling to represent the entire genetic variability. Hence, landraces are the most vulnerable to loss of genetic variation during conservation.

2.2 Obsolete Varieties

Obsolete varieties are the improved varieties of the recent past no longer under cultivation. These varieties are maintained as germplasm without any difficulty, as their characterization is already available and they are more genetically uniform than landraces. Apart from maintenance, their detailed characterization and genetic potential need to be evaluated and documented for use in current and future crop improvement programmes.

2.3 Wild Relatives

Wild relatives are nondomesticated species that share a common gene pool with crops and possess important gene sources for various biotic and abiotic stress tolerances. The wild relatives are grouped into three gene pools, viz., a primary gene pool (comprised of easily crossable wild relatives with cultivated species), a secondary gene pool (consisting of crossable wild relatives with cultivated species, but hybrids are usually sterile), and a tertiary gene pool (comprised of wild species which can be crossed with the cultivated species only through the use of modern techniques like embryo rescue and genetic transformation). The wild relatives are mostly conserved in their natural habitats in situ, hence always in danger of loss due to various reasons like construction of roads, dams, canals, industries, mining, and wars. Ex situ conservation of wild relatives is becoming more difficult owing to their adaptation to a particular environment, seed dormancy, and susceptibility to shattering and lodging. Moreover, the wild relatives in nature are always under the process of evolution in response to environmental stresses and accumulate important gene combinations required for survival in changing environments, which can be bypassed if conserved ex situ in seed banks.

2.4 Commercial Varieties

Commercial varieties are developed through scientific breeding methods and possess valuable features like improved yield, uniformity, and resistance to diseases, pests, and abiotic stresses. Due to their genetic uniformity and homozygosity, commercial varieties can be preserved, multiplied, and distributed to users with ease in seed banks and maintained through selfing or inbreeding, as the characteristic features are well documented.

2.5 Advanced Breeding Material/RILs

Breeding material in advanced stages, recombinant inbred lines, and parents of hybrids are important crop genetic resources that need to be conserved and used in future breeding programmes, molecular breeding, developing linkage maps, and identification of gene-specific markers. Genetic and cytogenetic stocks are also important for

Fig. 1 Genetic variation for seed coat colour and size in landraces of mungbean (*Vigna radiata* L. Wilczek)

advanced basic studies, which need to be conserved for future use, as their development is tedious and needs expertise, time, and resources.

3 Strategies to Promote Cultivation of Land Races in Specific Areas

Various strategies should be planned and implemented to promote conservation and utilization of land races, farmer varieties, wild relatives, and other traditional varieties which are used for special rituals, festivals, and as medicine (Fig. 2).

3.1 Promoting Land Races for Organic Farming

Landraces are the result of traditional cultivation practices that did not involve the use of fertilizers or pesticides, making them more suited for organic farming. These landraces of various crops can be promoted for organic farming, a $4 billion-a-year industry growing at an annual rate of 20% worldwide (UN FAO 1996). Organic products derived from landraces, with their superior nutritional qualities, improved packaging, and proper labelling, command higher prices in both national and international markets. Thus, the conservation of landraces can be encouraged by developing a specialized market catering to health-conscious elite consumers.

3.2 Community-Based Seed Banks

Community-based seed banks play a vital role in conserving and maintaining sufficient seed stocks of local varieties preferred by farmers. In India, 97 community-based seed banks are preserving 779 crop varieties across 10 crops, including 69% of landraces (538 indigenous crop varieties/landraces) (Nagubandi et al. 2021). While in Indonesia, community seed banks are managed by 10–20 farmers which incorporate income-generating schemes and training programmes on cultivating endangered traditional rice and other crop varieties (Soetomo 1992). Governments should promote such community seed banks by recognizing their members, providing financial and technical support, and safeguarding their efforts through farmers' rights.

3.3 Detailed Evaluation and Documentation of Medicinal and Nutritional Values of Landraces

Landraces of certain crops are traditionally known for their medicinal and nutritional values. Such traditional varieties need to be scientifically analysed using suitable check varieties to validate their medicinal and nutritional properties and should be registered to get recognition at national and international levels. Landraces with these unique traits can command premium prices in the market, enabling the development of commercial products. As a result, farmers may prefer cultivating such unique landraces despite their

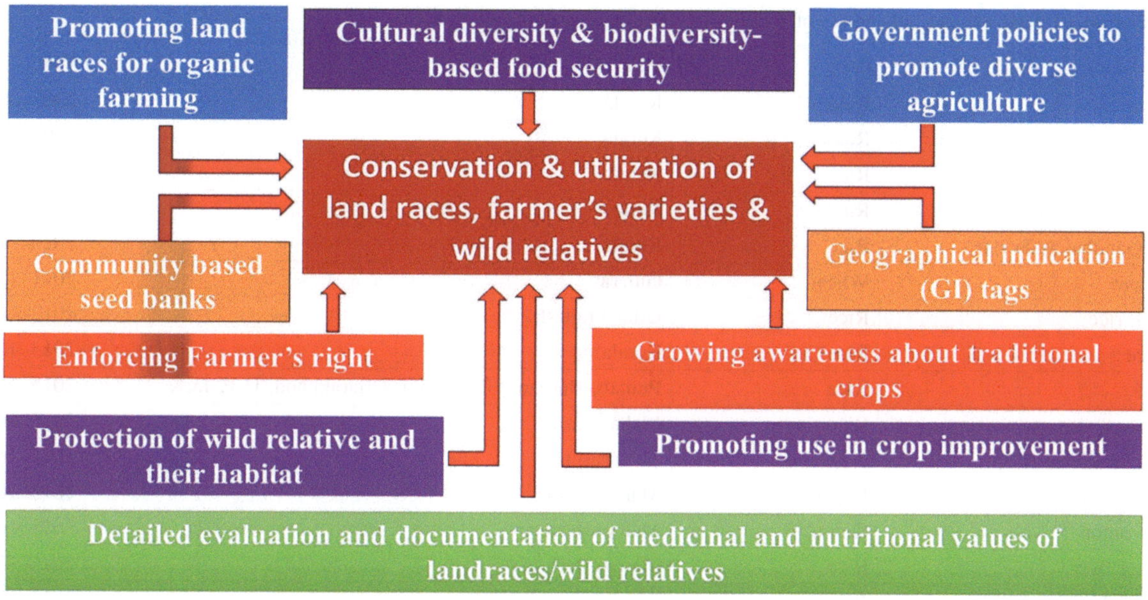

Fig. 2 Strategies to promote conservation and utilization of plant genetic resources

lower productivity compared to improved varieties. A notable recent example is the rice landrace variety 'Sanjivani', released for the state of Chhattisgarh, India. It is a selection from the rice landrace 'Layacha', which is traditionally used as a medicine for healing. More than 350 different phytochemicals were identified which can boost immunity and enhance antioxidant responses (Shalakha et al. 2023). In Peru, native potatoes are sold at higher prices due to their superior taste and prestige as gifts, which helps safeguard farmer's interest in cultivating these landraces (Brush 1991). The concept of farmer's markets selling the product of traditional varieties should be promoted in towns and cities all over the world as seen in farmer's markets in the USA and India (Hyderabad and Pune). Such marketing initiatives directly connect farmers with consumers, increasing demand for produce derived from traditional varieties.

3.4 Growing Awareness About Traditional Crops

In recent years, efforts have been made to raise awareness about the nutritional benefits of pulses and millets. The United Nations declared 2016 as the 'International Year of Pulses' and 2023 as the 'International Year of Millets', promoting the production and consumption of these crops. This initiative ultimately boosted the cultivation of traditional varieties of these previously neglected crops. The concept of 'natural food co-ops' in the U.S. has created a market for quinoa and blue corn which are now incorporated into cereals, chips and tacos. Similar target efforts are needed to promote the conservation, cultivation, and consumption of products derived from landraces.

3.5 Promotion of Land Races Through Geographical Indication (GI) Tags

Geographical indication (GI) is a legally protected name or sign to registered products that resembles to a definite geographical origin. GI protection prevents both direct and indirect commercial misuse and imitation of the product, thus safeguarding the interests of producers and consumers. In India, Geographical Indications of Goods (Registration and Protection) Act, 1999, was executed in 2003. The landraces of various crops, including spices, condiments, fruits, vegetables, and flowers, have been registered and awarded GI tags based on their place of origin or traditional cultivation. Among the GI-tagged crops, rice dominates, with several landraces receiving registration (Table 2). Rice landraces possess unique features like aroma, taste, and nutritional and medicinal value, which are expressed only under specific microclimate or niches. For instance, Ajara Ghansal, a traditional rice landrace of Ajara Tahsil in Kolhapur district of Maharashtra, develops its distinct quality, aroma and taste only in that specific location. After getting GI tag in 2015–16, Ajra Ghansal rice has gained higher market price both nationally and internationally (Desai et al. 2022). Similarly, GI recognition of other landraces of rice (Chinnor, Ambemohar, Boka rice, Jeeraphool, and basmati), wheat, onion, pigeon pea, French bean, and sorghum will encourage their conservation by promoting cultivation among farmers.

Table 2 Registered geographical indications of Indian landraces in various crops

GI	Crop	State	Year
Navara rice	Rice	Kerala	2007–08
Palakkadan Matta rice	Rice	Kerala	2007–08
Pokkali rice	Rice	Kerala	2008–09
Wayanad Jeerakasala rice	Rice	Kerala	2010–11
Wayanad Gandhakasala rice	Rice	Kerala	2010–11
Bhalia wheat	Wheat	Gujarat	2011–12
Kalanamak rice	Rice	Uttar Pradesh	2013–14
Kaipad rice	Rice	Kerala	2013–14
Basmati	Rice	Punjab, Haryana, HP, Delhi, Uttarakhand, U P, J&K	2015–16
Ajara Ghansal rice	Rice	Maharashtra	2015–16
Mangalwedha Jowar	Sorghum	Maharashtra	2015–16
Navapur Tur Dal	Pigeon pea	Maharashtra	2015–16
Lasalgaon onion	Onion	Maharashtra	2015–16
Ambemohar rice	Rice	Maharashtra	2016–17
Joha rice of Assam	Rice	Assam	2016–17
Gobindobhog rice	Rice	West Bengal	2017–18
Tulaipanji rice	Rice	West Bengal	2017–18
Katarni rice	Rice	Bihar	2017–18
Boka Chaul	Rice	Assam	2018–19
Jeeraphool	Rice	Chhattisgarh	2018–19
Chokuwa rice of Assam	Rice	Assam	2019–20
Chak-Hao	Rice	Manipur, Nagaland	2019–20
Balaghat Chinnor	Rice	Madhya Pradesh	2021–22
Munsyari Razma	French bean	Uttarakhand	2021–22
Alibag white onion	Onion	Maharashtra	2022–23
Attappady Thuvara	Pigeon pea	Kerala	2022–23
Tandur Redgram	Pigeon pea	Telangana	2022–23
Nagri Dubraj	Rice	Chhattisgarh	2022–23
Sharbati Gehu	Wheat	Madhya Pradesh	2022–23
Adamchini Chawal	Rice	Uttar Pradesh	2022–23
Bhandara Chinoor rice	Rice	Maharashtra	2023–24
Mushqbudji rice	Rice	Jammu and Kashmir	2023–24
Marcha rice	Rice	Bihar	2023–24
Bhaderwah Rajmash	French bean	Jammu and Kashmir	2023–24
Arunachal Pradesh Khaw Tai	Rice	Arunachal Pradesh	2023–24
Uttarakhand Lal Chawal	Rice	Uttarakhand	2023–24
Uttarakhand Kala Bhat	Rice	Uttarakhand	2023–24
Uttarakhand Pahari Toor Dal	Pigeon pea	Uttarakhand	2023–24
Kalonunia rice	Rice	West Bengal	2023–24
Koraput Kalajeera rice	Rice	Odisha	2023–24
Arunachal Pradesh Angnyat	Millet	Arunachal Pradesh	2023–24
Bundelkhand Kathiya Gehu	Wheat	UP, MP	2023–24
Borsuri Tur Dal	Pigeon pea	Maharashtra	2023–24
Dagdi Jowar of Jalna	Sorghum	Maharashtra	2023–24

4 Conserving Minor Crops Through Preserving Cultural Diversity and Biodiversity-Based Food Security

Some traditional varieties of minor crops are an integral part of local cuisines and are essential for cultural and religious rituals. As a result, they have been conserved for centuries in remote and tribal parts throughout the world. Cultural heritage and agriculture have been closely interwoven since ancient times, as farming was the primary means of livelihood. In Andean culture, different potato varieties are used in various rituals and culinary preparations, such as sancochado (boiled), huatia (over cooked, roasted), chuno (freeze-dried), and *moraya* (freeze-dried and washed in a stream) across different communities (Ortega 1997). In India, the aromatic rice landrace 'Vishnubhog' is traditionally used to prepare sweets offered to deities during Hindu festivals. Similarly, Mexican cuisine features 15 different dishes that require 15 distinct maize varieties (Brush 1991). Cultural diversity plays a crucial role in conserving such unique crop genetic resources, which otherwise would have faced extinction. Hence, promoting cultural diversity and traditional cuisines through systematic efforts is essential. Diversity-based food security enhances resilience against natural calamities in a changing environment. A diverse agricultural system provides farmers and consumers with greater options, reducing risks when specific crops fail due to adverse climatic conditions. Ensuring future food security requires biodiversity-focused policies encouraging traditional varieties, such as the European Union's 'Farm to Fork' strategy (EU biodiversity strategy for 2030).

5 Increasing the Importance of Crop Genetic Resources through Promoting Their Use in Crop Improvement

Conservation without utilization holds little meaning, as so much efforts, time, and resources are invested in conserving crop genetic resources. Therefore, the utilization of genetic variability for crop improvement is the most effective approach (Mooney 1992). A major challenge in utilizing germplasm for breeding is the limited seed distribution network. This can be addressed by strengthening seed production and distribution network. Coordinated efforts among scientists, NGOs, and local farmers are needed to ensure the availability of adapted landrace seeds in local markets. Farmers should also be trained in seed production programmes to maintain genetic purity for farmer-to-farmer distribution. Participatory breeding programme is another way of promoting germplasm exchange and knowledge sharing between scientists and farmers. This process leads to better-adapted varieties, as they are developed under target environmental conditions and tailored to user needs. Furthermore, farmer's right should be imposed to ensure the benefit sharing from the use of crop genetic resources, and recognizing the contributions of those who have conserved important germplasm (Halewood et al. 2020).

6 Government Policies to Promote Diverse Agriculture

Government policies should not discourage the cultivation of traditional crop varieties in favour of improved varieties. Instead, government programmes should offer equal credit, crop loans, and insurance, for both landrace cultivation and the adoption of new, improved varieties. Contract farming for traditional crops should be encouraged to create export potential for neglected crops with unique traits, such as quality, medicinal, and nutritional value. In addition, the traditional seed exchange between farmers should be encouraged by exempting these activities from government regulations related to variety release, seed certification, and plant breeder's rights.

7 Strategies to Protect Wild Relative and Their Habitat for in Situ Conservation

Wild genetic resources are mostly maintained in protected areas ensuring maximum genetic variability. However, their conservation is not prioritized due to their limited utility in crop improvement programmes. To develop better strategies, it is necessary to understand the basic requirements that can improve the efficiency of present germplasm conservation efforts in this protected area. The International Board for Plant Genetic Resources (IBPGR) has identified some of the paucities in the present in situ conservation practices, such as lack of adequate inventories and documentation, the small population size, which complicates preservation of allelic diversity and even the existence of the species, and the failure to consider minimum conservation requirements to maintain intra-specific diversity. Additionally, limited monitoring often is insufficient to guarantee effective conservation (IBPGR 1984). To address these issues, the protected areas should have well-defined conservation objectives for specific wild species or groups of wild species. This would enable the documentation and monitoring of existing genetic variability, ensuring an adequate population size and the proper area required for their efficient conservation without losing allelic diversity in endangered species. The molecular characterization based on whole-genome sequencing, and the identification of latent traits in wild species can generate interest among plant breeders for the utilization of wild germplasm in crop breeding programmes. The seeds of wild germplasm and genomic resources should be made easily reachable to users on a pay basis, with the revenue generated being reinvested into the conservation of wild germplasm. The free transfer of genes from wild to cultivated species is one of the major obstacles in the use of wild relatives. This can be addressed by intensifying research on embryo rescue, marker-assisted selection, and transgenic approaches to transfer economically valuable genes from wild to cultivated species without any linkage drag. The strategies should be implemented by prioritizing germplasm conservation, its utilization, and the sharing of benefits with the surrounding communities of protected areas.

8 Mitigating the Genetic Erosion During Ex Situ Conservation

Ex situ conservation involves botanical gardens, gene banks, seed banks, and cryopreservation, where genetic erosion can occur at various stages of germplasm conservation. The collection of germplasm is a primary step, and loss of genetic variation may happen if seed samples are collected insufficiently or in a non-scientific manner from the source natural population. It is advisable to collect fewer seeds from more sites rather than collecting more seeds from just a few sites, as this ensures a wider representation of genetic variability. Before collection, eco-geographic, sociological, and ethnological planning should identify the most desirable range of locations to sample. The size of the collected sample should be sufficiently large to store seeds for long periods without the need for regeneration before storage. The frequency of regeneration should be kept minimum to avoid or reduce the loss of alleles or the entire accession. The samples should be large enough for storage in both base and back-up collections, and germination tests should be conducted to ensure that the seeds meet minimum standards for storage. To minimize risks due to natural disasters, duplicate samples should be shared with both national and international seed banks, accompanied by comprehensive documentation and characterization. This will ensure the regaining of the complete collection in instance of total losses at the site of origin.

9 Gene Bank Infrastructure and Their Safety

The world is witnessing severe natural and man-made disasters at the beginning of twenty first century, incurring heavy economic losses and loss of life, with gene banks being no exception. Periodic assessment of gene bank infrastructure, their locations and the potential threats posed by earthquakes, cyclones, floods, and wars is crucial to minimize risks. For example, Nicaragua's earthquake, Honduras's hurricane, Philippines' typhoon and instances of civil unrest and (e.g. the civil war in Syria and Iraq, or the Israel-Palestine-Lebanon war) have all affected gene bank facilities. Notable examples of risk management include the successful transfer of 1400 valuable Iraqi varieties in 2003 to ICARDA in Syria, followed by relocation of ICARDA to Lebanon due to civil war and deposition of valuable germplasm for safety backup in other gene banks or the Svalbard Global Seed Vault to ensure safety backup, preventing a complete collapse of the gene bank in such crises (Westengen et al. 2013).

10 Svalbard Global Seed Vault

Conservation of germplasm is a costly affair due to the energy and manpower requirements needed for maintenance of infrastructure facilities such as cold storage and cryopreservation for long-term storage. The Svalbard Global Seed Vault in Norway is one of the global initiatives to conserve the germplasm collections by utilizing natural cold conditions (temperature ranging from − 30 to − 40 °C) which minimizes the energy requirements, thereby lowering expenditure. Further, the Vault is situated near the North Pole, making it relatively far from man-made disasters. This global resource is owned and run by the Ministry of Agriculture and Food on behalf of the Kingdom of Norway. Preparation and shipment of seeds from developing countries to the facility is funded by the Global Crop Diversity Trust. The Seed Vault acts as an insurance against both incremental and catastrophic losses of crop diversity from various gene banks running around the world. It functions as a backup gene bank, ensuring food security for future generations.

11 Use of Modern and Throughput Approaches for Documentation and Characterization of Germplasm

Modern and throughput approaches like plant phenomics, whole-genome sequencing, and marker-assisted selections could be utilized for documentation, maintenance, and characterization of germplasm at DNA level which can boost utilization and thereby conservation of PGR (Fig. 3).

11.1 Use of Molecular Markers in Maintenance of Germplasm

The maintenance of germplasm is crucial to maintain the genetic purity of each germplasm line, particularly the accessions possessing special traits such as disease resistance and quality traits. Genetic erosion of these genes or

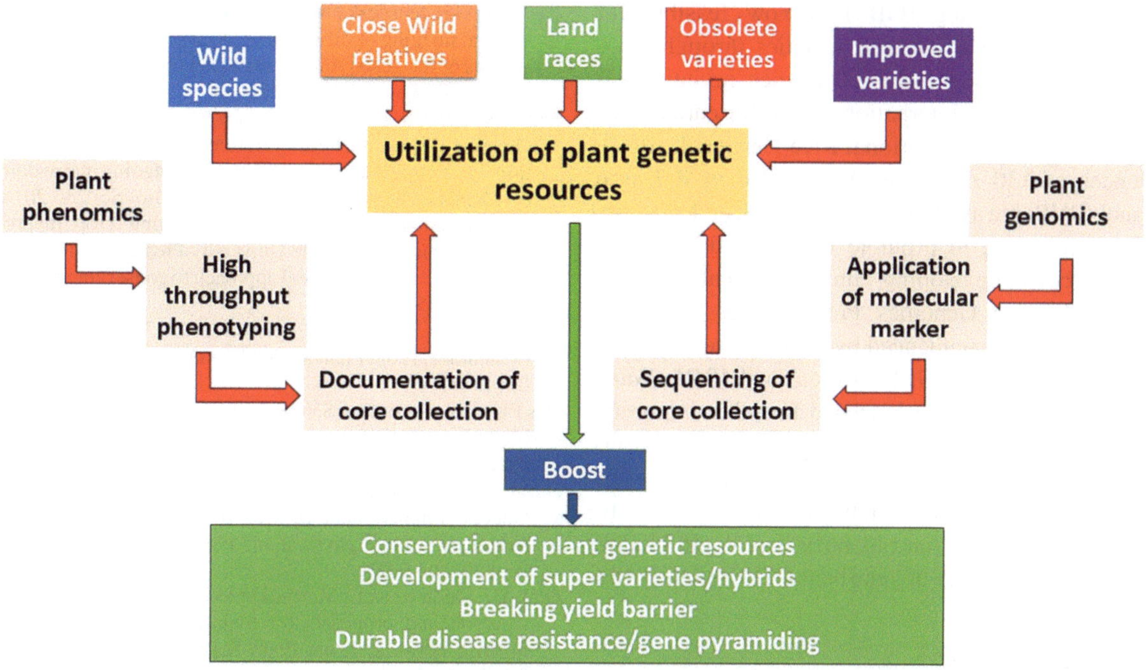

Fig. 3 Application of advance plant genomics and phenomics methods to boost conservation and utilization of plant genetic resources

alleles can occur during the regeneration process. Maintaining such germplasm lines becomes increasingly challenging, as screening a large collection for disease resistance is not always achievable due to the lack of adequate disease screening facilities. Flaws during maintenance and regeneration can lead to the loss of disease resistance genes due to natural mutations, outcrossing (particularly in cross-pollinated crops), and mechanical mixtures. In such scenarios, marker-assisted maintenance and regeneration of germplasm can be a game-changer. With the availability of flanking markers, tightly linked SSR and SNP markers, as well as genic markers associated with genes governing disease resistance and quality traits, it becomes possible to precisely track and preserve these valuable traits in germplasm collections.

11.2 Phenotyping and Method of Screening for Disease Reactions

Phenotyping a large number of accessions is a crucial but challenging task in gene banks, influencing the potential utilization of these resources. Plant phenomics, an emerging field in high-throughput technologies, is poised to overcome the bottleneck of phenotyping by enabling the precise mapping of complex traits. Convolutional Neural Networks (CNNs) is found to be useful in the detection of disease severity in infested plants (Sandhu et al. 2021). CNNs can be integrated into mobile-based apps for detecting foliar diseases like powdery mildew (Ramcharan et al. 2019). These models can segment diseased leaf images at the pixel level, achieving 96%-pixel accuracy (Lin et al. 2019). Various imaging techniques have been used for disease detection, including mobile cameras for *Cercospora* leaf spot (Hallau et al. 2017), UAV/Drone imaging for bacterial leaf blight in rice (Das et al. 2015), and imaging using robots for stripe rust resistance in wheat (Zhang et al. 2019). Additionally, techniques like RGB imaging for powdery mildew (Lin et al. 2019), hyperspectral and thermal imaging for *Alternaria* leaf spot (Baranowski et al. 2015), and hyperspectral imaging for wheat rust disease (Ashourloo et al. 2014), have been successfully used for phenotyping disease reactions. Automated high-throughput phenotyping platforms enable large-scale population evaluations, enhancing selection intensity and precision (Araus et al. 2018). These throughput approaches not only save time and resources but also improve data quality via precise phenotyping.

11.3 Sequencing of Core Collection to Promote the Utilization of Crop Genetic Resources

Genomic resources are becoming as important as crop genetic resources in accelerating genetic gains in major cereal, oilseed, and pulse crops. Whole-genome sequencing databases are available for almost all major crops, with abundance of genetic resources like SSRs and SNPs that can be utilized for tagging agronomically important genes.

Genotyping by sequencing (GBS) is now feasible for large core collections and released varieties through next-generation sequencing approaches. One such initiative has been implemented by the International Rice Research Institute (IRRI) through the International Rice Genebank (IRG). The sequences of 130 rice varieties included in Released Variety Panel (RVP) are available on the SNP-Seek Database, along with their seeds, to promote the utilization of rice genomic resources in developing climate-resilient varieties. Under the 3000 Rice Genomes Project, over 3000 rice genomes (3K-RG) were sequenced by the Chinese Academy of Agricultural Sciences, BGI-Shenzhen, and IRRI (Wang et al. 2018). Such initiatives are needed for other important crops to emphasize the importance of germplasm conservation. Genotyping of collected germplasm can also be utilized to identify duplicates, establish varietal limit, estimate population variability, and identify haplotypic blocks, thus enhancing the efficiency of gene banks (Díez et al. 2018).

12 Conclusion

The conservation of PGR is elemental to enduring agricultural productivity, ensuring food security, and mitigating the adverse effects of climate change. Despite significant advancements in conservation techniques, challenges such as genetic erosion, habitat loss, and the total displacement of traditional varieties with modern high-yielding cultivars threaten the genetic diversity of crops. It is essential to have an inclusive and combined approach of in situ and ex situ conservation methods for preserving the valuable PGR. Strategies such as community-based seed banks, promotion of landraces through GI tags, and application of advanced biotechnological tools can enhance conservation efforts. Moreover, the active participation of farmers, researchers, policymakers, and international collaborations is crucial in safeguarding PGR for future generations. Strengthening conservation policies, narrowing funding gaps, and leveraging modern genomic and phenotyping technologies will help maximize the utilization of PGRs in crop improvement. In conclusion, the sustainable conservation and utilization of PGRs will secure global food production and preserve the biodiversity that underpins resilient and adaptive agricultural systems.

References

Araus JL, Kefauver SC, Zaman-Allah M et al (2018) Translating high-throughput phenotyping into genetic gain. Trends Plant Sci 23:451–466

Ashourloo D, Mobasheri M, Huete A (2014) Evaluating the effect of different wheat rust disease symptoms on vegetation indices using hyperspectral measurements. Remote Sens (Basel) 6:5107–5123. https://doi.org/10.3390/rs6065107

Baranowski P, Jedryczka M, Mazurek W et al (2015) Hyperspectral and thermal imaging of oilseed rape (*Brassica napus*) response to fungal species of the genus *Alternaria*. PLoS One 10:e0122913. https://doi.org/10.1371/journal.pone.0122913

Bellon MR, Mastretta-Yanes A, Ponce-Mendoza A, Ortiz-Santamaría D, Oliveros-Galindo O, Perales H, Acevedo F, Sarukhan J (2018) Evolutionary and food supply implications of ongoing maize domestication by Mexican campesinos. Proc R Soc B Biol Sci 285:20181049

Brush SB (2004) Farmers' bounty: locating crop diversity in the contemporary world. Yale University Press, New Haven

Brush S (1991) A farmer-based approach to conserving crop germplasm. Econ Bot 45:153–165

Casañas F, Simó J, Casals J, Prohens J (2017) Toward an evolved concept of landrace. Front Plant Sci 8:145. https://doi.org/10.3389/fpls.2017.00145

Das PK, Laxman B, Rao SVCK et al (2015) Monitoring of bacterial leaf blight in rice using ground-based hyperspectral and LISS IV satellite data in Kurnool, Andhra Pradesh. India Int J Pest Manag 61:359–368

Desai S, Jadhav A, Ramteke A, Dhole V, Bapat V, Gaikwad N (2022) Genetic improvement of two Indian non-basmati aromatic rice landraces through physical and chemical mutagenesis. Int J Radiat Biol 98(1):83–89

Díez MJ, De la Rosa L, Martín I et al (2018) Plant gene banks: present situation and proposals for their improvement. The case of the Spanish network. Front Plant Sci 9:1794. https://doi.org/10.3389/fpls.2018.01794

Dyer GA, López-Feldman A, Yúnez-Naude A, Taylor JE (2014) Genetic erosion in maize's center of origin. Proc Natl Acad Sci USA 111:14094–14099

Ferrero-Serrano Á, Cantos C, Assmann SM (2019) The role of dwarfing traits in historical and modern agriculture with a focus on rice. Cold Spring Harb Perspect Biol 11(11):a034645. https://doi.org/10.1101/cshperspect.a034645

Gomes VJP, Pires CJ, Bajay MM et al (2020) Do the importations of crop products affect the genetic diversity from landraces? A study case in garlic (*Allium sativum* L.). Genet Resour Crop Evol 68:1199–1211

Halewood M, de Souza F, Dias B, Nnadozie K, Noriega I, Toledo A (2020) Including access and benefit sharing in the post-2020 global biodiversity framework. Technical Report, CGIAR, Rome, Italy

Hallau L, Neumann M, Klatt B et al (2017) Automated identification of sugar beet diseases using smartphones. Plant Pathol 67:399–410

IBPGR (1984) IBPGR regional Committee for Southeast Asia, five year plan of action, 1985–1989. International Board for Plant Genetic resources, Rome

ITPGRFA & FAO (2013) International treaty of plant genetic resources for food and agriculture. ITPGRFA & FAO, p 68. Available at: ftp://ftp.fao.org/docrep/fao/011/i0510e/i0510e.pdf

Khoury CK, Stephen B, Denise E, Costich et al. 2021 (2022) Crop genetic erosion: understanding and responding to loss of crop diversity. New Phytol 233:84–118. https://doi.org/10.1111/nph.17733

Lin K, Gong L, Huang Y et al (2019) Deep learning-based segmentation and quantification of cucumber powdery mildew using convolutional neural network. Front Plant Sci 10:155. https://doi.org/10.3389/fpls.2019.00155

Mooney PR (1992) Towards a folk revolution. In: Cooper D, Vellve R, Hobbelink H (eds) Growing diversity: genetic resources and local food security. Intermediate Technology Publications, London, pp 125–145

Nagubandi U, Siripurapu K, Laxmi B (2021) Mapping of the community-based seed banks in India. https://doi.org/10.13140/RG.2.2.24475.39200

Ortega R (1997) Peruvian in-situ conservation of Andean crops. In: Maxted N, Ford-Lloyd BV, Hawkes JG (eds) Plant genetic conservation: the in-situ approach. Chapman & Hall, London, pp 302–314

Ramcharan A, McCloskey P, Baranowski K et al (2019) A mobile based deep learning model for cassava disease diagnosis. Front Plant Sci 10:272

Sandhu KS, Mihalyov PD, Lewien MJ et al (2021) Combining genomic and phenomic information for predicting grain protein content and grain yield in spring wheat. Front Plant Sci 12:613300. https://doi.org/10.3389/fpls.2021.613300

Shalakha J, Chauhan A, Thada A et al (2023) Bioactive compounds in Layacha brown rice (*Oryza sativa* L.) improve immune responses in mice via activation of transcription factor Nrf2. Food Biosci 53:102785

Soetomo D (1992) Sowing community seed banks in Indonesia. In: Cooper D, Vellve R, Hobbelink H (eds) Growing diversity: genetic resources and local food security. Intermediate Technology Publications, London, pp 30–37

Trifonova AA, Dedova LV, Zuev EV, Goncharov NP, Kudryavtsev AM (2021) Comparative analysis of the gene pool structure of *Triticum aethiopicum* wheat accessions conserved ex situ and recollected in fields after 85 years. Biodivers Conserv 30:329–342

United Nations, Food and Agriculture Organization (1996) The state of the world's plant genetic resources for food and agriculture background documentation prepared for the international technical conference on plant genetic resources Leipzig, Germany, June 17–23 1996

Wang W, Ramil M, Zhiqiang H et al (2018) Genomic variation in 3,010 diverse accessions of Asian cultivated rice. Nature 557:43–49. https://doi.org/10.1038/s41586-018-0063-9

Westengen OT, Jeppson S, Guarino L (2013) Global ex-situ crop diversity conservation and the Svalbard global seed vault: assessing the current status. PLoS One 8(5):e64146. https://doi.org/10.1371/journal.pone.0064146

Zhang C, Pumphrey MO, Zhou J et al (2019) Development of an automated high-throughput phenotyping system for wheat evaluation in a controlled environment. Trans ASABE 62:61–74. https://doi.org/10.13031/trans.12856

Role of Seed Banks in the Conservation of Plant Diversity and Ecological Restoration

Anjani Kumar

Abstract

The numerous significant traits of plant species are improved by the incorporation of desired genes from germplasm. This germplasm is pronounced by several terms such as plant genetic resource, genetic resource, gene pool, and genetic stock. Therefore, conservation as well as preservation of these useful genetic stocks is a very important task in crop improvement programmes. However, the chief losses of this genetic material such as germplasm are possible because of gradual loss in it. The meaning of genetic erosion is the reduction of genetic diversity from various crop species, wild species, and wild relative in gene banks as well as seed banks. The conservation of germplasm or crop genetic resources is achieved mainly by two processes, viz. in situ conservation and ex situ conservation. A convenient procedure for conservation of a particular germplasm or crop genetic resource is a seed bank because of its significance. Seed bank is a place where plants are conserved as seeds known as seed bank. A seed bank comes under the classification of gene bank; consequently, the major role of a seed bank is to preserve genetic diversity to maintain ecological balance. The preservation of germplasm or crop genetic resources are appropriate in the form of seed because it requires smaller area than storage in the form of whole plant. There are currently more than a thousand seed banks that exist all over the world, depending on their type and size. The main objective for maintaining the seed bank is successful sustainable management and conservation of biodiversity through plant breeders developing disease-resistant, nutritional, drought-tolerant varieties in various crop species. Consequently, seed banks and gene bank have played a dynamic approach to the supporting in conservation as well as sustainable use of plants. This book chapter discusses every aspect related to storage of seed in gene bank or seed banks for the appropriate execution of genetic diversity and sustains biological balance.

Keywords

Seed bank · Gene bank · DNA bank · Conservation · Diversity

1 Introduction

The various crop species have the ability to survive for many decades in the form of seed in dry and cold surroundings. Due to this amazing capacity of survival, plants tolerate every type of surroundings and enable extensive dispersal of seed during different time and space as well as increasing the possibility of seed viability as well as creation of baby plant. The capacity of various seeds from different crops to preserve in storage conditions has been effectively done through seed bank. Gene banks have the ability to protect seed from being threatened by genetic vulnerability and help in the safeguarding with exchange of genetic stocks around the world. It has been reported that around 60 countries face severe genetic vulnerability, especially with *Phaseolus* and *Sechium* species, which are major problems for genetic erosion (FAO 2010). These banks are actual vital resources for the protection or preservation of numerous cultivated and wild species that are disturbed through various factors such as unbalanced ecology, pollution, overexploitation, alien species, and climate alteration. The main importance of seed banking over the in situ types of conservation such as gene sanctuaries is their cost-effective nature and it gives crucial resource to help endangered crop species to maintain ecological restoration. At present, the destruction rates of many

A. Kumar (✉)
Department of Genetics and Plant Breeding, Agriculture College Garhwa, BAU, Ranchi, Jharkhand, India
e-mail: anjani0039@gmail.com

N. Kumar (ed.), *Biotechnology in Plant Conservation*, Sustainable Landscape Planning and Natural Resources Management,
https://doi.org/10.1007/978-3-032-02605-7_5

crops and animal species are much higher than in the recent geographical past, leading to extinction—permanently loss of all species in their natural surroundings. A seed bank is an area where different crop species are conserved or preserved outside their natural home; therefore, it comes under general kind of ex situ type of conservation. Seed banks have special facilities that conserve the seed according to their required temperature. These processes help maintain the viability of different kinds of stored seeds and provide material for future research purpose. It has been reported that the conservation of domesticated crop species in seed bank has been practiced since ancient times (Plucknett et al. 1987). Various organizations conserve agricultural seed bank and maintain collection of nondomesticated crops, such as wild species and wild relative, including United States National Plant Germplasm System. It has been observed that various botanical gardens have initiated the construction of new seed banks for storing diverse crop-related germplasm. The best example for the creation of new botanical garden is Royal Botanic Gardens, Kew (Great Britain). Moreover, approximately 700 botanical gardens are available in all over the world to maintain germplasm of numerous wild species of medicinal plants, ornamental plant, and different kinds of crops (FAO 1996). There are many medicinal compounds occurred in agricultural crops for various functions such as providing defence against chronic disease. Therefore, medicinal plants are storeroom for secondary metabolites used in a wide range of therapeutic compounds, perfumeries, dyeing, and for enhancing flavour in many food products (Kumar et al. 2024). The major benefit of seed bank is the conservation of wild species alongside cultivated crop species, which can be directly used by humans for the purpose of fodder, medicinal plants, and trees. Nowadays, the concept of seed bank is more popular for the storage of various kinds of wild plant species which is supported by various regions. The first region for adopting this seed bank is to provide several valuable products like plant sample which is used by scientist and biologist for gaining different assets likewise innovative bases of remedies and nourishment. The second region for adopting this seed bank is to help in the scientific study of numerous wild species, because in these resources, some valuable genes are present that utilized in genetic improvement of crop. The last important role of seed bank is to protect plant genetic resources that are at risk of damage from harmful plant disease and insect pest. The land degradation in all over the world is caused by both anthropogenic activity and climate change; therefore, the existence of human being is decreased; hence, gene bank is very important to sustaining the biodiversity. The germplasm is also known in various terms such as gene pool, plant genetic resources, and genetic stock. Therefore, in gene pool system, the main work is to conserving these genetic stocks in two modes, viz. in situ and ex situ methods. However, the maintenance and storage of crop-related germplasm from outside to their original environment are known as ex situ mode of conservation. The conservation of plant genetic resources of several crops, wild relatives, and wild species in natural environment is called in situ mode of conservation. The selection is the main practice in plant breeding programme; therefore, plant breeders need to utilize agricultural genetic resources from farmers' fields and outside from natural habitat. The preservation, maintenance, and utilization of crop genomic resources are actual hectic tasks in plant breeding programme. Seed bank or gene bank is one of the types of biorepositories. Because it is a biological material repository that gathers, processes, preserves, and issues biospecimens to carry future scientific investigation. Conservation of living plants is done by field gene bank and also done through cutting from the plants as well as in vitro storage. Cryopreservation is also very important type of germplasm conservation, involving the preservation of genetic material at an artificially controlled temperature (− 196 °C) in liquid nitrogen. Seed bank plays a very important role to preserving or maintaining agricultural biodiversity in terms of crops, wild forms, wild relatives, and wild species of numerous plant species. The main work of seed bank is to preserve and conserve biodiversity; therefore, it comes under gene bank. There are so many plants that are used less frequently now, but these were used decades ago by humans; therefore, gene bank recommends an approach to safeguard historic, social, and traditional values. The collected seeds from several agricultural crop species are stored at fixed low temperature with less moisture level that is useful to protecting these materials from loss by several other factors. Therefore, these genetic materials can be conserved through methods such as field collections or in situ conservation. Seed bank works like a seed library that covers some useful information regarding crop species; therefore, this information is used for the creation of hereditarily modified forms of the remaining seeds and developed approaches to encounter crop harshness. When we talk about seed bank, there are approximately one thousand seed banks found in all over the world that depend on different types and sizes. Svalbard Global Seed Vault is the world's biggest and protected seed bank; it is situated in the north, 13 hundred kilometres away from the Arctic Circle, which started by the means of the Norwegian government in February 2008. When we talk about the largest seed bank in India, it is the Indian Seed Vault that is situated on high altitude in Ladakh, India, started in the year 2010. The abovementioned seed bank found in India is the second main seed bank in the world. The longevity and viability of the seed depend on the optimum conditions found in respective seed bank. When we store the seed in low-temperature conditions, seed comes under dormant conditions till resowing procedure. The one important thing about plant is that it can be stored in the form of seed because the size of seed is small and conquers minimum space. The storage of a huge variety of seeds makes this system possible. The conservation and preservation of genetic diversity in the ex situ mode require human activities such as the collection and maintenance of various genetic resources. Several plant species are preserved in their natural form in their living condition at botanical garden as well as in field gene banks for research purposes. The gene bank provides a raw genetic material for advancing the genetic traits of different agricultural crops. Therefore, the achievement of this organization has encouraged plant breeders to initiate breeding work as well as the conservation of germplasm. Currently, there are approximately 1750 gene banks or seed banks around the world that keep exclusive recognition evidence such as taxon, location, and collection periods. It has been reported that nearly 15 thousand to 20 thousand taxa are available in agriculture-based gene banks in the form of base collection and about 45 thousand to 55 thousand taxa are available in preservation-based seed banks. The overall improvement of agricultural crops is possible due to the contribution of various types of crop-related genetic stocks, including farmer varieties, landraces, and different wild species of cultivated plants. It has been revealed that accession collecting from wild species causes several difficulties such as collecting cost as well as the fact that their management in a seed bank is also a very difficult task; therefore, there is a requirement for further necessities other than seed-based conservation approaches (Pence et al. 2011; Walters, 2015; Walters et al. 2018). This book chapter discusses every aspect related to the prospect of seed banks for the preservation as well as the conservation of genetic diversity and for sustaining environmental balance through various ex situ conservation methods. At this point, I have also explained detailed information on the maintenance of crop genetic resources through DNA-based gene banks.

2 The Biological Overview of Seed Banking

Majorly, seeds may be divided into two main classes based on their desiccation tolerance: orthodox as well as recalcitrant. Orthodox seed is desiccation tolerant; however, recalcitrant seed shows desiccation sensitive. If we want to know the actual meaning of desiccation tolerance and desiccation sensitivity, it is the capacity of the seed to survive as well as recover after losing almost all of its cellular water without suffering irreversible damage. The above-mentioned trait (desiccation tolerant) is a common feature of plants and animals that survive in drought-like situations. There are some examples of living organisms that show desiccation tolerance, viz. bryophytes, resurrection plants, and orthodox seeds. The major feature of orthodox seed is the survival under the 7% water level and stands subsequent rehydration deprived of noteworthy harm of seed viability. The durability of orthodox types of seeds increases in expectable way when room temperature as well as the moisture content of the seed is decreased. The important features of orthodox types of seeds are that they may be stored at very low temperature for long durations. Here, I mentioned some important examples of orthodox seeds, such as seeds of maize, rice, wheat, carrots, beets, papaya, gram, lentils, soybeans, cotton, sunflowers, and all members of brassica family. Therefore, these desiccation-tolerant seeds have the ability to store seeds in all types of seed banking. Most of the seeds stand at low moisture content; however, others do not survive in water-deficit conditions. However, there are some significant seeds which are sensitive to drying, such as recalcitrant seeds. Although this is seen in every recalcitrant type of seed, these seeds cannot tolerate when internal water of seed is removed and also show a sign of seed viability loss. Traditional seed bank facilities are not suitable for the drying and storage of recalcitrant seeds. There are also some examples of tropical fruits that need to be stored between the temperatures suitable for the orthodox and recalcitrant types; storage at − 20 °C can be damaging. Tropical fruits may be defined as a varied category of fruits that belong to tropical regions, which are hot and humid areas concerning the Tropics of Cancer and Capricorn. These fruits are known for their exotic appearance, rich flavour, and nutritional benefits. Some examples of tropical fruits are mango, cashew, custard apple, papaya, watermelon, avocado, banana, and passion fruit. The desiccation tolerance is developed constantly throughout the development and ripening of orthodox types of seeds on their original base plant as a result of usual terminal occasion in their improvement. Few plant seeds have the capacity to gradually shift from a dehydration-intolerant to a dehydration-accepting condition. Normally, desiccation tolerance in any type of seed is developed during the early and the midpoint in the seed development process. The desiccation tolerance level of seed increases to a supreme level during the last phase of development; essentially, if earlier seeds give out, during this stage, possible seed durability also occurs its highest level. For this reason, it is advised to harvest the seed of different plants for storage, throughout the age of normal dispersal of seed, because this potential seed durability decays (Way, 2003). Most of the crop variety seeds are conserved according to their types, such as orthodox types of storage manners, and these plant species tolerate infrequent or periodic drought, viz. drought, desert, and grassland ecologies. In contrast, recalcitrant seeds are those seeds that prefer moist and hot surroundings where irregular or minor differences in climate conditions occur, such as tropical and aquatic environments. Some seeds fall under both categories, i.e. orthodox and recalcitrant types suitable for equally drought and moist environments. The expectation for the storage of different kinds of seeds, such as orthodox and recalcitrant types, is a challenging job. It has been reported that the capacity of orthodox seed is improved due to their several factors, such as numerous seeded berries, different capsules, several dry planted pods, siliques, and one-seeded fruit. Apart from this, entirely three groups for seed storage characteristics occur in various varieties of fruit types such as pods, capsules, and drupes. In relation to water content, there are some classes of varieties in which seed is picked at around 20% water level or below; this moisture content indicates orthodox nature; nevertheless, not all species that drop out seed at more than 60% moisture level show recalcitrant storage manners. The job of seed collectors is to collect different kinds of crop seeds and assess the quality of seeds because there are several biotic and abiotic factors that deteriorate the quality of seeds in different plant families, such as *Asteraceae* and *Poaceae*. The physical assessment for seed quality of different kinds of seeds is done at enormous gene banks with the help of X-ray machine; this provides a clear-cut image of different parts of seeds. However, a simple cut assessment for the different seeds of crop is typically effective at evaluating all the minutest seeds in the field condition. However, collecting different kinds of seeds is a very hardworking job because the dispersal arrangements of wild crops may be very movable; therefore, watchful seed sampling from at least 50 different plants randomly throughout the population is probable to covering 95% of the non-rare allelic variations existing in the supreme crop genotypes. The one significant advantage of conserved seed is to re-establish the base population, which is disturbed during the time of catastrophic loss, as well as most open-pollinated species, which are also characteristic of the biological diversity of the species. The major task of seed banks is to preserve this uniformity through working to appropriate gathering and duration criteria. Although the common role of a seed bank is to conserve essential data related to the collection as well as regeneration process and collect records from the conducted examination for the seed, testing, research, and proliferation were assumed with the hereditary material.

3 Detailed Analysis of Crop Genetic Resources

Crop genetic resources may be defined in various ways; this is the total genetic material or sum total of allelic constitution found in a particular plant species and wild species, known as crop hereditary resources or plant hereditary resources. There are various terms that are used to denote crop genetic resources, including gene pools, genetic stocks, and germplasm. Plant breeder started his experiment for improving genetic trait of crop by utilizing base material such as gene pool and germplasm. There are certain significant characteristics that occur in plant genetic resources. Genetic stock characterizes the whole genetic differences or variability existing in a plant species. A plant genetic resource consists of farmer variety, improved variety, obsolete variety, breeding lines, and different forms of wild species in various agricultural plants. Genetic stock is also pronounced for crop hereditary resources and it includes both cultivated plant and noncultivated plant that occurred in mountains, valley, etc. The collection of valuable genes in the form of genetic stocks is carried out from farmers' fields, different seed companies, valleys, mountains, forests, primary centre of origin, secondary centre of origin, natural park, etc. Genetic resources are the ample homes of valuable genes for beginning a crop improvement programme. Genetic resources may be indigenous, which are collected within the country, or exotic, which are gathered from other countries, such as outside India.

3.1 Different Kinds of Crop Genetic Resources

The different kinds of crop genetic resources consist of innumerable genotypic resources for agricultural plant species. I have mentioned here some very important kinds of plant resources which are mostly applied by respective plant breeder for the crop improvement. These crucial crop genetic resources are varieties that are grown by farmer to their field from very old times; such kind of varieties is popularly known as landraces. In addition to the above represented different kinds of varieties, there are also different categories of plant genetic resources available in different locations, such as obsolete genotypes, numerous mutant germplasm, and homozygous lines. Therefore, the above-depicted numerous types of crop genetic resources are also discussed below for clear understanding about the different terms because of right decision taken during the time of crop improvement. The diagrammatic representations of different kinds of genetic stocks are depicted in Fig. 1 for clear understanding.

3.1.1 Farmers Variety

Farmer's varieties are those varieties which are cultivated by the farmer to fulfil their regular requirement from very old time's generation to generations. These crucial varieties are not selected by any respective plant breeder as well as not adopting any advanced plant breeding methods for genetic improvement of this farmer varieties. The contribution for the evolvement of these kinds of genotypes totally goes to farmer. These varieties are also popularly called as landraces and have a broad genetic base; therefore, these varieties are capable of tolerating every kind of environment. Hence, it has very significant genes in their genome that is used in plant breeding programme for genetic advancement of numerous crop genetic species. However, a key limitation of these varieties is that they do not hybridize as easily as advanced breeding lines. Because of that, these materials are

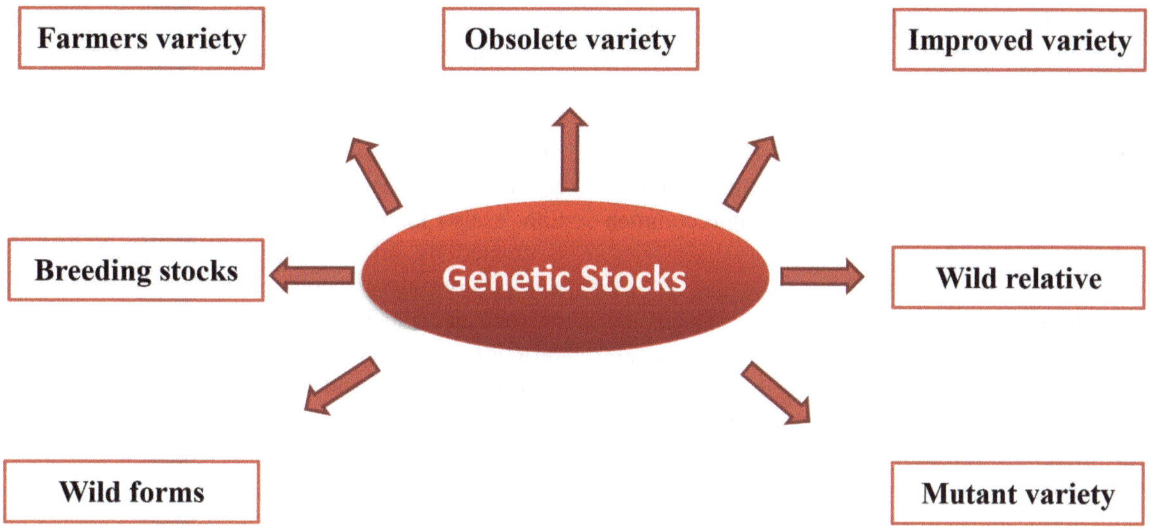

Fig. 1 Classification of genetic stocks

applied after pre-breeding procedures. Apart from this, the other very important features of farmer's varieties are genetic diversity. It has lots of variations seen in the population of farmers' varieties due to this genetic diversity; it has capacity to tolerate against biotic and abiotic factors. Farmers' varieties come from the process of existence of agriculture carried out through the farmers. The major demerits of these types of varieties are not phenotypically similar and also very low productivity.

N. I. Vavilov was the first person who collected farmers' varieties in crop paddy. The credit for the discovery of primary centre of origin goes to N. I. Vavilov. Therefore, this is all about the landraces or farmers genotypes which could help in any genetic advancement of different traits in agricultural plants.

3.1.2 Obsolete Variety

These varieties are basically outdated varieties cultivated by the farmers in very old times. Although these varieties are also produced from a systematic plant breeding procedures, these outdated cultivars are currently not in use. Nowadays, the main reason for not cultivating these kinds of varieties is the availability of many improved varieties. Currently, there are lots of improved varieties of different crop species replaced the obsolete cultivar. There are many systematic plant breeding methods that exist to increase the overall character of plants. There are numerous kinds of desirable genes that are present in obsolete kinds of genotypes because this makes popular and constitutes important part of genetic stock. The whole K-series varieties (K65, K68, and PB 591) of wheat are popular during the time of release but these are replaced by adopting high-yielding Mexican dwarf wheat. Recently, these above-mentioned varieties are no longer cultivated because of the existence of semidwarf high-yielding varieties. The only use of these varieties is for wheat-based research programme for the enhancement of superior grain.

3.1.3 Improved Variety

These are varieties which produced through particular systematic plant breeding procedures. The productivity of improved or modern varieties is higher as compared to obsolete and landrace cultivar. I have mentioned here one significant example of improved varieties, which is improved BPT-5204. This cultivar is produced by converting the BPT-5204 into improved BPT-5204 with the help of marker-assisted backcross method. The above-mentioned variety (improved BPT-5204) is resistant to bacterial leaf blight. Bacterial leaf blight is a major disease of rice and causes severe loss in productivity of rice grain. The phenotypic appearance of this improved variety is homogeneous or similar to obsolete and landrace cultivar. Improved varieties are used as a parent during hybridization procedures for transferring many desirable genes into the genomes of target genotypes. Because of that reason, these cultivars come under the base collection and also constitute an important part of gene pool. One main demerit of these genotypes is the narrow genetic base population, and due to this, the adaptability of this cultivar is less as compared to other kind of varieties.

3.1.4 Breeding Stocks

Breeding stocks are those materials which are not grown by farmers; basically, these cultivars are under breeding procedure. Although they are pre-released varieties produced from plant breeder, they contain beneficial genes in their genome. Therefore, they may be applied as a source of plant genetic resources. Breeding stocks consist of many types of genotypes such as mutant genotypes, lines developed from biotechnology, genetically modified crops, and inbred lines, which are the main categories of breeding stock. Therefore, the above information confirms that this breeding stock is not ready for supply to the farmer. However, these materials are tolerant to biotic and abiotic stress and are therefore considered during the time of breeding recommendations.

3.1.5 Wild Form Genotypes

There are many wild form genotypes that exist for several agricultural plant species. Wild forms are those genotypes which do not share mutual heritage with cultivated crop species but these are naturally founding plant species. Common pedigree record creates the main differences between wild form and wild relative of cultivated plants. Therefore, wild forms are those plants which have common ancestor record with the cultivated crop species; however, wild relative does not have these types of properties. Biotic and abiotic stress properties are the main features of these genotypes, because of that reason, they are highly utilized in plant breeding programme for enhancing stress-tolerant plant. These materials are easily used in hybridization work for gene transfer into cultivated plant. However, numerous wild forms of several cultivated plant species have been lost due to several reasons.

3.1.6 Wild Relative Genotypes

Wild relatives are those genotypes which have common pedigree record with crop plants during the time of evolution. However, wild forms are those genotypes which are not communal heritage with cultivated crop species but these are naturally founding plant species. Apart from this, wild relatives can easily cross with cultivated crops and gene transfer possible. There are lots of valuable genes occurred in their genome, and because of that, these play an important role during the time of hybridization. These materials have wider adaptability because of broad genetic base and rich several biotic and abiotic stress genes occurred in their genome. Wild relative genotypes are also used to increase the quality as well as the yield potential trait for so many important agricultural crops.

3.1.7 Mutant Variety

Mutant varieties are created from special types of plant breeding methods, like mutation breeding approaches. Special breeding approaches (mutation breeding) are generally followed when desired traits are not improved by general breeding procedures. Mutation breeding works are based on various types of mutagens such as physical and chemical mutagens. There are basically two types of mutation, viz. spontaneous mutation and artificial mutation. In case of spontaneous types of mutation, the frequency of mutation in any plant population is very less. Because of that reason, such kind of mutation work is not applied further. Therefore, artificial mutation plays a very vital role in creating genetic variation and producing varieties according to plant breeder's requirements. Artificial mutation is also known as induced mutation because mutation is created by scientist. Hence, these mutant varieties are also part of plant genetic resources or genetic stock and can be used as a source population. Sometimes, these mutant varieties are not considered for release by the state variety release committee and the central variety release committee. Therefore, these materials only come under gene pool system.

4 Germplasm Restoration

Germplasm collections have to be regenerated every few years since seeds of most species drop variability with the storage duration. Germplasm regeneration or rejuvenation consists of growing seeds of the various increases in the field and harvesting fresh seeds for further storage. Generally, an accession is regenerated when its regeneration falls below 95% of its regeneration at the start of storage. The frequency of regeneration depends mainly on the storage conditions and on the concerned crop species; in some crops, like onion and lettuce, it also depends considerably on the varieties involved. Generally, (1) about 50–100 plants should be grown per accession, but some workers may be satisfied with a smaller number and (2) regeneration should be done under climatic conditions similarly, if not identically, to those from which an accession was collected. The above precautions are essential to prevent genetic drift. Genetic drift represents change in the frequencies of genes as well as cultivars of a population exclusively due to change. Therefore, routine germination checks are done every 5–10 years to decide if regeneration of a sample has become necessary. It may be pointed out that ancient types of seeds with decreased capability should increase the frequency of natural mutation. The process of rejuvenation and multiplication should be utilized to eliminate viral and other infectious diseases from the accession.

5 Action Taken During Germplasm Conservation

There are many actions taken during the time of different activities operated for protecting plant genetic resources for use in breeding works. The protection of plant genetic resources or genetic stocks is significantly challenging and

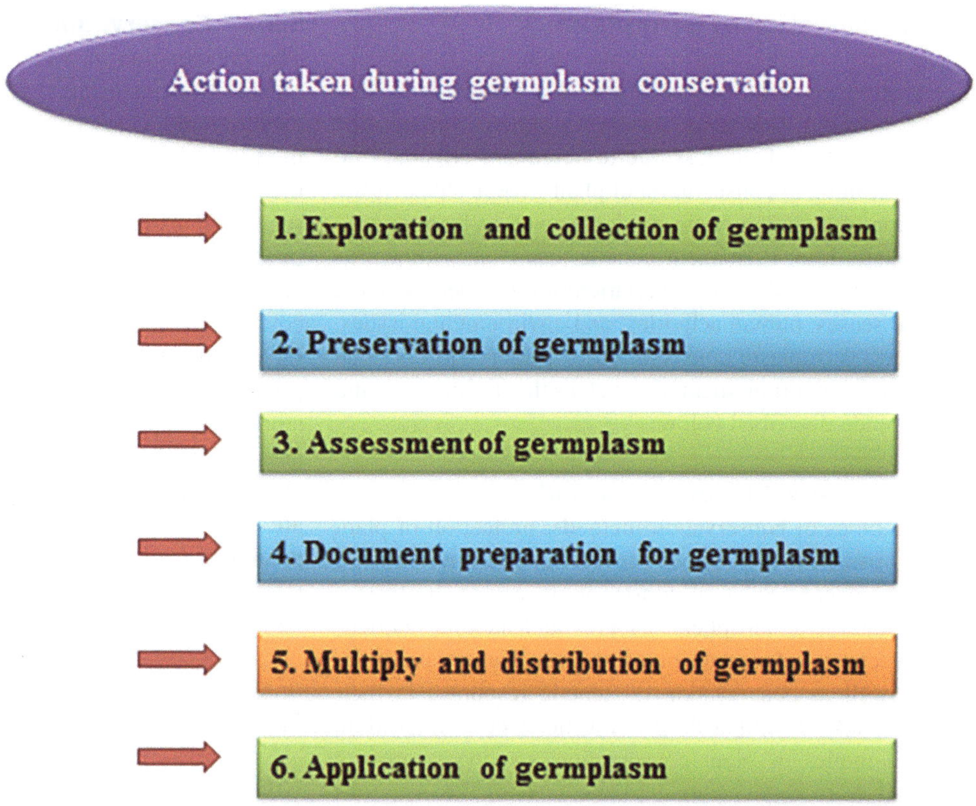

Fig. 2 Activities for collection of plant genetic resources

important task. Scientists are required to give proper concentration during the time of plant genetic resource-related activities. Therefore, some important actions to be taken during the time of plant genetic resources are depicted in Fig. 2, and each of them is described in different subheadings below.

5.1 Exploration and Collection of Germplasm

It is a trip that is organized for the collection of numerous plant genetic resources by plant breeder, selecting all genetic diversity from different sources, as well as storing at one location. Exploration and collection of germplasm are very important procedures because many scientific principles are used during this event. There are various regions where these steps are followed, such as farmers' field, any private seed companies, mountains, and centre of diversity, which are the main places for tapping the genetic variability and putting it into safe places. During the process of collection for germplasm, various steps are kept in mind, viz. collection places, collection for crucial genetic stock, organizations for gathering to genetic stock, collection methods, sampling procedures, and size of different samples.

Advantages and Disadvantages

Here are some very important points for advantages and disadvantages for exploration and gathering of plant genetic resources:

Advantages

1. With the help of collection and exploration, all genetic diversity is stored in a single habitation.
2. It helps protect the plant genetic diversity which is disturbed by hereditary erosion.
3. Occasionally, we acquire some very crucial genetic stocks during the time of collection.
4. It also helps protect genetic stocks from permanent genetic loss.

Disadvantages

1. The important demerit of germplasm collection is that, during the time of collection, the new diseases and insect pests enter in our country with plant genetic resources.
2. Germplasm gathering is a very hectic work.
3. Occasionally, many germplasm collectors are killed by dangerous wild animals during germplasm collection.
4. There are difficulties in transportation during the time of gathering plant genetic resources.

5.2 Preservation of Germplasm

Preservation of plant genetic stock is a very crucial step in seed banking, therefore conserving genetic variation of agricultural plants from loss due to hereditary erosion. The overall meaning of conservation of plant genetic stocks is the protection of it from genetic erosion as well as from permanent genetic loss. Permanent genetic loss is also known as extinction, where all genetic variation of crop species is lost due to diseases, human activities, and environmental issues. Therefore, germplasm conservation or preservation is done through two important methods, such as the in situ mode and ex situ mode described below:

5.2.1 Conservation by in Situ Mode

In these types of conservation methods, protection of plant genetic resources is possible in their natural environment. Therefore, these modes of conservation require the formation of natural or biosphere reserves and national gardens for protecting endangered plant species. Hence, preservation as well as conservation of genetic resources in their original places is called the in situ mode of conservation. In these modes of plant conservation, genotypes are not disturbed from their original places and all the activities are done on their original places. So, the above-mentioned objective is fulfilled by the process of adopting such infrastructure, like gene sanctuary, national garden, and biosphere reserve. A gene sanctuary is an area which is protected from human activities so that the genetic diversity is not disturbed by human interference. This is mainly seen in the primary centre of origin for any crop species. The maximum genetic diversity observed in the primary centre of origin is known as a microcenter. There are many gene sanctuaries that are constructed by the National Bureau of Plant Genetic Resources, New Delhi, for conserving genotypes of citrus, sugarcane, mango, and paddy. In India, the first gene sanctuary area was established in Garo Mountains of Assam for the conservation of citrus wild species. The gene sanctuary for coffee was established in 1984 at Ethiopia. The main merits of gene sanctuaries are as follows: (a) they protect genetic diversity from human activities; (b) they operate natural selection and evolution; (c) chances of genetic diversity loss due to ex situ mode are not operated. Some important demerits regarding gene sanctuary are as follows: (a) overall genetic diversity is not conserved at single place and (b) establishment as well as maintenance of infrastructure for gene sanctuary is a very challenging job.

5.2.2 Conservation by Ex Situ Mode

The preservation as well as conservation of genetic stock outside from its original place is known as conservation by ex situ mode. Conservation of genetic resources through the mode of ex situ pattern has following advantages: (a) in this method, protection of all genotypes, wild relatives, and wild species in a single place, so entire genetic diversity is found in same place; (b) simple maintenance of genetic diversity; and (c) conservation of genotypes by this method is very cost-effective. Therefore, preservation of genetic diversity by means of ex situ conversion using seed is the appropriate approach for storage. There are several reasons for storing seed as a germplasm: it requires very less space and maintenance is easy. Because of that reason, conservation occurs in the form of seed as a germplasm. There are three storage conditions available for conserving genetic resources as seeds: short-term, medium-term, and long-term storage circumstances. There are so many materials used for conserving seed such as plastic tray, glass tray etc.

5.3 Assessment of Germplasm

After the collection of germplasm, the most important activities for plant genetic resources are to evaluate the germplasm in terms of phenotypic as well as genotypic parameters. There are also assessment processes for biological, chemical, and physiological parameters. By the process of assessment of plant genetic resources, we know the actual potential of genotypes in terms of disease-resistant, insect-resistant, and abiotic stress-tolerant genotypes. During the time of evaluation of germplasm, a team of scientists from different specializations of agriculture is required for proper assessment of plant genetic resources. Lists of character parameters are available during the process of evaluation. Currently, there are numerous character parameters are available for analysing genetic stocks for different agricultural crops. The assessments of plant genetic resources are conducted at three different locations, such as farmer's field, greenhouse, and laboratory.

5.4 Document Preparation for Germplasm

This is a very crucial step of plant genetic resources because it requires lots of attention for the collection and well-organized sharing of evidence about genetic stock. In the

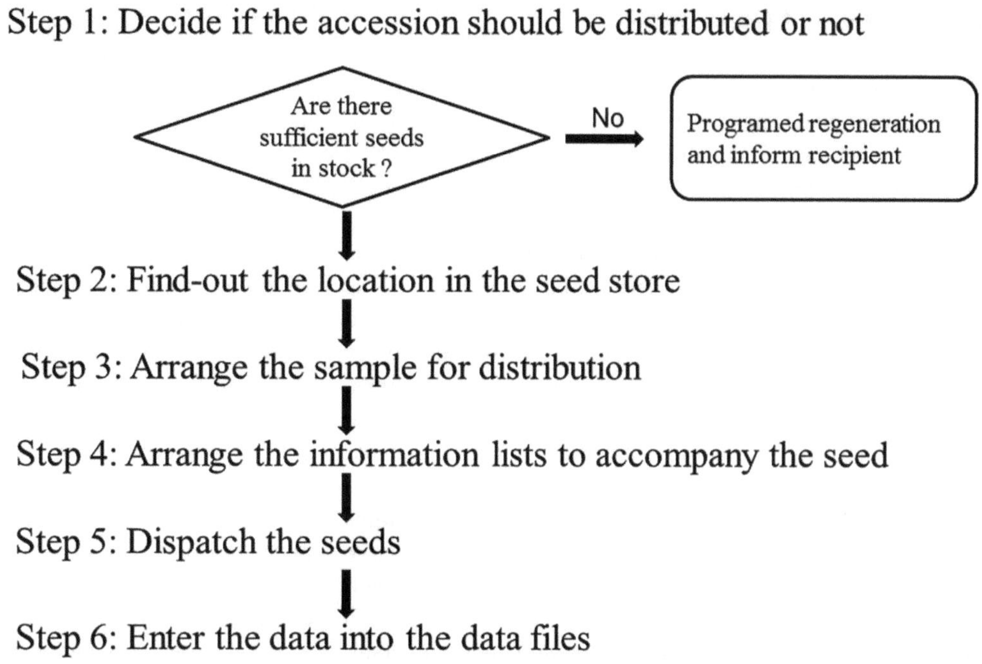

Fig. 3 Flow diagram for the process of germplasm distribution

documentation process, there are so many procedures that are involved, such as identification, collection, storage, management, and sharing of information about plant genetic material. This is based on computerized as well as manual operations conducted for doing above-depicted points. Information system is also used as documentation for plant genetic resources. The small sample sizes of plant genetic resources are handled by the manual process; however, when dealing with a large number of plant accessions, there is a requirement for a well-computerized system.

5.5 Multiply and Distribution of Germplasm

The plant genetic resources are used to improve the genetic trait of various crop species and for the improvement of crop requirement of distribution of genetic material. Gene bank is responsible for multiplication as well as the supply of genetic material to plant breeder for genetic improvement of crop. The plant genetic resources are also supplied to the person who engaged with crop improvement programme. In the process of plant genetic stock distribution, there is a requirement to properly maintain the record for the supply of genetic stock. The distribution of many kinds of plant genetic resources depends on the demand of research worker and respective plant breeder. Plant genetic resources are distributed to scientists from active collection and working collection in the form of seeds. Plant genetic resources are never distributed from the base type of collection. One very important consideration taken during the time of distribution of seed is seed viability and germination percentage. The supply of seed must be in good condition, such as maintaining genetic purity, physical purity, seed free from inert matter, and weed seed. The supply of different kinds of germplasm to the researchers and scientists is depicted in Fig. 3.

5.6 Application of Germplasm

The plant genetic resources or genetic stock are used for the improvement of crop in terms of yield and other traits. The genetic stock of various crop species can be utilized in various ways, such as being used as a cultivar and also applied as parent during the process of hybridization. Plant genetic stock also constitutes major component of gene pool. Plant genetic resources are used to transfer many valuable genes into the genome of crop species. The disease-resistant and insect-resistant genes are also provided by this enormous genetic diversity such as genetic stock.

6 Overview of Gene Bank

It is an organization where genetic materials are stored in active conditions. These banks store genetic materials in various forms, like seed, different plant parts, and DNA, which is the main source of conserving genetic stock. 'Germplasm bank' is another term that can be used instead of 'gene bank'. The large number of genetic diversity is also conserved by means of gene bank. There are various types of gene banks occurred for the protection of large number of genetic resources. Each of the different types of gene banks is discussed in detail under subheadings. Seed gene bank, field gene bank, DNA bank, and stem gene bank are the main categories that fall under gene bank. These organizations are known as biorepositories and used to protect plant genetic resources by storing living things or parts of them, such as seeds. They are used to conserve the hereditary information of threatened species, as well as major crop species and cultivars. These organization works as a very important task to maintain genetic diversity and genetic improvement of crop. Farmers use gene banks to develop new crop varieties that can tolerate changing climates and environments. The collected material of plant genetic resources is generally preserved at 0 °C temperature. In seed gene bank, the plant genetic materials are also stored at − 196 °C during the time of cryopreservation. There are various gene banks constructed for the cultivation of plant genetic resources in living condition, providing constant temperature and also applying artificial nutrient medium. The artificial nutrient medium is provided during the time of preservation of plant genetic resources in the laboratory-based gene bank. There are approximately 80 thousand species and nearly 7.4 million plant genetic resources are conserved at 1750 gene banks and 2500 botanical gardens in all over the world (Anishetty et al. 2016). There are so many reasons why our genetic biodiversity is disturbed; the one of the main reasons is land degradation. Land degradation is one of the very important reasons why genetic diversity is disturbed, and this is created through environmental change and several human-based activities. Therefore, due to the above-mentioned reasons, humankind is also affected by changing climatic conditions. Hence, gene banks are also very essential to maintain ecological balance by conserving several species of plants, and through these procedures, the risk of climatic conditions is reduced. The above different types of mentioned reasons are managed by the process of adopting environmental restoration system. Therefore, ecological restoration is the crucial step for improving our biodiversity in all aspects. The United Nations General Assembly decided to choose United Nations Decade on Ecology Renovation in the years 2021–2030. The convention of biological diversity has aim to conserve the degraded 15% ecology by the year 2020 (Wambugu et al. 2023). The demand for ex situ types of plant conservation has increased because the genetic diversity of several crop species is facing challenges (Godefroid et al. 2010). Seed banks are the very practical organizations used to save numerous plant species. Therefore, these organizations do a very important job in the protection of crop varieties which are naturally extinct (Abeli et al. 2020). This is also reported that the collection of plant genetic resources for conservation in seed banks saves the plant diversity and regains different crop species. The recovery of endangered crop species is possible by restoring in situ-based plant genetic resource conservation (León-Lobos et al. 2012). Gene bank is one of the types of biorepositories because it is a biological material repository that gathers, processes, preserves, and issues biospecimens to carry future scientific investigation. Sometimes, genetic material is stored using cryopreservation-based preservation at very low temperature (− 196 °C) and stored in a non-natural measured nutrient medium. These organizations protect and store the hereditary material of various cultivated genotypes and their wild genotypes for maintaining agricultural biodiversity. A number of important gene banks for various crops in India, as well as international or global gene banks, are mentioned in Tables 1 and 2.

6.1 Requisite for a Gene Bank

Gene banks are maintained so that they can supply viable and useful germplasm to users, viz. breeders, agronomists, entomologists, plant pathologists, geneticists, crop ecologists, soil scientists, production specialists, and growers. In order to be able to achieve this objective, a gene bank must fulfil the following requirements.

(a) The accession maintained in the gene bank should provide comprehensive and representative coverage of germplasm of the concerned species.
(b) It should have dependable and cost-effective preservation facilities.
(c) It should generate sufficient seed stocks for distribution and exchange.
(d) It should publish and make available easy and freely accessible information on the accessions maintained in the bank.
(e) It should enjoy adequate financial and technical support.
(f) The staff of the gene bank should be technically capable and, more particularly, willing to perform their assigned duties.

Table 1 Major gene bank of important crops in India

S. No.	Name of research institute	Name of crop	Place of research institute
1	Directorate of Wheat Research (DWR)	Wheat	Karnal
2	Central Rice Research Institute (CRRI)	Rice	Cuttack
3	Indian Institute of Maize Research (IIMR)	Maize	Ludhiana
4	National Research Centre for Sorghum (NRCS)	Sorghum	Hyderabad
5	Indian Institute for Pulse Research (IIPR)	Pulses	Kanpur
6	National Research Centre for Groundnut (NRCG)	Groundnut	Junagarh
7	National Research Centre for Grapes (NRCG)	Grapes	Pune
8	National Research Centre for Citrus (NRCC)	Citrus	Nagpur
9	Indian Institute for Horticultural Research (IIHR)	Horticultural crops	Bangalore
10	Indian Grassland and Fodder Research Institute (IGFRI)	Forage crops	Jhansi
11	Central Tuber Crop Research Institute (CTRI)	Tobacco	Trivandrum
12	Central Potato Research Institute (CPRI)	Potato	Shimla
13	Central Tobacco Research Institute (CTRI)	Tobacco	Rajahmundry
14	Central Institute for Cotton Research (CICR)	Cotton	Nagpur
15	Sugarcane Breeding Institute (SBI)	Sugarcane	Coimbatore
16	National Research Centre for Soybean (NRCS)	Soybean	Indore
17	Directorate of Oilseed Research (DOR)	Oilseed	Hyderabad
18	Central Plantation Crops Research Institute (CPCRI)	Plantation crops	Kasaragod
19	National Research Centre for Litchi (NRCL)	Litchi	Muzaffarpur
20	Directorate of Medicinal and Aromatic Plants Research (DMAPR)	Therapeutic and fragrant crops	Anand

Table 2 Major gene bank of important crops in the world

S. No.	Name of research institute/centre	Crop species	Place
1	International Rice Research Institute (IRRI)	Rice	Philippines
2	Asian Vegetable Research and Development Centre (AVRDC)	Soybean, mungbean, sweet potato, and vegetables	Taiwan
3	International Centre for Agriculture Research in Dry land Areas (ICARDA)	Durum wheat, beans, and barley	Syria
4	International Crop Research Institute for Semi Arid Tropics (ICRISAT)	Camel crop, bajra, arhar, peanut	India
5	International Potato Centre (CIP)	Potato	Peru
6	International Institute for Tropical Agriculture (IITA)	Lobia, *Glycine* spp., lima bean, cassava, *Ipomoea* spp.	Nigeria
7	International Centre for Tropical Agriculture (CIAT)	Cassava, beans, rice, and maize	Columbia
8	International Centre for Improvement of Wheat and Maize (CIMMYT)	Maize and wheat	Mexico
9	International Council for Research in Agroforestry (ICRAF)	Tree-like cocoa, coffee, rubber, and oil palm	Nairobi
10	United States Department of Agriculture (USDA)	Sugarcane	Florida

6.2 Losses of Genetic Diversity by Genetic Erosion

The losses in plant diversity occur in all the gene banks due to lack or deficiency of one or more of the following: refrigerated store rooms, field space, trained personnel, and administrative and financial support. Repeated regeneration in a small plot also leads to undesirable effects in terms of genetic identity and population structure. Genetic erosion is more frequent in gene banks located in tropical regions. Factors that lead to genetic erosion are as follows: discontinuity in programmes and personnel, human neglect, shifts in programmed direction or methodology of maintenance, poor storage facilities, disappearance of some records, and lack of periodic monitoring of seed viability. Hereditary loss is the reduction of the different species of crops and wild relatives that can happen in gene banks and other populations. Genetic erosion can happen quickly or gradually. Small populations are more susceptible to genetic erosion than larger populations. The different smart techniques adopting for cultivation of agricultural crops are the main reasons for degradation of various alleles of crop species. Therefore, these losses of different alleles of species are possible due to the adopting modern techniques of agriculture and replacing primitive varieties by cultivating high yielding cultivar. The other reasons for degradation of different genotype are due to loss of plant genetic diversity by gradual loss of plant population through the effect of advanced plant breeding procedures. It has been reported that the loss of genetic diversity occurs due to human intervention and takes place at three different levels: plant, variety, and allele (Rogers 2004). The main reason for loss in genetic diversity among crop species is the biotic and abiotic factors and genetic uniformity. The gradual loss of genetic diversity in agricultural crops occurs due to the pressures on the maintainable managing of domesticated genetic stock (Brown and Brubaker 2002). Genetic erosion is mainly possible by the process of human-related activities in very faster frequencies for larger area than spontaneous or natural way of genetic erosion. It has been reported that the association between the size of the plant population and the degradation of plant genetic resources has been well studied and calculated (Wright 1931). Genetic drift is one of the very important phenomena of genetics, in which gene and genotype frequency changes in small population due to improper handling of samples. Larger plant population has low risk for changing gene and genotype frequency by genetic drift. It has been seen that plant population with very small generation times shows loss of plant variations in real period (Frankham et al. 2004). It has been reported that practical research was also carried out for knowing the association of plant population size with variation of genotypes (Ellstrand and Elam 1993). Therefore, this association has been calculated on different species, and the same type of result was found as mentioned above, and numerous reviews show that limited and rare species have less diversity than general plant species (Cole 2003; Gitzendanner and Soltis 2000; Hamrick and Godt 1990). This is very crucial to point out for clear understanding of the association between genetic variation and population size compared to species. Genetic drift is the second main reason which reduces the genetic variation of plant genetic resources. It has been reported that smaller plant populations have a high risk of degradation for plant genetic resources due to inbreeding or selfing (Frankham et al. 2004). Genetic erosion is a very serious problem for plant genetic resources, in which lots of valuable alleles are lost. Therefore, numerous factors are responsible for causing genetic erosion in gene bank. Some very important reasons that cause genetic erosion in ex situ include improper storage condition, error during mixing of seeds, biological alteration of seed, poor management practices, limitation of financial support, accidental damage by human, several biotic and abiotic factors, and a lack of well skilled workers.

6.3 Classification of Gene Bank

Gene banks are collections of plant genetic resources that are used to conserve genetic material, which is useful for maintaining genetic restoration. This infrastructure provides a suitable place where plant genetic resources can be preserved in a viable state. These banks store genetic materials in various forms, such as seed, different plant parts, and DNA, which is the main source of conserving genetic stock. Gene banks are mainly classified under the following headings.

6.3.1 Seed Banks
6.3.2 Field Banks
6.3.3 Meristem Banks
6.3.4 Shoot Tip Banks
6.3.5 Cell and Organelle banks
6.3.6 DNA Banks.

The above revealed different types of gene banks are discussed below.

6.3.1 Seed Banks

Seed bank is a very important type of gene bank where plant genetic resources are stored as seed form. These plant parts are very easy and an appropriate for conservation because they take very smaller place and do not use high maintenance as compared to complete genotypes. Although agricultural crops seeds may not be conserved in very low temperature in seed bank, only orthodox types of seeds are conserved at very low temperature without reducing viability. There are three types of conservation adopted for a seed bank: short-term seed bank, medium-term seed bank, and long-term seed bank. The working collection is generally used for those plant genetic resources which are conserved for 3–5 years at 5 °C. This collection is also known as the active collection. In case of medium-term collection, plant genetic resources are conserved for 10–15 years at 0 °C. The other very crucial type of seed bank is the base collection, in which plant genetic resources are preserved for up to 100 years at − 18 to − 20 °C. The different important merits of seed bank are as follows: (a) Small area or places are required for conservation of large number of plant genetic resources. (b) Handling as well as maintenance of plant genetic resources in seed bank is easy. (c) Neat and clean places are provided for conservation of plant genetic resources; therefore, there is no risk of attack of insect and pest. Out of the advantages of seed bank, there are also some demerits recognized in seed bank. (a) Recalcitrant seeds cannot be conserved in a seed bank because of their perishable nature. (b) Electricity is a major problem in seed bank because after power cut, the viability is lost. (c) Seed viability must be checked time to time. Multiplication of seed is a very important step to get fresh or new seed in a seed bank. The durability and viability of the seed depend on the good facilities of the seed bank. In a seed bank, seeds are stored at low temperature so that dormancy is induced and after that dormancy breaks during resowing. The storage of a huge variety of seeds is made possible through this system.

6.3.2 Field Banks

In these types of germplasm conservation, different plant species are preserved in an area in living condition or standing position. The term 'plant gene bank' is also used in place of 'field gene bank'. This is different from seed gene bank because plants are stored and conserved by sowing plant propagule in particular area and proper agricultural practices adopted for maintaining in living condition. However, this is not the in situ types of germplasm conservation like gene sanctuary, where plant genetic resources are protected by human activities. These types of plant genetic resources come under ex situ types of conservation. In ex situ mode of germplasm conservation, plant genetic stocks are preserved outside from natural habitat. Basically, recalcitrant seed-producing plants as well as those plants which is not producing seeds are conserved by these gene bank. Vegetatively propagated crops are also conserved through the plant gene bank conservation. Generally, plant gene bank is maintained for mainly working collection of alive germplasm for research purpose. Some very important locations around the world for maintaining germplasm by field gene bank are mentioned in Table 3.

The main important merits and demerits of plant gene bank are listed below:

Merits

1. Plant gene bank provides chances for the evaluation of various genetic traits phenotypically.
2. Plant genetic resources in these banks are used directly in plant breeding procedures.

Table 3 Different plant species and their field bank in world

S. No.	Location	Plant species
1	India	Coconut
2	Indonesia	Coconut and different perennial plants
3	Philippines	Banana
4	Malaysia	Oil palm

Demerits

1. It is not conserving the whole plant genetic diversity simultaneously.
2. Plant genetic resources in these banks are high risk of insect pest, and the maintenance of plant genetic resources in these banks is very expensive.

6.3.3 Meristem Banks

In these types of plant genetic resources, vegetatively propagated crops are preserved by the process of meristem gene bank method. Horticultural plants are basically preserved or conserved through the process of these methods. There are two procedures applied for the conservation of plant genetic resources by these methods, viz. slow growth culture and cryopreservation. Cryopreservation is generally used for the preservation of plant genetic resources for long duration at a temperature of -196^0C in liquid nitrogen. However, all the biological and physiological activities of plants are stopped at that temperature.

6.3.4 Shoot Tip Gene Banks

Shoot tip gene bank is basically used for the preservation of apical shoot or growing tips of plant in liquid nitrogen. In these types of conservation, vegetatively propagated plant species are used for the preservation in liquid nitrogen. Shoot tips contain the apical meristem; therefore, they have the potential to generate organs and new cells. Shoot tip gene banks can conserve large amount of plant genetic resources in very small places, which confirms food security. During the process of shoot tip culture, subculturing is done to every 6 months to 3 years. The upper most part of the plants is isolated and then sterilized by surface sterilization and then preserved in liquid nitrogen.

6.3.5 Cell and Organelle Gene Banks

The genetic material is preserved using liquid nitrogen by the process of cell and organelle gene banks. There are numerous types of genetic material used for preservation such as zygotic embryo, somatic embryo, and apex part of shoot. In these types of conservation, endangered plant species are preserved and protected from permanent loss of genetic material. The important advantages of this conservation are storage of plant genetic resources in long duration; however, seeds or tissue cannot be preserved by this process.

6.3.6 DNA Banks

This is most popular types of gene bank due to preservation of hereditary material for long duration. DNA bank is generally used for various types of research purposes such as disease prediction, individual identity, and analysis of genetic disorder. In these banks, isolated DNA molecules are

Table 4 Major heredity bank of the world

S. No.	Heredity bank	Location	Websites
1	Australian Plant DNA Bank, Centre for Plant Conservation Genetics, Southern Cross University	Australia	http://www.dnabank.com.au/index_files/index.php
2	BGBM DNA Bank, Botanic Garden and Botanical Museum	Germany	http://www.bgbm.org/bgbm/research/dna/
3	DNA Bank Brazilian Flora Species, Rio de Janeiro Botanic Garden	Brazil	http://www.jbrj.gov.br/pesquisa/div_molecular/bancodna/index.htm
4	DNA Bank at Kirstenbosch, South African National Biodiversity Institute	South Africa	http://www.nbi.ac.za/research/dnabank.htm
5	IRRI, DNA Bank, International Rice Research Institute	Philippines	http://www.irri.org/GRC/GRChome/Home.htm
6	Missouri Botanic Garden DNA Bank	USA	http://www.mobot.org/MOBOT/research/diversity/dna_banking.htm
7	National Herbarium Netherlands DNA Bank	Netherlands	http://www.nationaalherbarium.nl/taskforcemolecular/dna_bank.htm
8	NIAS DNA Bank, National Institute of Agrobiological Sciences	Japan	http://www.dna.affrc.go.jp/
9	Plant DNA Bank Korea (PDBK), Graduate School of Biotechnology	Korea	http://www.pdbk.korea.ac.kr/
10	Royal Botanic Garden Edinburgh DNA Bank	Scotland	http://www.rbge.org.uk/rbge/web/science/research/
11	Royal Botanic Garden Kew DNA Bank	England	http://www.rbgkew.org.uk/data/dnaBank/
12	TCD DNA Bank, Department of Botany, School of Natural Sciences	Ireland	http://www.dnabank.bot.tcd.ie
13	Tropical Plant DNA Bank, Fairchild Tropical Botanic Garden and Florida International University	USA	http://www.ftg.org/research

Modified and adapted from Hodkinson et al. (2007)

stored by using several types of vectors such as plasmid, cosmid, and phagemid and conserved as pure DNA form. The above-stored DNA material is evaluated, and the desired segment of these molecules is used to produce genetically modified organism. This new technique is very helpful to preserve the endangered plan genetic resources which are highly used to transfer the valuable genes into target genotypes. Although the conservation as well as preservation of plant genetic resources by these types of conservation takes high rates of attention. This method is very costly and sophisticated. DNA bank is manly constructed for preservation of numerous rare plant species which are used to develop resistant genotypes against so many stresses. When we study the botanical prospect of research, the alteration of genomic DNA is regular exercise and causes a large impact on genetics, breeding, molecular biology, and biodiversity assessment. The isolation of genetic material is very easy from most of the plant species and conserved for long duration without any alteration. It has been reported that numerous scientists all over the world adopt to DNA bank for conservation of agricultural crops and also give information about biological research and provide suitable preservation technologies (Kelleher et al. 2005). DNA bank helps to solve various taxonomic complexes of different plant species and calculated plant diversity (Carolan et al. 2006). Recently various agricultural crops have sequenced their genome with the help of pure isolated DNA and genetic variations of different agricultural crops are also evaluated. Therefore, the above work is done in well manner with the help of highly sophisticated molecular biology laboratory. The first genome map was completed in *Arabidopsis thaliana* in the year 2000 and their genome size is 125 mega base pair. The genetic variation of many plant species is evaluated with the help of DNA sequencing. Characterization as well as usefulness of biodiversity is possible due to this novel conservation method. Currently, there is very minimum plant DNA bank available in all over the world. Here, I have mentioned some very important plant DNA bank links that are available for the entire researchers working in the field of biological sciences. Therefore, the useful internet link is provided in Table 4. This is not very hectic list of plant DNA bank, but some crucial lists are mentioned.

7 Conclusion

Seed banks are very significant resources for the preservation or conservation of numerous cultivated and wild species. These crop species, wild species, wild forms, and wild relatives come under crop genetic stocks. The major damage to these crop heredity stocks is due to genetic erosion. Therefore, genetic loss occurs in all the seed banks and gene banks because of deficiencies in refrigerated storerooms, field space, trained personnel, and administrative and financial support. Genetic erosion is the reduction of various important agricultural species in same species in gene bank and populations. Genetic erosion can happen quickly or gradually. Small populations of crop species are facing more genetic loss than larger plant populations. To summarize, this book chapter provides detailed information regarding conservation and different activities of crop hereditary stocks and how these crop hereditary stocks are important to sustaining the biodiversity. Plant breeders need a good plant population where maximum genetic diversity is available. In this chapter, I have also discussed the DNA bank currently highly utilized for the preservation of germplasm in highly sophisticated conditions. In these banks, isolated DNA molecules are stored by using several types of vectors such as plasmid, cosmid, and phagemid and conserved as pure DNA form.

Acknowledgements The author of this chapter is very thankful to Agriculture College Garhwa under Birsa Agricultural University, Ranchi, for support and kind collaboration during the writing of the manuscript.

References

Abeli T, Dalrymple S, Godefroid S, Mondoni A, Müller JV, Rossi G, Orsenigo S (2020) Ex situ collections and their potential for the restoration of extinct plants. Conserv Biol 34(2):303–313

Anishetty NM, Rao VR, Sivaraj N, Babu BS, Sunil N, Varaprasad KS (2016) Gene banks: Management of genetic erosion in Ex Situ collections. Indian J Plant Genet Resour 29(03):268–271

Brown AHD, Brubaker CL (2002) Indicators for sustainable management of plant genetic resources: how well are we doing? In: Managing plant genetic diversity. Proceedings of an international conference, Kuala Lumpur, Malaysia, 12–16 June 2000. CABI Publishing, Wallingford UK, pp 249–262

Carolan JC, Hook IL, Chase MW, Kadereit JW, Hodkinson TR (2006) Phylogenetics of Papaver and related genera based on DNA sequences from ITS nuclear ribosomal DNA and plastid trnL intron and trnL–F intergenic spacers. Ann Bot 98(1):141–155

Cole CT (2003) Genetic variation in rare and common plants. Annu Rev Ecol Evol Syst 34(1):213–237

Ellstrand NC, Elam DR (1993) Population genetic consequences of small population size: implications for plant conservation. Annu Rev Ecol Syst 24(1):217–242

FAO (1996) Report on the state of the world's plant genetic resources for food and agriculture. FAO, Rome

FAO (2010) The second report on the state of the world's plant genetic resources. Food and Agriculture Organization. United Nations. Rome (Italy). Available at: http://www.fao.org/agriculture/crops/core-themes/theme/seeds-pgr/sow/sow2/ country-reports/en/. Accessed 12 Oct 2011

Frankham R, Ballou JD, Briscoe DA (2004) A primer of conservation genetics. Cambridge University Press, Cambridge, UK

Gitzendanner MA, Soltis PS (2000) Patterns of genetic variation in rare and widespread plant congeners. Am J Bot 87(6):783–792

Godefroid S, Van de Vyver A, Vanderborght T (2010) Germination capacity and viability of threatened species collections in seed banks. Biodivers Conserv 19:1365–1383

Hamrick JL, Godt MW (1990) Allozyme diversity in plant species, pp 43–63

Hodkinson TR, Waldren S, Parnell JA, Kelleher CT, Salamin K, Salamin N (2007) DNA banking for plant breeding, biotechnology and biodiversity evaluation. J Plant Res 120:17–29

Kelleher CT, Hodkinson TR, Douglas GC, Kelly DL (2005) Species distinction in Irish populations of *Quercus petraea* and *Q. robur*: morphological versus molecular analyses. Ann Botany 96(7):1237–1246

Kumar A, Bhamini K, Singh RK, Singh DN (2024) Production of plant-derived natural compounds through hairy root culture. In: Biosynthesis of natural products in plants: bioengineering in post-genomics era. Springer Nature Singapore, Singapore, pp 213–239

León-Lobos P, Way M, Aranda PD, Lima-Junior M (2012) The role of ex situ seed banks in the conservation of plant diversity and in ecological restoration in Latin America. Plant Ecolog Divers 5(2):245–258

Pence VC, Plair BL, Charls SM, Clark JR, Taylor DD (2011) Micropropagation, cryopreservation, and outplanting of the Cumberland sandwort *Minuartia cumberlandensis*. J Kentucky Acad Sci 72(2):91–99

Plucknett DL, Smith NJH, Williams JT, Anishetty NM (1987) Genebanks and the World's Food Princeton: Princeton University

Rogers DL (2004) Genetic erosion: no longer just an agricultural issue. Native Plants J 5(2):112–122

Walters C (2015) Gene banking seeds from natural populations. Nat Areas J 35(1):98–105

Walters C, Richards CM, Volk GM (2018) Genebank conservation of germplasm collected from wild species. In: North American crop wild relatives, vol 1. Conservation Strategies, pp 245–280

Wambugu PW, Nyamongo DO, Kirwa EC (2023) Role of seed banks in supporting ecosystem and biodiversity conservation and restoration. Diversity 15(8):896

Way M (2003) Collecting seeds from non-domesticated plants for long-term conservation. In: Smith RD, Dickie JB, Linington SH, Pritchard HW, Probert RJ (eds) Seed conservation: turning science into practice. Royal Botanic Gardens Kew, Kew (UK), pp 163–201

Wright S (1931) Evolution in Mendelian populations. Genetics 16:97–159

Biotechnological Strategies for Rare and Endangered Plant Conservation and Improvements

Shivani Bhutani, Manali Singh, and Nisha Dinkar

Abstract

The several kinds of in vitro instruments and techniques that are employed to preserve the genetic diversity of endangered and uncommon plant species worldwide have been covered in this article. Species that are at risk of going eradicated due to the swift loss of genetic variety and habitat are mostly found in places with limited resources, in underdeveloped regions of the world, in hotspots for biodiversity worldwide, and in island nations. Since only a few tiny populations, or perhaps just a few individuals, of these endemic species are left in the wild, they are special. This compilation of articles explains the direct use of cryopreservation and in vitro techniques, as well as the efforts being made in many places to prevent the extinction of plants. These contributions also offer advice on rare plant propagation, covering methods for large-scale propagation, preservation, and reintroduction.

Keywords

India · In vitro · Flowering plants · Conservation · Endangered

1 Introduction

Even though these floristically significant places have a remarkable concentration of endemic species, they are also facing threats, which increases the prevalence of vulnerable plant species. Therefore, to overcome the limitations of the same and preserve the threatened flora, quick micropropagation methods must be developed (Nagesh et al. 2010). The preservation can be carried out by applying in situ and/or ex situ methods, but due to intrinsic or extrinsic variables, the in situ method may not be enough to preserve such vast bioresources of endangered flora. Therefore, the use of in vitro techniques might be beneficial for the recovery and restoration of plant species that are endangered (Swamy et al. 2018).

Research on genetic diversity is crucial for uncommon and endangered plant species in their natural environments. In order to ensure their long-term survival and maintain their genetic diversity, it will be useful in developing management plans. According to Sreekumar and Renuka (2006), genetic diversity is influenced by a variety of geographical circumstances, which is why it exists in nature as subspecies, races, or ecotypes.

The distribution of plant species has an impact on species evolution in their natural habitat. Both short-term and long-term plant species survival in the wild depend on genetic variety (Schonewald-Cox et al. 1983; Lande 1988).

Growing a population by introducing new populations to replace species that have vanished from the community is essential to preserving the genetic variety of a population. However, a variety of environmental variables can disrupt a plant species' native habitat, which can result in a decrease in genetic diversity.

The evolution of endangered species depends on gene flow to provide adequate genetic variability both within and across populations. Additionally, environmental changes lead to novel selection pressures (Barrett and Kohn 1991).

S. Bhutani
Department of Biotechnology, Invertis University, Bareilly, India

M. Singh (✉)
Department of Life Sciences, Parul Institute of Applied Sciences, Parul University, Vadodara, Gujarat, India
e-mail: manalisingh15@gmail.com

N. Dinkar
Department of Biotechnology, Khandelwal College of Management Science and Technology, Kalapur, Bareilly, Uttar Pradesh, India

N. Kumar (ed.), *Biotechnology in Plant Conservation*, Sustainable Landscape Planning and Natural Resources Management,
https://doi.org/10.1007/978-3-032-02605-7_6

Since genetic drift can cause a decrease in genetic variance, which impacts the survival of natural populations, the loss of alleles leads to a decrease in genetic diversity, which may be caused by declining population sizes (Reed and Frankham 2003). Geographical proximity and ecological similarities may allow for gene flow across populations; nevertheless, low migration levels may result in less genetic variation and some pressure from natural selection.

2 Role of Molecular Marker Technology in Conservation of Diversity

2.1 Biochemical Markers

Isozymes are distinct forms of the same enzyme that perform the same or comparable tasks. They are effective instruments for researching genetic differences in both within and between animal and plant communities (Mondini et al. 2009). Clone identification and clonal structure analysis of plant populations have been accomplished with success using isozyme electrophoresis (Johanson et al. 1996; Lehmann 1997).

2.2 Phytochemical Markers

The development of contemporary methods like nuclear magnetic resonance (NMR) spectroscopy and mass spectrometry (MS) in conjunction with separation techniques made it easier to identify and clarify the structure of molecules. Individual species within a genus can be distinguished using metabolomics (Heinrich 2008; Tianniam et al. 2010), and certain species' regional origins can be distinguished using metabolomics (Kang et al. 2008).

2.3 DNA-Based Markers

These markers are useful for identifying plant species and their cultivars (Kiran et al. 2010). However, by identifying the putative genes that regulate significant plant features, molecular linkage maps are being effectively applied to improve several crop species. Certain DNA markers use polymorphism to measure genetic diversity. Multiple phenotypic variants of a trait occurring simultaneously can be referred to as polymorphism. The following lists the possible markers that are being used to assess the genetic diversity of a plant genome (Fig. 1).

2.3.1 Random Amplified Polymorphic DNA (RAPD)

Due to the primers' arbitrary sequence, RAPDs can be created quickly and easily because the method is cheap and simple to use (Williams et al. 1990).

2.3.2 Simple Sequence Repeat (SSR)

The eukaryotic genome contains these tandemly repeated DNA sequences. DNA polymerase regression during SSR region replication is the cause of this marker's length fluctuation (Levinson and Gutman 1987). It is estimated that this marker occurs once every 6–7 kb in the plant genome (Cardle et al. 2000). Consequently, there is a significant degree of allelic variability in these markers. SSR markers' revelation of allelic polymorphism and codominant nature has yielded comprehensive data on the genetic structure and gene flow of natural plant populations. In eukaryotic genomes where there is a dearth of publicly available DNA sequence data, the genomic library or SSR-enriched library is crucial for locating SSRs. Genetic diversity in a variety of crops, including sloe (*Prunus spinosa* L.), cherry plum (*Prunus cerasifera* Ehrh.), *Prunus domestica* L., *Zea mays*, *Apium graveolens*, and *Psathyrostachys huashanica*, has been evaluated using SSR markers.

2.3.3 Inter Simple Sequence Repeat (ISSR)

The ISSR marker is applied to assess genetic diversity in plant species with low sequence information (Zietkiewicz et al. 1994; Al-Qurainy 2010).

2.3.4 Restriction Fragment Length Polymorphism (RFLP)

Different restriction fragments are produced during digestion due to variations in the enzyme's restriction site, which is the basis for the RFLP principle. Since DNA fragments from every homologous chromosome are recognized, it can identify the coupling phase of DNA molecules because it is a codominant marker.

2.3.5 Amplified Fragment Length Polymorphism (AFLP)

This method works especially well for assessing genetic diversity in plant species for which there is no previous genetic data. It is a modified version of the RFLP that amplifies the restriction fragments produced when restriction endonucleases digest genomic DNA using polymerase chain reaction (PCR). More polymorphism would result from greater variety in the restriction location. But for many plant species, including.

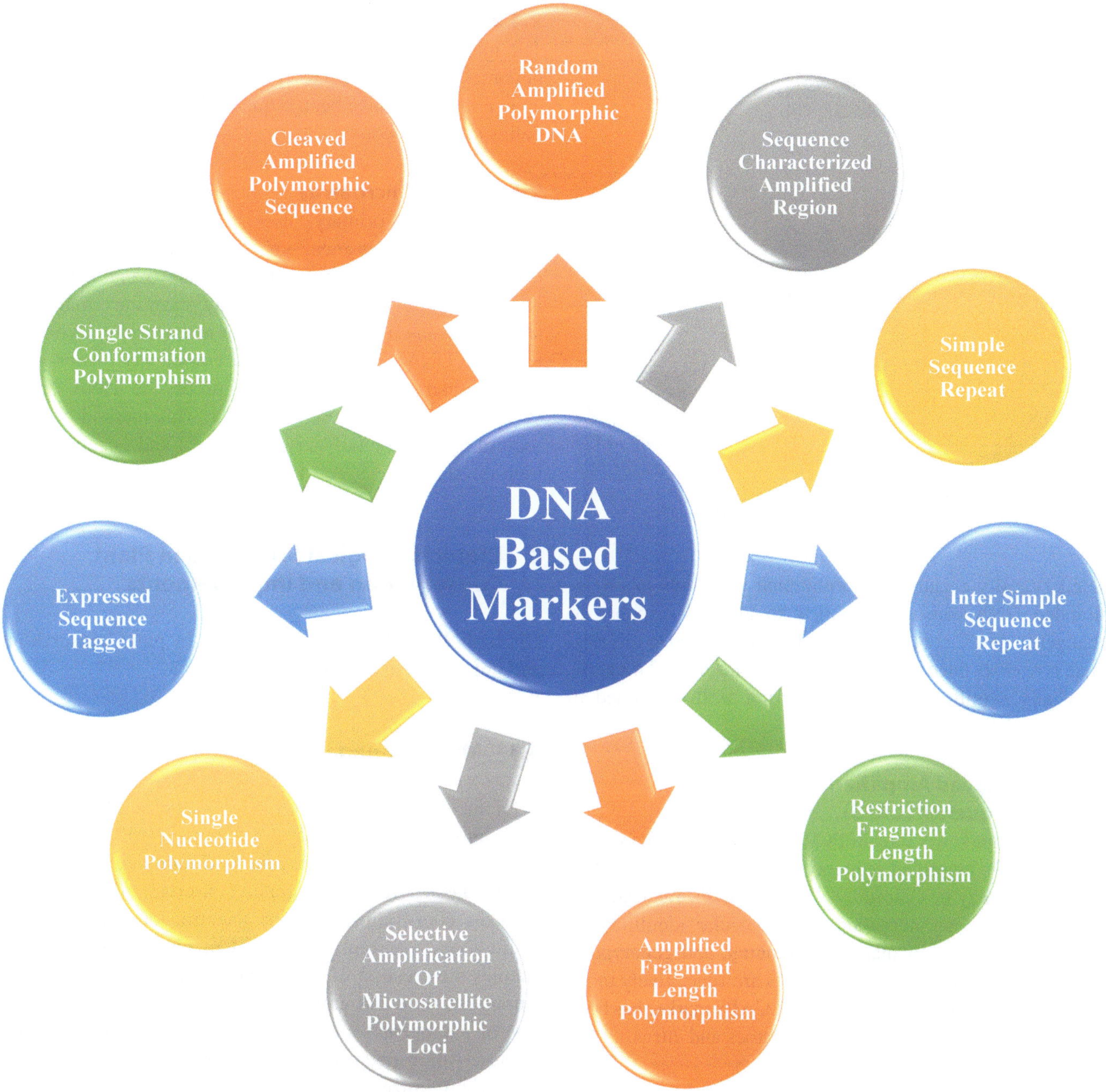

Fig. 1 Different types of DNA-based molecular markers

2.3.6 Selective Amplification of Microsatellite Polymorphic Loci (SAMPL)

It is a variation of the AFLP technique which is applied to the study of eukaryotic genomes' highly different microsatellite regions.

2.3.7 Single Nucleotide Polymorphism (SNP)

The most prevalent variable sequences in the plant genome are single nucleotide polymorphisms (SNPs), which are frequently employed as molecular genetic markers.

2.3.8 Expressed Sequence Tag (EST)

Additionally, an apparently distinct EST aids in the isolation of novel genes pertaining to advantageous agronomic features. Once produced, they can be used to map the synteny of functional genes in different related organisms and to clone particular genes of interest.

2.3.9 Single-Strand Conformation Polymorphism (SSCP)

This method is applied to both polymorphism and point mutation identification. This method makes it simple to identify

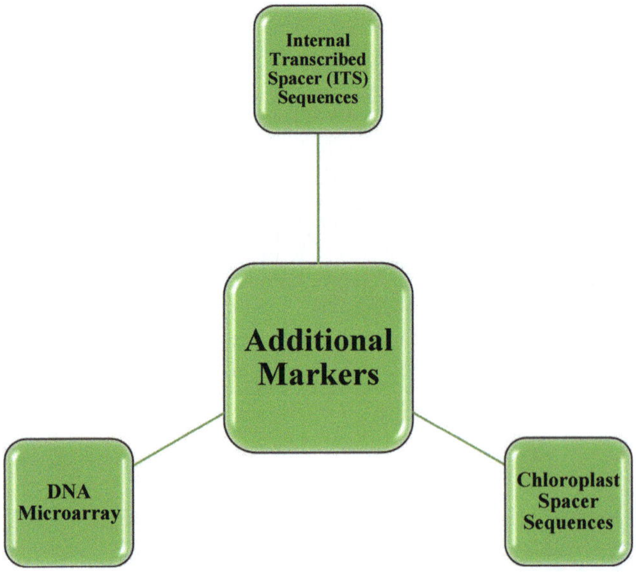

Fig. 2 Different types of additional molecular markers

heterozygosity because conformational variations in single-stranded DNA cause its mobility to alter in response to changes in its GC content. Although it can be used to discriminate progenies once appropriate primers are created for agronomically significant features, it is not well developed in plants.

2.3.10 Cleaved Amplified Polymorphic Sequence (CAPS)

The CAPS marker is often a PCR-based marker where only one of the two amplified sequences has a restriction site; this discrepancy may be caused by an SNP marker. Accordingly, CAPS markers are necessary for molecular identification research, positional or map-based cloning, and genotyping. Similar to sweet potatoes, strawberries are vegetatively propagated polyploid plants. Similarly, the CAPS marker has been used to identify cultivars of sweet potatoes and citrus.

2.4 Additional Markers

Different types of additional markers are presented in Fig. 2.

2.4.1 Internal Transcribed Spacer (ITS) Sequences

Based on sequence variation, ITS markers are used to identify different plant species. These sequences become preferred markers in evolutionary investigations at various taxonomic levels because they evolve at different rates.

2.4.2 Chloroplast Spacer Sequences

The majority of angiosperms have circular chloroplast (cp) genomes that range in size from 120 to 160 kb. 22 of the 132 cp genomes that have been sequenced so far come from date palms and monocots.

2.4.3 DNA Microarray

Currently, the primary uses of microarrays are the simultaneous assessment of gene expression and the discovery and genotyping of mutations using DNA variation analysis. A DNA microarray examines many loci simultaneously using gene chips or microarrays. Hybridization of cDNA or RNA reduces complexity. After using a fluorescently tagged probe to probe the DNA on the slide, polymorphisms are looked for. This method does not require agarose or polyacrylamide gels and is reliable and high throughput.

3 Biotechnological Strategies in Plant Conservation and Improvements

Different Biotechnological strategies in plant conservation and improvements are presented in Fig. 3.

3.1 Micropropagation

A biotechnological method for conserving plant species, including stubborn seeds of commercially significant crops, is plant micropropagation. Combining ex situ and in situ species into an integrated conservation program that involves cooperation between scientific or plant breeding institutions is currently the trend in plant conservation. It enables the creation of many plant species in a constrained amount of time and area. It is used to create cloned plants free of pathogens for use in forestry and agriculture. From a conservation perspective, the micropropagation approach should minimize somatic differences in the regenerated plantlets by minimizing the number of subculturing and axillary bud or shoot tip cultures. Since cells are given successive stimuli for healthy growth during in vitro culture, several plantlets will eventually arise. However, somatic variants can arise as a result of insertion, deletion, or substitution in a cell's gene pool. For micropropagation to be successful, the explants must be carefully chosen. The production of calluses or suspension cell culture typically results in somaclonal differences, which lowers the regenerated plants' economic value.

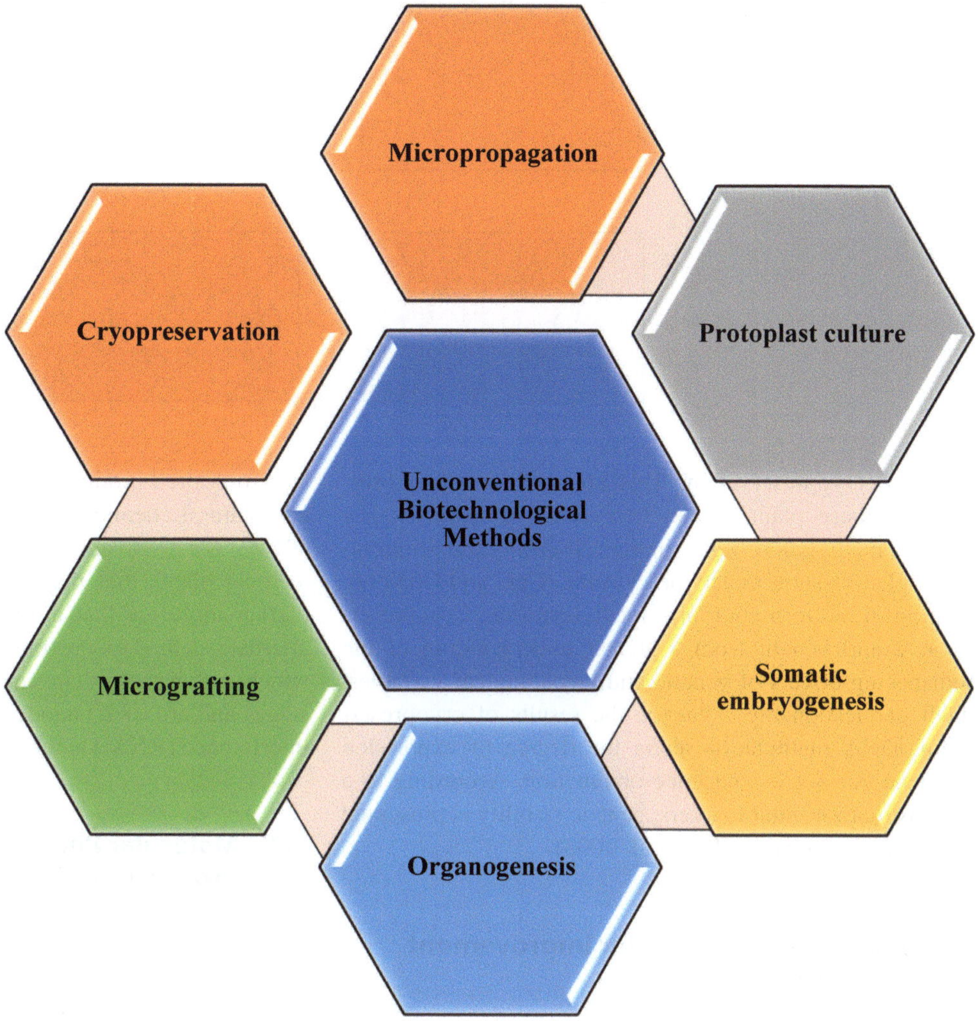

Fig. 3 Different biotechnological strategies in plant conservation and improvements

3.2 Protoplast Culture

In this process, the protoplasts of two plant species are allowed to fuse together with the help of a fusogen which helps in the generation of hybrid plants.

3.3 Somatic Embryogenesis

It is the process by which a single or several cells grow into a somatic embryo. Different steps of somatic embryogenesis are depicted in Fig. 4.

3.4 Organogenesis

The process of developing roots, shoots, and buds with the help of plant growth hormones is called organogenesis. New information about the interactions between auxin and cytokinin during plant organogenesis has recently surfaced, showing a connection between these hormones' distribution and de novo organogenesis.

3.5 Micrografting

The process of cutting and grafting the stem cuttings of two different plants and placing them together by wrapping them in a poly so that no external infection occurs is known as grafting.

3.6 Cryopreservation

For crucial preservation of seeds, this method is employed. These species can spread by grafting or cutting, but they cannot generate actual seeds. For different plant species, different cryopreservation techniques are employed. Because the hyper freezing rate prevents intracellular freezing, the cooling rate is also a crucial component of

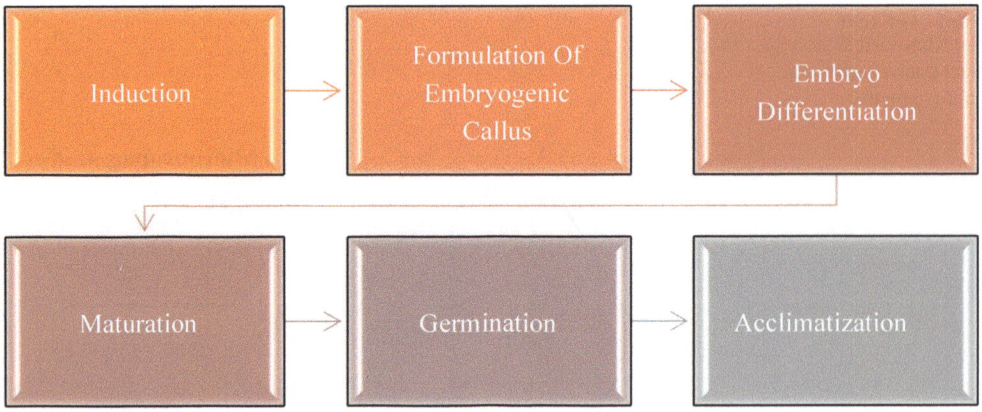

Fig. 4 Steps in somatic embryogenesis

successful cryopreservation. This approach to the preservation of these plant species has certain drawbacks. For example, methylation begins in the cell during the preservation of plant tissues, mostly on cytosine (CpG and CpXpG), resulting in the formation of 5-methylcytosine. Cryopreservation would benefit from a greater understanding of the predisposing effects of genetic and physiological variables, which unquestionably influence the results of cryopreservation. DNA methylation alters the degree of expression of a gene but not its sequence or function. According to a study using molecular markers, genetic stability is preserved under cryopreservation (Liu et al. 2008).

4 Genetic Manipulation for Improvement of Yield and Quality Traits

4.1 *Agrobacterium Tumefaciens*-Mediated Transformation

Plant selection, analysis, multiplication, and improvement all depend on biotechnological technologies (Khan et al. 2009). Combinatorial biosynthesis is a technique used to create bioactive molecules in plants by combining genes from several organisms. Root-inducing plasmid, also known as Ri plasmid with T-DNA, is an infectious plasmid found in *Agrobacterium rhizogenes*. The "hairy root disease" is caused by the plasmid infecting plant cells and causing the formation of hairy structures that resemble roots. In a hormone-free medium, the neoplastic (cancerous) roots that are created after infection have a high mass/medium ratio due to their rapid development rate and genetic and biochemical stability.

4.2 RNAi Technology

By reducing enzyme activity, the RNA interference (RNAi) approach can either channel metabolites into pathways that lead to desired products or prevent the synthesis of undesirable natural chemicals. As of right now, RNA interference (RNAi) is a promising approach that is being used to produce medical plants that may have characteristics for a certain marker molecule that is required to eradicate diseases. An alternate method for preventing the function of these enzymes, which are expressed in various organs and developmental stages and are each encoded by a multigene family, is RNA interference (RNAi) technology.

4.3 Molecular Pharming for High Production of Bioactive Compounds

Molecular pharming is used for the enhanced production of pharmaceutically active metabolites, which are very essential for the formulation of drugs and decoction of potent natural bioactive compounds.

5 Conclusion

Biotechnological methods are essential for the preservation and enhancement of rare and endangered plant species that are valuable both medicinally and economically. The major bottleneck is with the plants which are not available although the season and their conventional method of growth and agriculture is not possible due to seed dormancy or sometimes lack of germplasm. For such crucial plants, biotechnological approaches offer a major breakthrough for plantation and propagation, paving a new way in the field of agriculture.

References

Al-Qurainy F (2010) Application of inter simple sequence repeat (ISSR marker) to detect genotoxic effect of heavy metals on *Eruca sativa* (L) Afr. J Biotechnol 9(4):467–474

Barrett SCH, Kohn J (1991) The genetic and evolutionary consequences of small population size in plant: implications for conservation. In:

Falk D, Holsinger KE (eds) Genetics and conservation of rare plants. Oxford University Press, pp 3–30

Cardle L, Ramsay L, Milbourne D, Macaulay M, Marshall D, Waugh R (2000) Computational and experimental characterization of physically clustered simple sequence repeats in plants. Genetics 156:847–854

Heinrich M (2008) Ethnopharmacy and natural product research multidisciplinary opportunities for research in the metabolomic age. Phytochem Lett 1:1–5

Johanson BO, Jonsdottir IS, Cronbcrg N (1996) Clonal diversity and allozyme variation in populations of the arctic sedge *Carex bigclowii* (Cyperaceae). J Ecol 84:449–459

Kang J, Lee S, Kang S, Kwon HN, Park JH, Kwon SW, Park S (2008) NMR-based metabolomics approach for the differentiation of ginseng (*Panax ginseng*) roots from different origins. Arch Pharm Res 31:330–336

Khan S, Mirja KJ, Tayyab Md, Abdin MZ (2009) RAPD profile for authentication of medicinal plant *Glycirrhiza glabra* L. MAPSB 3(1):48–51

Kiran U, Khan S, Mirza KJ, Ram M, Abdin MZ (2010) SCAR markers: a potential tool for authentication of herbal drugs. Fitoterapia 81:969–976

Lande R (1988) Genetics and demography in biological conservation. Science 241:1455–1460

Lehmann C (1997) Clonal diversity of populations of *Caiamagrostis epigejos* in relation to environmental stress and habitat heterogeneity. Ecography 20:483–490

Levinson G, Gutman GA (1987) Slipped-strand mispairing: a major mechanism for DNA sequence evolution. Mol Biol Evol 4:203–221

Liu YG, Liu LX, Wang L, Gao AY (2008) Determination of genetic stability in surviving apple shoots following cryopreservation by vitrification. CryoLetters 29:7–14

Mondini L, Noorani A, Pagnotta MA (2009) Assessing plant genetic diversity by molecular tools. Diversity 1:19–35

Nagesh KS, Shanthamma C, Pullaiah T (2010) Somatic embryogenesis and plant regeneration from callus cultures of *Curculigo orchioides* Gaertn. Indian J Biotech (9):408–413. Retrieved from: http://nopr.niscpr.res.in/handle/123456789/10440

Reed DH, Frankham R (2003) Correlation between fitness and genetic diversity. Conserv Biol 17(1):230–237

Schonewald-Cox CM, Chambers SM, MacBryde B, Thomas WL (1983) Genetics and conservation: a reference for managing wild animals and plant populations. Benjamin/Cummings, Menlo Park, CA

Sreekumar VB, Renuka C (2006) Assessment of genetic diversity in *Calamus thwaitesii* BECC. (Arecaceae) using RAPD markers. Biochem Syst Ecol 34:397–405

Swamy MK, Paramashivaiah S, Hiremath L, Akhtar MS, Sinniah UR (2018) Micropropagation and conservation of selected endangered anticancer medicinal plants from the Western Ghats of India. In: Anticancer plants: natural products and biotechnological implements, pp 481–505. https://doi.org/10.1007/978-981-10-8064-7_20

Tianniam S, Bamba T, Fukusaki E (2010) Pyrolysis GC–MSbased metabolite fingerprinting for quality evaluation of commercial *Angelica acutiloba* roots. J Biosci Bioeng 109(1):89–93

Williams JGK, Kubelik AR, Livak KJ, Rafalski JA, Tingey SV (1990) Nucl Acids Res 18:6531–6535

Zietkiewicz E, Rafalski A, Labuda D (1994) Genome fingerprinting by simple sequence repeat (SSR)-anchored polymerase chain reaction amplification. Genomic 20:176–183

Plant Genetic Resources Conservation by Cryopreservation

K. Kavinaya Shri, A. Pavithra, Vivek Pazhamalai, S. S. Meenambiga, and S. Ivo Romauld

Abstract

Cryopreservation is one of the significant biotechnological approaches that has conserved plant genetic resources destroyed by habitat loss, climate change, and overexploitation. The method ensures long-term preservation of plant tissues, cells, and seeds at ultra-low temperatures that halt biological degradation and keep genetic materials safe for further use. Cryopreservation is a method that does not assimilate the characteristics of any other conservation methods, for its genetic materials are maintained in a stable form and hence apt to preserve rare and endangered species. It also assists with in situ conservation when it controls diseases and pests and acts as a gene bank, increasing diversity in reproductive techniques. Cryopreservation has recently been significant for sustainable food security by conserving the crop genetic diversity required to adapt to the agricultural challenges posed by climate change. Cryopreservation is one of the critical components for future directions, which will further support the preservation of genetic resources necessary for ecosystem restoration and innovative agriculture. This chapter highlights cryopreservation techniques, applications, and their vital role in plant genetic resource conservation, biodiversity, and sustainable resource management.

Keywords

Cryopreservation techniques · Sustainable agriculture · Genetic stability · Long-term storage of plant materials · Conservation of rare plant species · And food security

K. Kavinaya Shri · A. Pavithra · V. Pazhamalai (✉) · S. S. Meenambiga · S. Ivo Romauld
Department of Bio-engineering, School of Engineering, Vels Institute of Science, Technology and Advanced Studies, Chennai, Tamil Nadu, India
e-mail: vivek.se@velsuniv.ac.in

1 Introduction

Cryopreservation, a technique for preserving biological material at ultra-low temperatures (usually − 196 °C), has emerged as a crucial tool for plant conservation, especially for species that are endangered or resistant to conventional conservation methods (Shankar et al. 2024). In the eighteenth century, this method first appeared. The development of cellular and molecular responses to osmotic stress has made the procedure evolve to date, accompanied by cryoprotective agents that protect the cellular structure during freezing, resulting in much better survival post-thawing of the thawed cells. As a result, long-term plant genetic resources will be provided with an effective survival strategy. This will contribute to species survival under risks such as climate change and habitat destruction (Gubaidullin et al. 2024). The widest applicability is one of the remarkable features of cryopreservation. It has been successfully applied to various plant species, including palms, orchids, and other rare or economically important plants. This makes it versatile and highlights its role in biodiversity conservation. By conserving genetic material, cryopreservation not only preserves unique plant species but also supports research into crop improvement, resilience to environmental stresses, and restoration of ecosystems. It enables the conservationist and the researcher to bank plant material such as seed, embryo, or even shoot tips at ultracold temperatures, so a lot of genetic diversity in crops can be kept as a way of assuring sustainable agriculture and worldwide food security (Shankar et al. 2024).

Despite these and other benefits, cryopreservation is not without problems: technical challenges remain in the form of the necessity for optimization of protocols on diverse species and minimizing cytotoxicities of cryoprotectants. Cryoprotectants are fundamental in preventing ice crystals from forming during freezing, which can be toxic to plant

N. Kumar (ed.), *Biotechnology in Plant Conservation*, Sustainable Landscape Planning and Natural Resources Management,
https://doi.org/10.1007/978-3-032-02605-7_7

cells at high concentrations. Complexity in cryopreservation methods, which require strict regulations on cooling and warming processes, demands further research on innovation. The need, therefore, is to integrate the efforts of the international players in research, conservation, and policy-making to streamline current techniques with less toxic cryoprotectants and to increase ease of access to cryopreservation technology. By overcoming these obstacles, cryopreservation can play an even more prominent role in ensuring the long-term sustainability of plant biodiversity in the face of escalating environmental challenges (Nguyen et al. 2023).

1.1 Overview of Cryopreservation of Plant Conservation

Two major solutions have been established to overcome these issues and ensure biodiversity preservation: in situ and ex situ. The former seeks to protect species in their natural habitats, whereas the latter seeks to preserve species outside their natural habitats. Plant ex situ conservation includes preserving plant specimens in a variety of collections, including botanical gardens, germplasm banks (which preserve seeds, pollen, tissues, and organs), and field collections of vegetatively propagated crop species.

Cryopreservation is one of the most modern methods of ex situ biodiversity protection. The term refers to the process in which biological material is frozen and preserved at below-freezing temperatures. It maintains plant material at very low temperatures, namely liquid at − 196 °C and vapor over liquid at − 150 °C, using liquid nitrogen (LN). There are several advantages of using LN, including low cost, chemical inertness, accessibility, and a high level of independence from power. In cryogenic conditions, the reduction or cessation of kinetic energy and molecular motion in biological systems weakens transport and enzyme processes, leading to a slower aging process (Białoskórska et al. 2024).

However, the cryopreservation procedure has to be adapted to the specific species and type of explant being preserved. This is because an effective mechanism needs to be established for the prevention of ice crystals from accretion in the water in cells. The formation and development of ice crystals intracellularly rupture cell membranes, leading to loss of function and ultimately death. Recent developments in mammalian systems have contributed to the development of cryopreservation technologies for plant cells and organs.

For many species, especially those of economic interest, techniques of cryopreservation have been developed. However, differences in tolerance to cryopreservation have been found, and evidence of this includes the regeneration ability after cryopreservation. Such findings indicate that molecular-level pathways exist for freezing tolerance and metabolic restoration upon thawing (Shankar et al. 2024).

2 Principles of Cryopreservation

Cryopreservation is the preservation of biological materials at temperatures far below the freezing point of water, thereby freezing all of the cellular metabolic activities. This theoretical basis for cryopreservation is essentially vitrification, a process where an otherwise liquid can be effectively turned into glass without even forming ice crystals. This is important because the presence of ice crystals causes mechanical damage to cell membranes and results in cell death. To prevent this, cryoprotectants are used. Cryoprotectants are chemicals that decrease the freezing and supercooling points of water, hence limiting ice crystal formation. DMSO, glycerol, ethylene glycol, and sucrose are examples of cryoprotectants (Nausch and Buyel 2021; Whaley et al. 2021). Such agents must penetrate the cells well and have minimal toxicity for their effectiveness.

There are several steps to the cryopreservation process, which include preparation of the biological material, addition of cryoprotectants, cooling, storage, and thawing. The rate of cooling is very crucial and depends on the type of biological material. The slow-freezing methods cool the material at 0.5–5 °C per minute for controlled ice formation and the least damage (Nausch and Buyel 2021). Vitrification, on the other hand, involves rapid chilling that would avoid ice formation, thereby making it a glass-like solid state with preserved cellular structures. Success in cryopreservation lies in careful optimization of every one of these steps so that the biological material remains viable at the time of thawing (Kim and Popova 2023).

However, further elaborating this would say that cryopreservation is not only an advanced technique for preservation but also a strategy for the maintenance of the integrity of genetic material from samples over long periods of time. The method ensures the survival of cells under extreme conditions of liquid nitrogen storage by incorporating cryoprotectants. The cooling protocol to be used will either be slow freezing or vitrification, depending on the specific requirements of the biological material, further emphasizing that tailored protocols are necessary to ensure maximum effectiveness and success in cryopreservation (Kim and Popova 2023). These principles underscore a critical balance between scientific precision and practical application in the realm of biological conservation.

3 Cryopreservation Techniques for Plant Conservation

Techniques that can be developed include traditional slow freezing, along with cryopreservation procedures. The methodologies used here are vitrification, droplet vitrification, encapsulation dehydration, and cryo-plates (V-Cryo

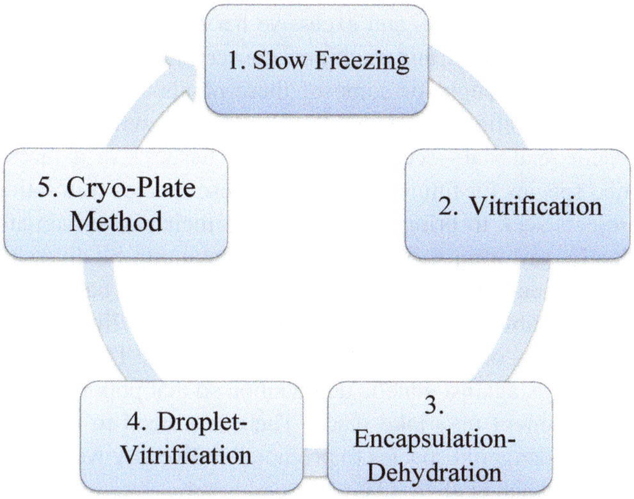

Fig. 1 Cryopreservation techniques

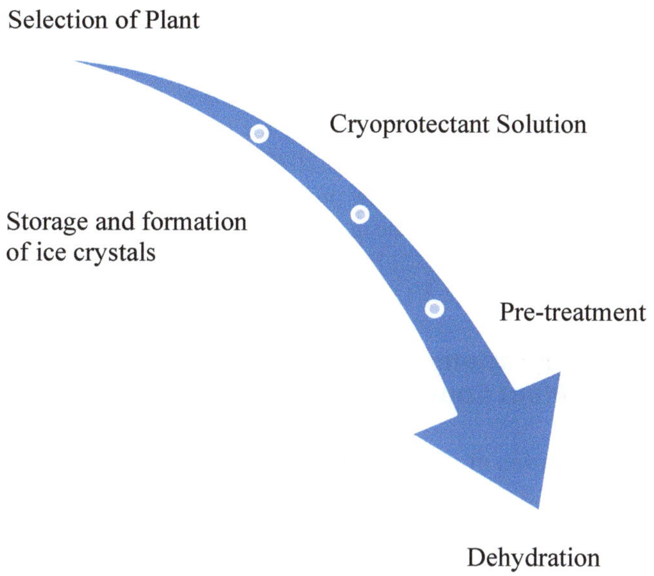

Fig. 2 Vitrification steps

and D-Cryo plates). The classification of cryopreservation techniques for plant conservation is shown in Fig. 1.

Slow Freezing

The method of slow freezing is used to preserve the cells, where the temperature gradually lowers to dehydrate and inhibit intracellular ice from forming. This is usually performed on seeds, somatic embryos, and other tissues but can only be applied to intolerant species of slow cooling. Controlled freezing equipment is employed in the slow cooling process, which is intended to reduce damage to cells from the generation of ice crystals. It applies to species with moderate to high cryotolerance. For slow cooling to be at its most effective, the rate of freezing and the cryoprotectant concentration must be optimized to decrease cryo-injury. There is the presence of cryoprotectants, which include such options as dimethyl sulfoxide and glycerol (Kakaei and Rehman 2023).

Vitrification

Rapid cooling of plant tissues in cryoprotectants. The cooling can be with an ice concentration but without crystalline ice formation; this would mean the production of a glass-like state. Shooting tips, embryos, or refractory seeds are also used as techniques. Depending on the applied cryoprotectant solution and particular pre-treatment, along with the dehydration method. Vitrification has been useful in allowing the long-term preservation of many species, especially those that have refractory seeds and cannot be preserved using conventional methods. Vitrification preserves cell integrity and viability through the rapid freezing of tissues, thereby preventing the formation of ice crystals (Mohammad et al. 2023). The steps for vitrification are shown in Fig. 2.

Encapsulation Dehydration

Encapsulation dehydration is a method whereby plant tissues are encapsulated in alginate beads, dehydrated, and frozen. This technique is well suited to tropical and subtropical species. The encapsulation process provides mechanical protection of the plant tissues, with dehydration reducing osmotic stress at the time of freezing. It has been proven to result in very high rates of regrowth after thawing and is useful in species that have recalcitrant seeds. Alginate beads help in providing structural integrity to the tissues, hence making them more resistant to freeze-thawing processes (Matsumoto 2017; Ruta et al. 2020).

Droplet Vitrification

It is an advanced method wherein plant samples are implanted in microscopic cryoprotectant droplets on aluminum foil strips and rapidly cooled in liquid nitrogen. Such uniform cooling produces very minimal thermal stress to tissues. The technique is appropriate for delicate tissues, like shoot tips and somatic embryos. It can be applied to the conservation of rare crops and endangered species. Droplet vitrification has a higher cooling rate, which is ideal to reach the vitrification phase in the absence of ice crystals (Kakaei and Rehman 2023; Matsumoto 2017).

Cryo-Plate Method

The V-Cryo and D-Cryo plate methods utilize special plates prepared to harbor the plant tissues for preservation using the method of vitrification. This improves cooling homogeneity and efficacy; hence, it can accommodate the preservation at high throughput. The dual-sided D-Cryo plate technique offers one side with a different rate of cooling, thereby ensuring optimal preservation conditions for every tissue. Improvements in cryopreservation technology make it possible to use such technology for more recalcitrant and vegetative species (Ruta et al. 2020). Improved consistency and control of the cooling process have also elevated the success rates of cryo-plate methods for the most difficult plant species (Matsumoto 2017).

4 Applications of Cryopreservation in Plant Conservation

Conservation of Genetic Resources

Cryopreservation is important for conserving plant genetic diversity, which is necessary for ecosystems and agriculture. Genetic material from rare, threatened, or economically important plant species is conserved in cryobanks at very low temperatures, like liquid nitrogen. This technique can help preserve seeds, embryos, and tissues for a long time, saving them from being lost due to habitat destruction, climate change, and other threats. Cryopreservation is particularly useful for plants with recalcitrant seeds that cannot tolerate dehydration, species with low seed viability, and endangered plants that are hard to propagate by conventional methods. The material can afterward be used for research and restoration efforts or reintroductions, thereby conserving biodiversity. Recent reviews place emphasis on the role that cryopreservation plays in the development of gene banks intended to conserve these vital plant resources (Tirado-Perez and Sandoval-Cancino 2022).

Propagation of Vegetatively Propagated Crops

Most of the major crops are reproduced vegetatively, which also implies that they do not produce viable seeds or possess genetic homogeneity through seed multiplication. Vegetative tissues like shoot tips, meristems, or somatic embryos can be cryopreserved to preserve clonal traits of crops. This is very important in conserving crop varieties, which are prone to diseases, pests, or environmental stress. In addition, this ensures the protection of commercially valuable types. The genetic material of such crops can be preserved in cryobanks and revived for breeding, research, or replanting purposes to ensure food security and agricultural diversity. Cryopreservation is a reliable approach to overcome the challenges of vegetative propagation and ensure the genetic integrity of these crops (Tirado-Pérez and Sandoval-Cancino 2022; Engels and Ebert 2021).

Restoration of Threatened Species

Cryopreservation is one of the most fundamental ways to conserve endangered or already extinct plant species. With habitat loss, climate, and excessive harvesting, some plants fall into extinction, and traditional conservation strategies just cannot be enough for some of these plants. This is where cryopreservation comes in, allowing for the saving of viable genetic material—seeds, embryos, or tissues—in a threatened species for future restoration. These species restoration projects seek to bring back species to their native habitats, thereby restoring ecosystems and maintaining biodiversity. Plant regeneration from cryopreserved material has proven quite valuable for species with problematic seeds, limited seed availability, or specialized ecological niches. This method safeguards against genetic degradation so that population re-establishment may take place. There is a need to cryopreserve endangered species to promote biodiversity worldwide (Amankwaah et al. 2022).

5 Role of Cryopreservation in Sustainable Agriculture and Food Security

Cryopreservation is vital for preserving plant genetic resources, ensuring biodiversity, and building resistance against climate change. It promotes sustainable agriculture by allowing crop improvement and guarantees global food security through genetic preservation. The role of cryopreservation is explained below and is shown in Fig. 3.

Conservation of Genetic Diversity

Cryopreservation is thus an important strategy for conserving genetic variety, which is critical in producing resilient crop variants. The preservation of seeds, tissues, and cells from many plant species allows for the preservation of a diverse range of genetic material that can be utilized to produce superior crops that are more resistant to diseases, pests, and other environmental challenges. Genebank ensures that valuable characteristics do not go extinct and can be added when necessary. Crop genetic diversity conservation through cryobanking supports crop engineering aimed at crops that thrive in diverse environmental conditions for a longer period and contribute to a sustainable agricultural industry (Tirado-Perez and Sandoval-Cancino 2022).

Long-Term Seed Storage

Cryopreservation provides an opportunity for long-term storage of seeds without a notable decline in viability. This is especially crucial for crops with short shelf life or seeds that are difficult to store under normal circumstances. Cryopreservation preserves seed viability throughout time, ensuring a steady supply of high-quality seeds to plant in future seasons and thereby contributing to food security. This method also saves seeds for use during disasters or in locations where typical seed preservation methods are not effective (Matsumoto 2017).

Fig. 3 Role of cryopreservation

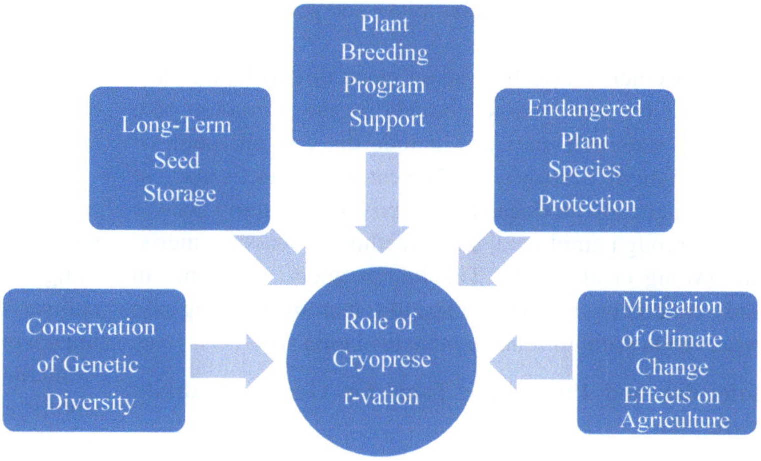

Plant Breeding Program Support

Cryopreservation can greatly facilitate the storage and transfer of germplasm; thus, it is essential for plant breeding programs. The availability of this genetic material allows for a new crop variety with desirable attributes, such as increased productivity, nutritional value, or resistance to abiotic stresses caused by drought and salinity. This speeds up breeding and helps in the creation of crops that can deal with changing environmental conditions. Through providing sound genetic material, cryopreservation allows for the continuation of research and innovation in plant breeding (Engel and Ebert 2021).

Endangered Plant Species Protection

Cryopreservation maintains the genetics of endangered plant species for their survival and reintroduction into natural habitats. This guarantees biodiversity and environmental stability, both being important in sustainable agriculture (Engels and Ebert 2021; Tirado-Pérez and Sandoval-Cancino 2022). The survival of endangered species through cryopreservation also guarantees the availability of resources for future research and restoration, which will increase the resilience of ecosystems and agricultural landscapes.

Mitigation of Climate Change Effects on Agriculture

Cryopreservation has the benefit of minimizing the effects of climate change on agriculture by conserving plant genetic resources that otherwise would be lost due to environmental changes. The researcher can then develop crops to better adapt to future scenarios of climate change with ensured food security in global warming. It ensures readiness and response to challenges facing the agricultural sector brought about by climate change. This proactive approach of conservation allows the supply of food security approach in subsequent generations (Praveen and Sharma 2019).

6 Genetic Stability and Long-Term Storage of Cryopreserved Plant Materials

Cryopreservation is an innovative method for preserving plant genetic resources, in which the viability and genetic integrity of these can be preserved at ultra-low temperatures for any duration.

Genetic Stability in Cryopreservation

Genetic stability ensures that the DNA sequence or epigenetic makeup of an organism will not be altered in cryogenic temperatures. Thus, the functionality and reproductive capabilities of preserved plant material are maintained intact. The metabolic activity in the cells will not be induced at these cryogenic temperatures, and thus, there will be no emergence of mutations, rearrangements, or any other alterations of genetics that may be injurious to the storage material.

The samples of a large number of plant species cryopreserved by using molecular tools, including SSR, AFLP, and RAPD, were proven to retain their genetic integrity. For instance, no detectable differences were revealed from fresh samples during cryopreservation of potato and banana shoot tips, thereby reiterating the efficiency of these techniques in maintaining genetic fidelity (Tikendra et al. 2019; González-Arnao et al. 2022). In addition, assessments of epigenetic stability by methylation-sensitive markers have supported the maintenance of critical regulatory patterns that ensure stable phenotypic expression after recovery.

Cryoprotective Agents in Maintaining Genetic Integrity

Cryoprotective agents (CPAs) are extremely important in protecting plant tissues from damage due to freezing and thawing. DMSO, glycerol, and sugars such as sucrose and trehalose prevent the formation of intracellular ice crystals, thus reducing the risk of cellular injury and genetic instability by alleviating osmotic stress.

Advancements have thus taken place in developing new techniques for vitrification methods by the formation of a glass-like state that does not include the crystallization of ice, which might destroy cells. The current advances made in such have already produced high survival concerning genetic fidelity through employing some methods of cassava meristems (Wang et al. 2021). The improvement will come in handy concerning recalcitrant species, thereby allowing safe conservation with minimal genetic deviation risks.

Molecular and Phenotypic Tools for Genetic Stability Evaluation

Molecular biology has enhanced the possibility of accurate measurement of genetic stability in cryopreserved materials. The throughput sequencing, methylation study, and transcriptomics of DNA provide a detailed assessment of the effects of cryopreservation on DNA and gene expression. For instance, next-generation sequencing of cryopreserved rice embryonic calli reported that the original genetic sequence remained intact, thus establishing that the techniques currently used are reliable.

Along with molecular evaluations, phenotypic analyses play a significant role in validating stability. Growth characteristics and morphology of cryopreserved and fresh samples are cross-compared to ensure the results are identical. In recent studies on the cryopreservation of bananas and yams, recovered plants demonstrated stability in both genotype and phenotype.

Longevity and Viability in Cryogenic Storage

Cryogenic temperatures completely stop biological aging. Thus, cryopreservation is the best technique for genetic resource storage for long durations. Optimally optimized protocols would allow materials to be viable and genetically intact for decades and even centuries. For example, wheat seeds preserved in liquid nitrogen for more than 25 years showed high rates of germination and genetically stable profiles upon recovery.

Preserved materials are highly dependent on the accurate control of freezing and thawing rates for viability. Slow freezing and rapid warming minimize the risk of ice crystal formation and prevent recrystallization that could potentially compromise cell integrity. More advanced monitoring systems ensure that the storage conditions remain constant, providing additional reliability in cryopreservation for extended periods (Bryant et al. 2022).

Mitigating Oxidative Stress and Optimizing Genetic Fidelity

The most significant problem during cryopreservation is the generation of reactive oxygen species, which takes place in the freezing and thawing process. These can cause DNA damage and also alter the cellular membranes. To counter this problem, antioxidants such as ascorbic acid and glutathione have been added to the cryopreservation protocols, thereby reducing oxidative damage and maintaining genetic integrity.

Cellular responses to cryopreservation have been identified using proteomic and transcriptomic studies, and the most significant genes and proteins that take part in stress mitigation have been pointed out. Based on these molecular pathways, researchers are designing specific protocols for improving cryopreservation for a wide variety of plant species (Marcantonini et al. 2022). Major steps involved in Genetic stability and long-term preservation through cryopreservation are illustrated in Fig. 4.

7 Cryopreservation and In Situ Conservation: Controlling Diseases and Pests and Increasing Diversity in Reproductive Techniques

Cryopreservation and in situ conservation are two major components of plant preservation. They are not only strong mechanisms for protecting such rare and endangered species, but they also help in managing diseases, pests, and genetic diversity in natural ecosystems. Given the rising threats from habitat destruction, climate change, and emergent plant pathogens, the integration of cryopreservation with in situ conservation strategies is increasingly acknowledged as a transformative approach to building ecological resilience for the long term (Kim and Popova 2024).

Cryopreservation as a Disease and Pest Management Tool

Cryopreservation controls diseases and pests in plants through sophisticated means of preserving pathogen-free tissues from plant species. Techniques, for instance, in the form of vitrification, encapsulation dehydration, and droplet freezing, are very helpful to separate and conserve plant materials that do not carry systemic infections. This material does not support harmful agents, which spread these noxious agents that devastate both agricultural crops and the wild ecosystem.

Especially useful for economically and ecologically significant crops, the ability of cryopreservation to eradicate pathogens has proven to be of immense value. For example, through the cryopreservation of meristems, it has been possible to remove viral infections from crops like cassava, potatoes, and bananas, thereby guaranteeing a constant supply of healthy and productive germplasm. Cryogenic techniques have also helped in the preservation of forest species. The pathogen-free plant materials are preserved through cryogenic techniques, thus preventing invasive diseases from entering their natural habitats. This is important in protecting wild plant populations from the risk of outbreaks but

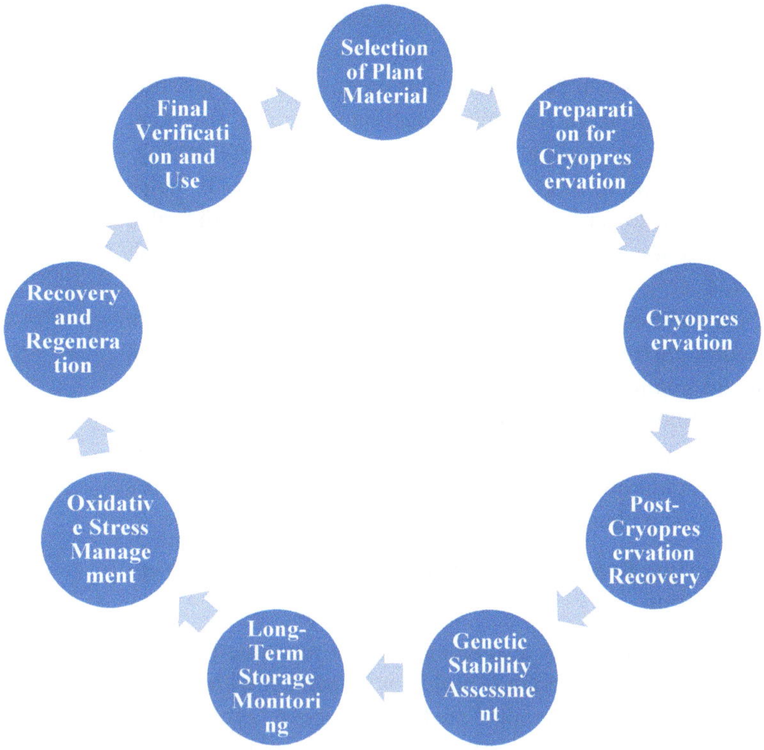

Fig. 4 Major steps in genetic stability and long-term preservation through cryopreservation

also enhances ecological stability through the maintenance of healthy and productive keystone species. Using cryopreserved, disease-free plant tissues means restoration of degraded habitats with genetic integrity, lower risk of recurrence in the disease, and, most importantly, reduces chemical treatments as part of practice to be used in sustainable agriculture in saving biodiversity in regions being threatened by invasive pests (Benelli et al. 2021).

Cryopreserved plant material is therefore a significant resource for in situ conservation, which is useful in source provision of disease-resistant genotypes for re-establishing natural habitats. Reintroductions, therefore, safeguard wild plant populations against the threat of epidemic outbreaks and stabilize ecosystems through the continued health and productivity of keystone species. Degraded habitats can be restored through the use of cryopreserved, disease-free plant tissues. This has the added advantage of ensuring genetic integrity and minimizing the likelihood of recurrence of disease.

Improving Genetic Diversity Through Cryopreservation

Preservation of genetic diversity through cryopreservation is often considered for species having complex reproductive systems or very limited natural propagation. It is also the most trusted method for the genetic conservation of recalcitrant-seed plants—those seeds that are intolerant to desiccation and conventional seed storage methods.

Rare alleles and unique genotypes can be cryopreserved in seeds, embryos, pollen, and somatic tissues. These resources are then critical input into ecological restoration projects and breeding programs. For instance, the possibility of cryopreserving pollen makes it possible to conserve male gametes from genetically diverse populations for controlled hybridization years or even decades later. This is particularly important for plant species that are experiencing genetic bottlenecks due to habitat fragmentation or population decline.

Equal weight has cryopreserved germplasm laid for increasing in situ population diversity. The regeneration of plants having maximum genetic fidelity through somatic embryogenesis, as well as shoot-tip cryopreservation, allows the restoration or return to the wild species under reproductive stress. Here, reproductive ones are included because of poor seed viability and insufficient gene flow. That implies restoration, sustainability, and, therefore, natural adaptation ability would follow natural laws (Białoskorska et al. 2024).

Integrating Cryopreserved Resources into In Situ Conservation

The in situ conservation strategy integrating cryopreservation techniques strengthens the ecosystem's resilience while supporting recovery efforts for species. On reintroduction into natural habitats, cryopreserved plant materials moderate the impacts of environmental changes in terms of habitat loss and variability in climate. Cryopreservation serves as a genetic reservoir to ensure the availability of diverse germplasm able to adapt to changing conditions in the ecological environment.

This is of special relevance to threatened species where prompt conservation intervention is essential. Cryopreserved tissues can be restored and propagated into planting material that facilitates reintroduction programs for plant populations as well as restoration of the general ecological interaction that the plants are supporting. It helps contribute toward general ecosystem recovery and also builds associated fauna and flora resilience (Zhang et al. 2014).

In addition, preserving plant material with features such as drought resistance or insect resistance will ensure that in situ populations are better prepared to cope with adverse environmental conditions. This direction fits well within the mainstream global conservation goals of integrated efforts aimed at mitigating the accelerating rates of biodiversity loss and climate change challenges.

Improving Reproductive Technologies Through Cryopreservation

Cryopreservation has also revolutionized reproductive technologies and has overcome problems associated with species conservation with complicated reproductive cycles. The cryogenic storage of zygotic embryos and somatic tissues may help bypass problems related to asynchronous flowering and low seed production. Resources thus preserved cryogenically can be utilized for the reproduction of clonal or sexually propagated plants, thus safeguarding the species with unusual reproductive features (Popova et al. 2015).

The germplasm of wild crop relatives that have been cryopreserved plays an important role in agricultural breeding programs to introduce desirable traits into the cultivated varieties. This ensures that, through cryopreservation, genetic diversity will be maintained over long periods, supporting sustained breeding activities, enriching the genetic base of crops, and conserving their wild relatives. This is a double-edged role that not only enhances agricultural productivity but also ensures the conservation of crop wild relatives for global food security. Cryopreservation and in situ conservation methods are illustrated in Fig. 5.

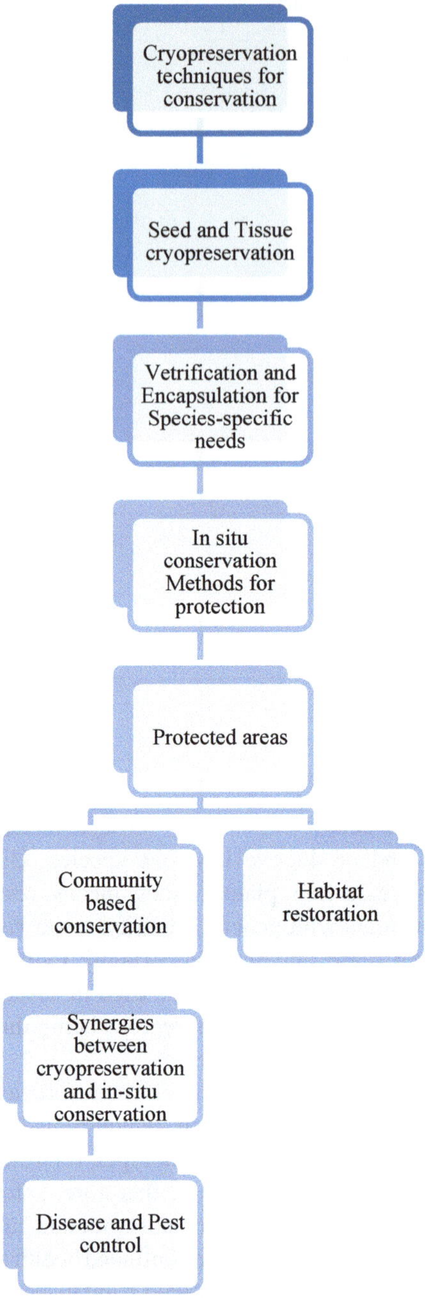

Fig. 5 Cryopreservation and in situ conservation

8 Challenges and Limitations of Cryopreservation

Cryopreservation, though indispensable in the conservation of plant genetic resources, is still faced with many biological, technical, logistical, and economic challenges. These must be overcome to improve the technique's application and ensure that conserved materials have long-term viability while making the technique useful in a much wider range of plant species.

8.1 Biological and Technical Limitations

- **Species-Specific Cryopreservation Needs**
  Cryopreservation is not applicable to all plant species. There is a big difference in the response of species to freezing, which primarily relates to their biological characteristics. For example, orthodox seeds, such as Arabidopsis thaliana, that can tolerate drying and freezing are easier to conserve. More challenging are recalcitrant seeds, such as Mangifera indica, which are much more sensitive to drying

and freezing. These have greater moisture content, leading to the formation of ice crystals at the time of freezing that can rupture cell membranes and, thus, lead to irreversible damage to the cell structure. Cryopreservation of recalcitrant seeds is often a very precise protocol, and results continue to be inconsistent (Parihar et al. 2023). Further research is required for the identification of better methods for species that have not yet been well studied.

- **Cryo-Injury and Cellular Damage**

  Cryo-injury refers to the physical and chemical damage that occurs to plant tissues when exposed to low temperatures. Plant cells become more susceptible to damage from the formation of ice within the cytoplasm, which often perforates membranes and even leads to cellular collapse. Even though techniques of rapid cooling, such as vitrification, have been developed to minimize the formation of ice crystals by transforming the water inside cells into an amorphous state, cryo-injury persists in many tissues, including embryos and meristems. The injury is especially difficult during the cooling and thawing phases, where even a small deviation in temperature may cause cell damage. Another major problem is the thermal shock that occurs during thawing, which can lead to the disruption of cellular integrity and impair recovery after thawing (Benelli 2021) (Fig. 6).

- **Cryoprotectant Toxicity**

  Cryoprotectants help prevent ice formation and maintain the integrity of cells during the freezing process. Glycerol, DMSO, and ethylene glycol are three of the most commonly used cryoprotectants. While the above substances are effective in the prevention of ice damage, using them in excess is usually toxic to plant cells. The most important challenge is to identify the right concentrations and exposure times that would minimize both cryo-injury and toxicity. It has been observed that higher concentrations of DMSO could cause membrane disruption and enhance osmotic stress, thus resulting in a lower regeneration potential (Kim and Popova 2024). Similarly, glycerol exhibits toxic osmotic effects at certain concentrations, so it is very important to calibrate appropriately to avoid cytotoxicity.

- **Optimizing Cooling and Thawing Protocols**

  The two critical steps that involve stringent management to effectively obviate cryo-injury are cooling and thawing. Different optimum cooling rates need to be applied for different plant tissues; rapid cooling that may be as fast as 10°C/min is suitable for some species, while others require slower cooling rates of about 0.3°C/min to avoid ice crystal formation. Equally important is the process of thawing; thawing too fast can result in thermal shock, whereas a very slow thaw will expose the material to damaging temperatures for a longer period of time. So, it is species-specific cooling and thawing protocols that may indeed improve success in cryopreservation (Chang and Zhao 2021).

8.2 Logistical and Economic Constraints

- **Infrastructure and Cost**

  This resource-intensive approach requires cryogenic facilities, liquid nitrogen storage tanks and freezers, and a number of refrigeration systems that allow low temperatures of storage. Such considerable requirements for infrastructure, along with experienced personnel to operate correctly, can be costly to the underdeveloped countries' institutes, as well as the underdeveloped research institutes of the developed or developing nations themselves. According to research, cryopreservation infrastructure requires a relatively large amount of initial investment. Besides, the recurring expenses in maintaining cryogenic conditions restrain the scalability of the method, especially for settings with low resources.

- **Continuous Resource Requirements**

  Cryopreserved material relies on a liquid nitrogen supply and a stable source of power. Interruptions in these resources, for example, caused by a natural disaster or power outage, might compromise the integrity of the preserved genetic material. It is also a challenge in the aspect of long-term storage stability; this requires some amount of money, particularly in unstable regions. For example, cryogenic storage systems become less reliable in regions of frequent political unrest or economic instability and thus result in a heavy loss of valuable genetic resources (Nagel et al. 2024).

8.3 Knowledge Gaps and Accessibility

- **Limited Protocols for Diverse Species**

  Despite the well-established protocols of cryopreservation for several species of plants, mainly model organisms and economically important crops, there are many other species, including rare and endangered species and species with economic value, that still require individualized cryopreservation protocols. In 2020, the study indicates that, for less studied species, cryopreservation databases are very few, thereby hampering the targeted

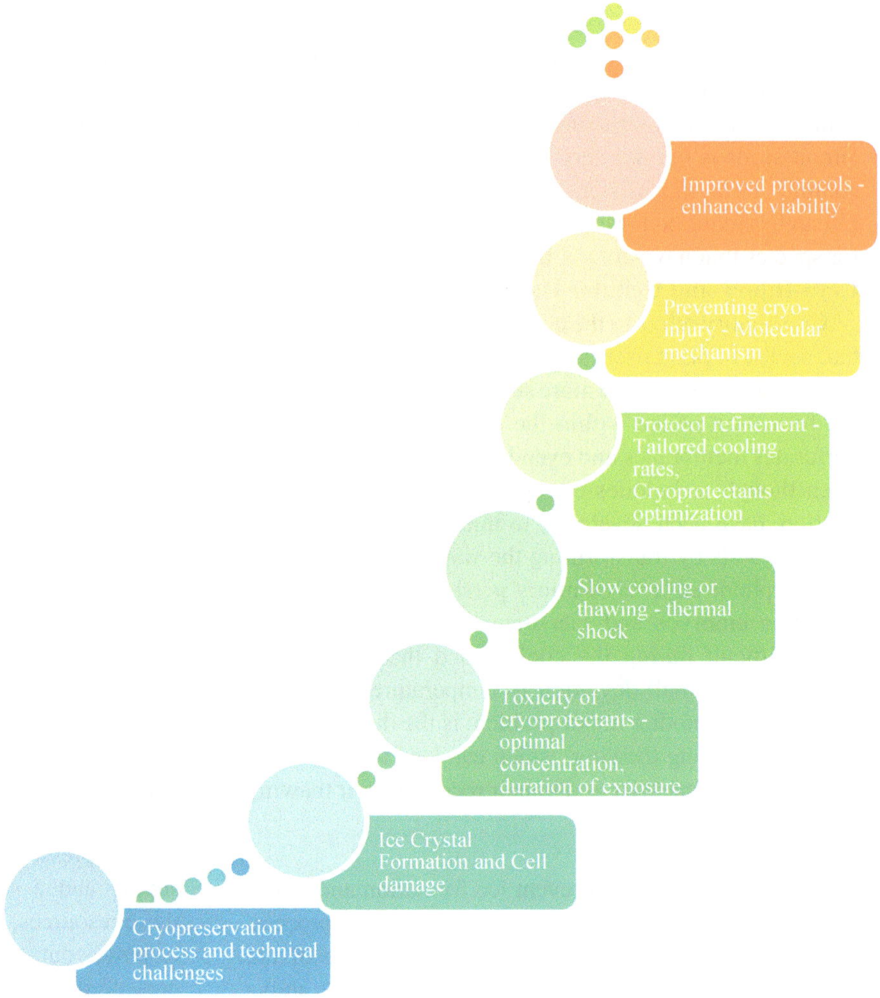

Fig. 6 Cryo-injury and technical constraints in cryopreservation

development of conservation strategies. In some species, even for the complex reproductive biology or the uniqueness in the genetic character, protocols have yet to be developed. This research gap limits the applicability of cryopreservation techniques to a much larger area of plant biodiversity, which may include crops critical to food security in the context of climate change.

- **Intellectual Property and Biopiracy**

  The other area of challenge that continues to face cryopreservation involves intellectual property rights and biopiracy. Many biodiversity-rich countries, particularly in the tropics, remain concerned about the exploitation of their plant species by external forces. This protocol has been of paramount importance in addressing such concerns through the fair and equitable sharing of genetic resources. However, the negotiation for access and benefit-sharing agreements is still complex and hinders collaborative cryopreservation efforts. National laws on access to genetic resources and the protection of Indigenous knowledge remain a source of complexity in international collaborations, hence limiting the potential for sharing valuable genetic resources (Kalaiselvi et al. 2017).

9 Future Directions in Cryopreservation Research and Applications

Although limited, cryopreservation does promise much with the prospects of overcoming many hurdles faced by biodiversity conservation. Enhancements in technologies, genomics, and molecular biology are likely to enable future innovations in the scope and efficacy of cryopreservation techniques as widely applied across species and for significantly increased success.

Advancing Cryopreservation Protocols

- **Universal Protocols for a Wider Range of Species**
  The development of universal cryopreservation protocols is the major aim in the field. Another promising approach is through vitrification, where intracellular water is quickly turned into an amorphous state to prevent the formation of damaging ice crystals. In a study by González et al., it has been proven that vitrification increases the survival rate of more fragile tissues because it reduces cryoprotectant toxicity. Additionally, methods such as encapsulation dehydration, in which plant tissues are immersed in a protective gel before freezing, are under further development for species whose tissues are of intricate architecture. These new developments should allow cryopreservation access to species previously refractory to preservation (Costa-Borges et al. 2024).
- **Nanotechnology**

  Nanotechnology is rapidly becoming an enabling area in cryopreservation science. Nanomaterials may help to increase the effectiveness of cryoprotectants by increasing penetration in plant cells and improving the protective ability of the latter. Studies conducted indicate that nanoencapsulation of cryoprotectants can reduce toxicity with increased protective efficiency. Real-time monitoring of the cryopreservation process using nanosensors also offers opportunities for monitoring cellular response to freezing and thawing, with such information helping in fine-tuning species-specific protocols (Tariq et al. 2020).

Expanding Applications

The value of cryopreservation has also started being appreciated outside the context of biodiversity conservation. Germplasm can be used in agriculture for crop development with climate change resilience, such as drought resistance. The ability to store climate-resilient crop varieties for long periods enables breeders to access the genetic diversity for future crop improvement (Benson et al. 2020). Similarly, the pharmaceutical industry also uses cryopreservation to store medicinal plant species long term, thus preserving their bioactive compounds for use in drug development. Advances in cryopreservation can also enable biotechnological production of biofuels and bioplastics that help make the new landscape of energy more sustainable and renewable (Tapiwa 2019).

Collaborative Networks and Data Sharing

Global collaborative networks will play a crucial role in addressing current challenges in cryopreservation. Successful international efforts at creating shared cryobanks include the Global Crop Diversity Trust and the Millennium Seed Bank Partnership. Adding such efforts to more species that also happen to have high economic or ecological value will enhance global outcomes in conservation. Open-access cryopreservation databases will enable the sharing of data and protocols among researchers in this field, thus improving standards and efficiency in cryopreservation throughout the globe (Khaydukova et al. 2024).

Ethical and Policy Considerations

Cryopreservation advances are still faced with significant ethical and policy challenges in contemporary times. Ensuring equitable access to cryopreserved genetic resources is fundamental to the prevention of biopiracy and guaranteeing the benefits of the technology for the Indigenous communities and the developing world. International agreements such as the Nagoya Protocol are vital to setting in place fair access to genetic resources and sharing the benefits arising from them (Mertes 2024). Genetic modification-related ethical issues and the application of CRISPR technologies to improve cryopreservation outcomes require careful consideration to ensure that the benefits of cryopreservation are distributed equitably across all sectors of society.

10 Conclusion

Cryopreservation is a vital tool for the conservation of plant genetic resources, possessing special benefits in protecting biodiversity and agrobiodiversity. Freezing the plant genetic material at ultra-low temperature avoids biological deterioration, and one can store the precious genetic resources in a long-term manner. It is especially needed for the preservation of the rare, endangered, and economically important species to prevent them from habitat destruction and climate change. Despite these promises, widespread adoption is complicated by several technical, logistical, and economic challenges. Technical factors are species-specific variations in the response to cryogenic conditions; complexities of cryoprotectants and cooling protocols have made standardization challenging. Access can also be limited due to the high costs of infrastructure, as well as the necessity for continuous, specialized storage systems, particularly in resource-poor areas. However, ongoing research into refining cryopreservation techniques—such as improving vitrification methods and cryoprotectant efficiency—holds the potential to overcome these challenges. Future perspectives on cryopreservation will depend on the integration of new emerging technologies such as nanotechnology and synthetic biology, which promise to raise the efficiency and accessibility of plant preservation. Thus, researchers, conservationists, and policymakers must work together to remove these barriers and make the technique sustainable. In this context, cryopreservation will serve as a backbone for the conservation of biodiversity while contributing to sustainable agriculture and ecosystem recovery.

In conclusion, cryopreservation is an essential solution for the preservation of plant genetic resources, food security, and the mitigation of climate change impacts. Despite the challenges, continued advancements in cryobiology and technology will enhance its utility and accessibility, positioning it as a key strategy in global conservation efforts. By protecting genetic diversity, cryopreservation helps safeguard the future of agriculture and ecosystems, playing a pivotal role in ensuring a resilient and sustainable world.

References

Amankwaah VA, Prempeh RNA, Quain MD (2022) Plant cryopreservation importance, approaches and future trends. In: Cryopreservation-applications and challenges. IntechOpen. https://doi.org/10.5772/intechopen.108806

Benelli C (2021) Plant cryopreservation: a look at the present and the future. Plants 10(12):2744. https://doi.org/10.3390/plants10122744

Benson JD (2020) Mathematical modeling and optimization of cryopreservation in single cells. In: Cryopreservation and freeze-drying protocols. Springer, New York, US, pp 129–172

Bialoskorska M, Rucinska A, Boczkowska M (2024) Molecular mechanisms underlying freezing tolerance in plants: implications for cryopreservation. Int J Mol Sci 25(18):10110. https://doi.org/10.3390/ijms251810110

Bryant SJ, Awad MN, Elbourne A, Christofferson AJ, Martin AV, Meftahi N, Drummond CJ, Greaves TL, Bryant G (2022) Deep eutectic solvents as cryoprotective agents for mammalian cells. J Mater Chem B 10(24):4546–4560. https://doi.org/10.1039/D2TB00573E

Chang T, Zhao G (2021) Ice inhibition for cryopreservation: materials, strategies, and challenges. Adv Sci 8(6):2002425. https://doi.org/10.1002/advs.202002425

Costa-Borges N, Coelo A, Cohen J, Gallardo M, Flores-Saiffe Farias A, Mestres E, Acacio M, Matia Q, Castello C, Calderón G, Chavez-Badiola A (2024) O-295 fast vitrification and warming protocols demonstrate similar efficiencies to a standard method and a substantial reduction in execution times. Hum Reprod 39(Supplement_1):108–352. https://doi.org/10.1093/humrep/deae108.352

Engels JM, Ebert AW (2021) A critical review of the current global ex situ conservation system for plant agrobiodiversity. I. History of the development of the global system in the context of the political/legal framework and its major conservation components. Plants 10(8):1557. https://doi.org/10.3390/plants10081557

González-Arnao MT, Cruz-Cruz CA, Hernández-Ramírez F, Alejandre-Rosas JA, Hernández-Romero AC (2022) Assessment of vegetative growth and genetic integrity of vanilla planifolia regenerants after cryopreservation. Plants 11(13):1630. https://doi.org/10.3390/plants11131630

Gubaidullin N, Balnur R, Kali D, Tussipkan S, Manabayeva S (2024) A promising strategy for conservation of endemic plant Euonymus koopmannii. https://doi.org/10.13057/biodiv/d250733

Kakaei M, Rehman FU (2023) Cryopreservation procedure technique in medicinal plants. Agrotech Ind Crop 3(1):23–29. https://doi.org/10.22126/atic.2023.8978.1088

Kalaiselvi R, Rajasekar M, Gomathi S (2017) Cryopreservation of plant materials—a review. IJCS 5(5):560–564

Khaydukova IV, Ivannikova VM, Zhidkov DA, Belikov NV, Peshkova MA, Timashev PS, Tsiganov DI, Pushkarev AV (2024) Current state and challenges of tissue and organ cryopreservation in biobanking. Int J Mol Sci 25(20):11124. https://doi.org/10.3390/ijms252011124

Kim HH, Popova E (2023) Unifying principles of cryopreservation protocols for new plant materials based on alternative cryoprotective agents (CPAs) and a systematic approach. Cryo Lett 44(1):1–12. https://doi.org/10.54680/fr23110110112

Kim HH, Popova E (2024) Improvement and innovation of cryopreservation and in vitro methods in plant resource protection. Biology 13(9):741. https://doi.org/10.3390/biology13090741

Marcantonini G, Bartolini D, Zatini L, Costa S, Passerini M, Rende M, Luca G, Basta G, Murdolo G, Calafiore R, Galli F (2022) Natural cryoprotective and cytoprotective agents in cryopreservation: a focus on melatonin. Molecules 27(10):3254. https://doi.org/10.3390/molecules27103254

Matsumoto T (2017) Cryopreservation of plant genetic resources: conventional and new methods. Rev Agric Sci 5:13–20. https://doi.org/10.7831/ras.5.13

Mertes H (2024) Ethical issues related to cryopreserved reproductive specimen. in cryopreservation in assisted reproduction: a practitioner's guide to methods, management and organization, pp 407–414. Springer International Publishing, Cham. https://doi.org/10.1007/978-3-031-58214-1_41

Mohammad Y, Amini J, Benson D (2023) Technologies for vitrification based cryopreservation. Bioengineering 10(5):508. https://doi.org/10.3390/bioengineering10050508

Nagel M, Pence V, Ballesteros D, Lambardi M, Popova E, Panis B (2024) Plant cryopreservation: principles, applications, and challenges of banking plant diversity at ultralow temperatures. Annual Rev Plant Biol 75. https://doi.org/10.1146/annurev-arplant-070623-103551

Nausch H, Buyel JF (2021) Cryopreservation of plant cell cultures—diverse practices and protocols. New Biotechnol 62:86–95. https://doi.org/10.1016/j.nbt.2021.02.002

Nguyen VA, Nguyen PT, Le MA, Bazrafshan A, Sisunandar S, Kalaipandian S, Adkins SW, Nguyen QT (2023) A practical framework for the cryopreservation of palm species

Parihar A, Kumar A, Panda U, Khan R, Parihar DS, Khan R (2023) Cryopreservation: a comprehensive overview, challenges, and future perspectives. Adv Biol 7(6):2200285. https://doi.org/10.1002/adbi.202200285

Popova E, Shukla M, Kim HH, Saxena PK (2015) Plant cryopreservation for biotechnology and breeding. In: Advances in Plant breeding strategies: breeding, biotechnology and molecular tools, pp 63–93. https://doi.org/10.1007/978-3-319-22521-0_3

Praveen B, Sharma P (2019) A review of literature on climate change and its impacts on agriculture productivity. J Public Aff 19(4):e1960. https://doi.org/10.1002/pa.1960

Ruta C, Lambardi M, Ozudogru EA (2020) Biobanking of vegetable genetic resources by in vitro conservation and cryopreservation. Biodivers Conserv 29:3495–3532. https://doi.org/10.1007/s10531-020-02051-0

Shankar V, Kumar KC, Puneeth GM, Deepak DA, Jeevan N, Sathish K (2024) Synergizing cryopreservation, thermal analysis, and in vitro regeneration: strategies for long-term plant germplasm preservation. Int J Adv Biochem Res 8(8S)1224–1235. https://doi.org/10.33545/26174693.2024.v8.i8Sp.2014

Tapiwa KA (2019) A review on the effectiveness of cryopreservation as a germplasm management option. JOJ Wild Biodivers 1:80–83

Tariq A, Ilyas S, Naz S (2020) Nanotechnology and plant tissue culture. Nanoagronomy 23–35. https://doi.org/10.1007/978-3-030-41275-3

Tikendra L, Koijam AS, Nongdam P (2019) Molecular markers based genetic fidelity assessment of micropropagated *Dendrobium chrysotoxum* Lindl. Meta Gene 20:100562. https://doi.org/10.1016/j.mgene.2019.100562

Tirado-Pérez B, Sandoval-Cancino G (2022) Cryopreservation of plant genetic resources: a legacy for humanity. Afr J Biotech 21(2):55–63. https://doi.org/10.5897/AJB2021.17379

Wang MR, Bi W, Shukla MR, Ren L, Hamborg Z, Blystad DR, Saxena PK, Wang QC (2021) Epigenetic and genetic integrity, metabolic stability, and field performance of cryopreserved plants. Plants 10(9):1889. https://doi.org/10.3390/plants10091889

Whaley D, Damyar K, Witek RP, Mendoza A, Alexander M, Lakey JR (2021) Cryopreservation: an overview of principles and cell-specific considerations. Cell Transplant 30:0963689721999617. https://doi.org/10.1177/0963689721999617

Zhang J, Xin X, Yin G, Lu X, Chen X (2014) In vitro conservation and cryopreservation in national genebank of China. Acta Hortic 1039:309–317

Gene Bank: Conservation of Genetic Resources in the Genomic Era

Md. Sajedur Rahman, Md. Zahidur Rahman, Niloy Gain, Mst. Meherun Nesa, Mohammad Zahidul Islam, Deen Mohammad Deepo, and Jamilur Rahman

Abstract

The conservation of genetic resources is vital for preserving biodiversity, enhancing agricultural resilience, and ensuring environmental sustainability. In the genomic era, gene banks have evolved from traditional repositories of physical genetic materials into advanced hubs that integrate cutting-edge genomic tools. These tools, such as genome sequencing, gene editing, and bioinformatics, enable precise characterization, efficient storage, and optimized use of genetic resources. Digital gene banks further elevate this role by incorporating technologies like cloud computing, blockchain, and artificial intelligence, which ensure secure, scalable, and accessible genetic data management. Gene banks play a crucial role in addressing challenges like genetic erosion, habitat loss, and climate change by protecting genetic diversity and providing adaptive traits for crop improvement and ecosystem restoration. The fusion of genomic data with traditional practices has facilitated global collaboration, accelerating breeding programs and ensuring the development of climate-resilient, high-yielding crop varieties. However, significant challenges remain, including high infrastructure costs, data security concerns, and issues related to interoperability. Overcoming these obstacles requires robust policies, international cooperation, and sustainable funding. By leveraging technological innovations and fostering global partnerships, gene banks are pivotal in conserving genetic heritage, promoting ecological stability, and supporting sustainable development. As we navigate the genomic era, they will continue to ensure the long-term preservation of biodiversity and its critical contributions to both human well-being and environmental health. This chapter will explore gene banking, its tools, strategies, digitization, the challenges involved, and their solutions, as well as the future of this technology in ensuring biodiversity conservation for a sustainable ecosystem for future generations.

Keywords

Genetic resource conservation · Gene banks · Genomic era · Digital gene banks · Biodiversity preservation

Md. S. Rahman · Md. Z. Rahman · N. Gain · Mst. M. Nesa · J. Rahman (✉)
Department of Genetics and Plant Breeding, Sher-e-Bangla Agricultural University, Dhaka 1207, Bangladesh
e-mail: jamilur@sau.edu.bd
URL: https://sau.edu.bd/teacher/157

Md. S. Rahman
e-mail: sajedurgpb@gmail.com

Md. Z. Rahman
e-mail: jahidurrahman2013@gmail.com

N. Gain
e-mail: kbdniloygain@gmail.com

M. Z. Islam
Genetic Resources and Seed Division, Bangladesh Rice Research Institute, Gazipur 1701, Bangladesh
e-mail: zahid.grs@gmail.com

D. M. Deepo
Department of Horticultural Science, Kyungpook National University, Daegu 41566, Korea

Md. S. Rahman · Md. Z. Rahman · N. Gain
Research and Development, Lal Teer Seed Limited, Gazipur 1704, Bangladesh

1 Introduction

The preservation of genetic diversity is an essential endeavor for safeguarding our agricultural resilience, environmental sustainability, and biological heritage. At the forefront of these efforts stand gene bank repositories specifically designed to conserve, document, and facilitate the utilization

N. Kumar (ed.), *Biotechnology in Plant Conservation*, Sustainable Landscape Planning and Natural Resources Management,
https://doi.org/10.1007/978-3-032-02605-7_8

of genetic materials across a wide array of species. These institutions play a vital role in ensuring that invaluable genetic resources remain accessible, not only for current needs but also for the unpredictable demands of future generations. As highlighted by the Food and Agriculture Organization (FAO), the scope of gene banks has progressively expanded over time, embracing advancements in biotechnology, data management, and, more recently, genomics (Commission on Genetic Resources for Food 2010). Against this dynamic backdrop, this chapter explores the significance of gene banks in the genomic era, delving into both the fundamental concept of genetic resource conservation and the transformational impact of emerging technologies.

Gene banks traditionally have served as centralized facilities where plant seeds, animal germplasm, and microbial strains are preserved under controlled conditions. The fundamental goal is to maintain the genetic integrity and viability of these resources for extended periods, enabling researchers, breeders, and other stakeholders to retrieve specimens when needed for breeding programs, restoration projects, and fundamental research (Engels et al. 2024). This approach is instrumental for mitigating the risks posed by genetic erosion, wherein valuable traits can be lost from the global gene pool due to environmental pressures or agricultural homogenization. Beyond merely acting as repositories, gene banks also provide critical services documenting lineage information, facilitating characterization of stored materials, and fostering collaboration across scientific and conservation-oriented communities. By preserving both widely used and underrepresented species, gene banks act as guardians of biodiversity, bridging the gap between research innovation and on-the-ground conservation practices. On the other hand, the concept of a digital gene bank (DGB) represents a transformative approach to conserving genetic resources, merging cutting-edge technologies with genetic data preservation to ensure the sustainability of biodiversity. Unlike traditional gene banks that store physical materials like seeds or DNA, DGBs leverage cloud computing, blockchain, artificial intelligence (AI), and big data analytics to store, manage, and analyze genetic information. These technologies facilitate secure, scalable, and accessible genetic repositories, enabling researchers, policymakers, and conservationists to utilize genetic data for research, crop improvement, and biodiversity conservation globally. While offering unparalleled benefits, including enhanced data analysis and global collaboration, challenges like high infrastructure costs, data security, and interoperability must be addressed to maximize their potential. This innovation is evident in platforms such as Genesis, the China National Gene Bank, and NCBI Gene Bank, which exemplify the application of advanced digital systems to revolutionize genetic resource management.

Despite their vital importance, gene banks and broader conservation efforts face a host of challenges that compromise long-term genetic resource availability. Climate change, for instance, not only alters habitats and species distributions but also exacerbates extreme weather events, threatening the viability of in situ conservation efforts. As habitats shrink or become fragmented due to agricultural expansion, urbanization, and industrial development, the risk of losing unique landraces and wild relatives intensifies. These pressures contribute significantly to the phenomenon of genetic erosion, an irreversible loss of genetic diversity within plant, animal, or microbial populations (Commission on Genetic Resources for Food 2010). Gene banks, although well-positioned to counteract these threats through ex situ preservation, must continually update their strategies to keep pace with ecological changes and emerging stressors. This necessity is emphasized in many studies, where the complex interplay between habitat fragmentation and declining genetic diversity is underscored, reinforcing the importance of proactive measures and robust conservation infrastructures. The advent of high-throughput sequencing, CRISPR gene editing, and advanced bioinformatics tools has precipitated a paradigm shift in how genetic resources are conserved, characterized, and utilized. In this genomic era, gene banks are no longer limited to simply storing physical specimens; they are increasingly seen as hubs for generating and disseminating comprehensive genomic data. By sequencing the stored accessions, researchers can unearth the underlying genetic variation that imparts stress tolerance, disease resistance, or quality traits—insights that can fast-track breeding programs and conservation strategies (Engels et al. 2024). Moreover, genomic databases linked to gene banks enable far-reaching collaboration, where scientists worldwide can access, analyze, and build upon shared genomic information. This convergence of physical and digital repositories heralds a more integrated and informed approach to conservation, where decisions about breeding, habitat restoration, and population management are guided by molecular-level evidence. These advancements are reshaping how gene banks set priorities, optimize collections, and engage in global efforts to safeguard genetic diversity for the benefit of society (Commission on Genetic Resources for Food 2010).

Advancements in genomic tools and technologies are revolutionizing gene banking and crop improvement. The narrowing genetic base of many crops highlights the importance of wild relatives as reservoirs of beneficial alleles for enhancing traits like yield, disease resistance, and nutritional quality. Tools such as high-throughput sequencing (HTS), marker technologies, and transcriptomics enable comprehensive genetic characterization and identification of desirable traits. HTS, including next-generation and third-generation sequencing, provides efficient genomic analyses, aiding in the discovery of structural variations and rare alleles in wild and domesticated species (Tripodi 2022). Marker technologies, especially single-nucleotide polymorphism (SNP) and genotyping-by-sequencing (GBS), facilitate genetic diversity analysis and marker-assisted selection for traits like drought and pest resistance. Transcriptomics offers insights into gene expression under various stress conditions, promoting the development of resilient crop varieties (Huang et al. 2024). Integration of omics approaches, genomics, transcriptomics, epigenomics, and phenomics bridges the gap between genetic resources and practical breeding applications (Huang et al. 2024). These tools enhance understanding of genetic diversity and adaptation mechanisms, enabling the development of crops tailored to diverse agro-ecological niches and climate resilience. Successful case studies, such as genomic sequencing in rice, barley, and wheat, demonstrate the transformative potential of these technologies in addressing global food security challenges (Alaux et al. 2018; McCouch et al. 2016). Despite challenges, the incorporation of advanced genomic data into gene banks strengthens their role as a cornerstone for biodiversity conservation and agricultural innovation. In the subsequent sections of this chapter, we delve deeper into the operational frameworks of gene banks, explore how genomic tools are revolutionizing conservation, and present case studies that illustrate the real-world implications of these innovations. By examining both the challenges and the opportunities posed by modern genomics, we underscore the critical relevance of gene banks in preserving our collective genetic heritage in an era of rapid environmental and technological change.

2 Conservation of Genetic Resources

In the context of climate change, the preservation of economically significant genetic resources, including landraces, wild species, breeding materials, and modern varieties, is crucial. This conservation of genetic resources is essential not only for sustaining agriculture but also for maintaining ecological balance and ensuring food security. However, two strategies, i.e., in situ and ex situ approaches, can be marked in a broad sense regarding the conservation of genetic resources. These strategies aim to safeguard valuable resources that are essential to ecosystem stability and resilience.

2.1 In Situ Conservation

In situ conservation is an active and dynamic strategy aimed at preserving genetic resources within their natural habitats, thereby supporting the natural processes of evolution and selection within ecosystems. This approach not only emphasizes the protection of individual species but also focuses on maintaining the overall health and balance of the ecosystems in which these species exist. Central to in situ conservation are strategies such as the establishment of national parks, wildlife sanctuaries, and biosphere reserves, which provide legal safeguards for species and their habitats. This holistic method goes beyond the preservation of plants and animals, extending to include microorganisms and their nonliving environmental components, ensuring biodiversity is conserved across multiple ecological levels. By addressing both biotic and abiotic factors, in situ conservation promotes the sustainable management and integrity of ecosystems, fostering the continued evolution and adaptation of species within their native environments.

2.1.1 National Parks

National parks can play a vital role in maintaining ecological balance and act as a carbon sink. It is a protected area that was established by the government to preserve landraces, wildlife, and genetic materials in their natural habitats. Yellowstone National Park is the world's first natural protected area established by the United States in 1872 (Mishra and Singh 2023). Yosemite National Park, Bhawal National Park, Grand Canyon National Park, Kaptai National Park, Lawachara National Park, Glacier National Park, and Acadia National Park are some other examples of national parks across the world. National parks serve as an anchor for larger ecosystems and can help to mitigate climate change. Subsequently, it can play a significant role in supporting the livelihood of impoverished communities residing in and around these regions. These communities often rely heavily on forest resources for their subsistence, and the sustainable management of these protected areas can contribute to their economic well-being while ensuring the conservation of biodiversity.

2.1.2 Wildlife Sanctuaries

A wildlife sanctuary plays a critical role in the conservation of biodiversity by providing protected habitats for various species and supporting both local and global conservation efforts. It is defined as a designated area where wildlife, including animals, plants, and microorganisms, along with their surrounding environments, are protected. Unlike national parks, wildlife sanctuaries may permit limited human activities such as resource collection and tourism while maintaining a focus on the conservation of individual species (Mishra and Singh 2023). Examples of wildlife sanctuaries include Rema-Kalenga Wildlife Sanctuary, Sundarbans Wildlife Sanctuary, Kassalong Reserve Forest, and Hail Haor Wildlife Sanctuary, and internationally, notable examples include Masai Mara, SanWild Wildlife Sanctuary, Bhadra Wildlife Sanctuary, Maenam Sanctuary, and the Sloth Sanctuary of Costa Rica. These sanctuaries serve as a "living gene bank" and thus play a crucial role in the sustainable utilization of natural resources by balancing conservation efforts with controlled human interaction.

2.1.3 Biosphere Reserves

The concept of biosphere reserves emerged with the landmark definition provided in 1970 during the UNESCO General Conference's proposal for the launch of the Man and the Biosphere (MAB) program (Mahanayak 2024). It is an area designed to balance biodiversity conservation with sustainable resource use. These serve as "living laboratories" where researchers and breeders can conduct experiments and manage genetic diversity alongside ecosystem processes. The UNESCO and Biosphere (MAB) program encourages the establishment and management of biosphere reserves globally (Mahanayak 2024). Biosphere reserves distinguished themselves as open rather than closed systems from most of the protected areas that were established earlier, and thus allowed for a dynamic interaction between conservation efforts and sustainable development within designated regions such as the Cold Desert, Nanda Devi, Dibru-Saikhowa, Great Nicobar, and Seshachalam Hills across the world. A biosphere reserve promotes a holistic approach to conservation that integrates ecological, social, and economic objectives. Some other in situ conservation approaches, namely Sacred Groves (Forest areas that are protected by local communities because of their religious or cultural value) and World Heritage Sites (The United Nations Educational, Scientific, and Cultural Organization (UNESCO) has designated a landmark or area as a World Heritage Site because of its significant cultural, historical or economic value) also play a vital role in the conservation of genetic resources.

2.2 Ex Situ Conservation

On the other hand, ex situ conservation entails the preservation of genetic resources, including animals, plants, organisms, or entire ecosystems, outside their natural habitats. This approach becomes essential when in situ conservation is not a feasible option due to various limitations, making it the only viable alternative for protecting species under severe threat. While ex situ conservation cannot entirely replicate the ecological advantages provided by in situ methods, it serves a vital function in addressing immediate risks to biodiversity. Moreover, it offers valuable opportunities for conducting detailed studies on species' biology, genetics, and breeding patterns. Additionally, ex situ conservation enables the reintroduction of species into their natural environments once the threats to their survival have been mitigated.

2.2.1 Botanical Gardens

The first Botanical Gardens Conservation Strategy, published in 1989, established a framework for enhancing the conservation role of botanical gardens during the 1990s (Mishra and Singh 2023). A botanical garden is a carefully managed space housing a diverse collection of plant species, typically labeled with their botanical names. These gardens provide visitors with opportunities to explore a wide range of plants, understand their ecosystems, and appreciate their importance to the environment. Additionally, botanical gardens play a crucial role in ex situ conservation by protecting plant species that face extinction in their natural habitats. This comprehensive strategy highlights the significance of botanical gardens as essential institutions for scientific research, public engagement, and the conservation of biodiversity.

2.2.2 Zoological Gardens

Zoological gardens, where wild animal species are enclosed, displayed to the public, and cared for under controlled conditions. These conservations play a significant role in advancing research, providing opportunities for recreation, and delivering educational experiences to the public. Most importantly, zoological gardens contribute to the ex situ conservation of animal genetic resources, offering a controlled environment for the preservation and study of species that are threatened or endangered in their natural habitats. Through these efforts, zoological gardens help to safeguard biodiversity, support species recovery programs, and foster greater public awareness and engagement in conservation issues.

2.2.3 Plant Herbarium

Plant genetic resources, encompassing crop plants, wild relatives, and landraces, can be effectively preserved in herbariums either as dried specimens or seed samples. This preservation method plays a crucial role in maintaining genetic variability, which is essential for the study of plant evolution, adaptation, and breeding. Herbariums serve as invaluable repositories for plant genetic diversity, offering researchers access to historical and contemporary plant materials for scientific inquiry, conservation planning, and the development of improved crop varieties.

2.2.4 Gene Banks

A gene bank serves as a form of ex situ conservation, enabling the long-term storage of genetic material, the preservation of rare and endangered species from extinction, and the provision of genetic diversity for future research and breeding programs. To optimize the conservation and management of genetic resources and maximize the utilization of genetic variation, five distinct types of gene banks have been established. These banks are categorized based on the conservation methods employed and the types of genetic material stored.

Seed Bank

Owing to the technological and infrastructural capabilities, a seed bank can contribute more directly to the conservation and restoration of ecosystems and biodiversity, especially by providing high-quality seed that is both ecologically and genetically appropriate. A useful and practical strategy for the preservation of plant genetic resources (PGRs) is the establishment of seed banks. Seed bank collections are particularly helpful in promoting the recovery of rare, endangered, extinct, and protected species. Seeds are the preferred form of germplasm due to their status as complete organisms rather than a gamete (Walters and Pence 2021). During their development, most seeds acquire desiccation tolerance as they accumulate food reserves within the embryo cells, making them particularly suitable for long-term storage and conservation. To ensure that only viable seeds are preserved, a fundamental procedure called seed viability testing is regularly carried out (Walters and Pence 2021). However, in a seed bank, seeds are preserved in several tiers, which are as follows.

- Tier I: Base collection—Long-term conservation for future generation (rarely used).
- Tier II: Active collection—For ongoing research and breeding programs, frequently used for regeneration, multiplication, distribution, evaluation, and documentation.
- Tier III: Duplicate collection (of base collection)—Backup in case of loss, which is housed for security in a different location from the corresponding base collection.

Field Gene Bank

Field gene banks serve as vital repositories for the conservation of plant genetic resources, particularly for species that cannot be conserved through traditional seed storage methods. It involves the collection of genetic material from different locations and placing it into a second site where it can be grown and monitored. The primary objective of field gene banks is to preserve as much genetic variation as possible within a limited sample size, ensuring the long-term conservation of valuable genetic traits. The plant genetic resources in field gene banks (FGBs) are preserved as living plants that grow and need to be maintained continuously, and they work as a "working sample" for experimental purposes. Several endangered economic tree crops have been identified and established within the field gene bank, including *Melicia excelsa*, *Triplochiton scleroxylon*, *Ceiba petandra*, *Entandrophagma cylindricum*, *Erythrina senegalensis*, *Pausinystalia johimbe*, species of *Khaya* and *Thaumatococcus danielli*, and so on (Walters and Pence 2021). Since some species have short-lived seeds (cocoa, coconut, oil palm, mango, jackfruit, etc.) that cannot be dried below 12% moisture content or have vegetatively propagated (banana, potato, sweet potato, taro, etc.), the field gene bank is the most commonly used method for conserving these genetic resources. Prominent examples of field gene banks include the National Fruit Collection in the United Kingdom, the Cassava Genebank in Colombia, the Sugarcane Genebank in Brazil, and the Potato Collection in Peru. These gene banks play a crucial role in the conservation and sustainable management of genetic resources and their potential use in addressing challenges such as food security, climate change, and biodiversity loss.

In Vitro Gene Bank

An in vitro gene bank involves the preservation of animal or plant genetic material, particularly for species facing specific challenges such as recalcitrant seeds, extinction in the wild, or sterile seeds. This method entails maintaining genetic material under controlled conditions outside of its natural habitat. This process involves two key steps: in vitro introduction and multiplication. In the in vitro introduction phase, explants are aseptically transferred from the field or screen house into a sterile culture container, ensuring a contamination-free environment for initial culture establishment. The multiplication phase follows, wherein the explants are provided with optimal environmental conditions, including appropriate light, temperature, and nutrients, to promote their development into small plantlets. This systematic approach is essential for the successful propagation and conservation of genetic material in an in vitro gene bank setting. In this process, a nutrient medium is utilized, which can be in liquid or gel form and is maintained under specific light and temperature conditions. The International Potato Center (CIP) in Lima, Peru, has the world's largest in vitro gene bank, dedicated to the conservation of potato genetic resources. In vitro conservation is primarily employed for the medium-term storage of genetic material and is particularly beneficial for the preservation of seedless plants and plants that reproduce asexually or those that need to be conserved as clones. This technique ensures the availability of genetic resources at any time for distribution, research, and breeding purposes. Furthermore, in vitro conservation techniques are continually evolving with advancements in tissue culture technology, cryopreservation, and genetic engineering, expanding the scope and efficacy of in vitro gene banks. This technique can enhance the capacity to conserve a wide range of genetic materials, making in vitro gene banks an indispensable component of comprehensive biodiversity conservation strategies.

Cryopreservation Bank

In recent years, cryopreservation has become a potential strategy as a part of ex situ conservation in enabling the long-term preservation of biological materials such as seeds, tissue, embryos, pollen, and other genetic material under

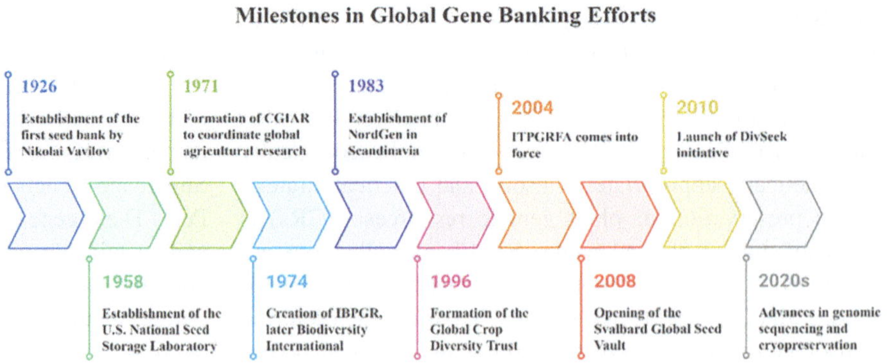

Fig. 1 Key Moments and Achievements in the Journey of Global Gene Banking Programs

ultra-freezing conditions. In this method, genetic material is preserved in liquid nitrogen (N_2) at an extremely low temperature (– 196 °C). A wide range of genetic resources can be cryopreserved without the risk of any epigenetic variation in the stored materials. Nowadays, vitrification is the most commonly used cryopreservation technique to preserve genetic material, especially oocytes, embryos, and ovarian tissue. Over hundreds of plant species have been successfully cryopreserved in this way**.** In this technique, samples are exposed to a high concentration of cryoprotectants and require rapid cooling to prevent the formation of ice crystals. Instead of freezing, the water within the sample transitions directly into a glass-like solid state. This eliminates the structural damage caused by ice crystals and facilitates long-term conservation of sensitive cells, tissues, and embryos.

DNA Bank

DNA bank has emerged as a new resource conservation strategy with a great opportunity for conservation and utilization of genetic diversity. It may be one of the best alternatives to conserve these genetic materials, which are difficult to preserve or are at the stage of extinction. DNA can be extracted from the majority of species, and once purified, it can be stored for long-term conservation. As the genetic component of an organism is encapsulated within its DNA, this method allows for the study of genetic variation within and between species through DNA sequencing. Several attempts have been made to coordinate a global network of DNA banks. Several notable DNA banks include the NIAS DNA bank, RBG Kew DNA bank, Genethon's DNA and Cell Bank, and the Wisconsin State Laboratory of Hygiene DNA bank. With the goal of establishing a global network among all DNA banks in the world, DNA bank-Net was created in the late 1980s (Fig. 1). This initiative seeks to facilitate the sharing and conservation of genetic material, supporting international efforts to preserve biodiversity and enhance the application of genetic resources in research and conservation.

3 Core Functions of Gene Bank

Gene banks play a crucial role in conserving genetic diversity, fostering research innovation, and ensuring food security through the implementation of effective management practices. By systematically collecting and preserving genetic resources, gene banks facilitate the sustainable use of these resources and make them available to researchers. The following key functions highlight the essential contributions of gene banks.

A. **Collection**

The main objective of genetic materials collection is to collect the maximum amount of genetic variation in the smallest number of samples. There are mainly three reasons for collecting genetic materials: loss of genetic diversity, gap filling, and need based on future use. Since it is a very costly and time-consuming process, reviewing the past collection of resources is necessary before conducting any collection mission. Resource exchange among different repositories across the world will be effective in this process. The collection procedure involves identifying the species, varieties, or population to be collected and choosing the collecting location based on species distribution, population size, and ecology. Landscape genomics can play a significant role in ex situ conservation by integrating genetic and environmental data to guide the management and preservation of genetic resources. It provides a science-driven framework for ex situ conservation. By analyzing the spatial pattern of genetic diversity across landscapes, it helps in identifying regions with unique or high levels of genetic diversity.

B. **Characterization**

Characterization focuses on the screening of germplasm for resistance to biotic and abiotic stresses alongside the evaluation of grain food quality. After the collection of genetic material, proper characterization is essential to ensure traceability, effective management, and subsequent use of genetic resources. It involves identifying and characterizing the genetic materials based on their agronomical and morphological features. It aims to estimate the extent of variation present within the germplasm collection. These activities were carried out in collaboration with scientists from various disciplines. Through comprehensive field and laboratory screening, followed by rigorous purification processes, a wide range of sources with desirable traits was identified in the assembled germplasm. Proper characterization provides information and facilitates utilization for breeders, researchers, and conservationists.

C. **Evaluation**

Unlocking the practical potential of genetic resources is essential for guiding their optimal use in sustainable agriculture and biodiversity conservation. The evaluation of genetic materials involves a comprehensive analysis of various characteristics, including genotypic, phenotypic, chemical, and biological traits. This process is crucial for identifying resources based on their quantitative properties, such as adaptability, performance, and storage quality, as well as their potential for further research. By thoroughly assessing these attributes, it is possible to determine the value of genetic resources for specific applications in breeding, conservation, and agricultural productivity, thereby ensuring their effective and sustainable use.

D. **Conservation**

Proper preservation of genetic materials is essential to ensure their availability for future utilization. The choice of conservation strategy is largely determined by the specific characteristics of the genetic material being conserved, as well as the intended purpose of conservation. Different strategies may be employed based on factors such as the biological nature of the species, its reproductive traits, and the environmental conditions required for its long-term preservation. Tailoring the conservation approach to these factors ensures the effective safeguarding of genetic resources for future research, breeding, and biodiversity conservation.

E. **Distribution**

Gene banks have become a critical source of biodiversity, providing invaluable genetic resources to researchers worldwide. The primary goal of preserving germplasm in gene banks is twofold: to restore biodiversity that has been lost in natural habitats and to enhance crop quality through breeding and related research (Walters and Pence 2021). The distribution process of genetic materials involves the systematic provision of these resources to researchers, breeders, farmers, or organizations for various purposes, including research, breeding, and agricultural applications. In response to requests from germplasm users, gene banks follow a Germplasm Distribution Protocol, which facilitates the efficient supply of genetic resources or working samples to ensure the continued advancement of agricultural and conservation efforts. Furthermore, gene banks serve as critical partners in international collaborations, facilitating the exchange of genetic material across borders to address shared challenges in biodiversity conservation and agricultural development (Walters and Pence 2021). Through these collaborations, gene banks not only contribute to the conservation of global biodiversity but also support the development of innovative solutions to address the food security challenges of the future.

4 Genomic Tools in Gene Banking

Over the recent years, the genetic base of the crop species has been narrowed down, becoming a bottleneck for successful crop improvement programs. Wild relatives act as reservoirs of noble alleles of many major and minor crop species, which can be incorporated into modern breeding programs in developing superior crop cultivars showcasing higher yield, disease resistance, and nutritional qualities. However, the characterization and identification of genes of interest for a particular trait or traits through screening of large accessions is a difficult task. Genomic approaches now open a new avenue in crop characterization that can be used as potential genetic stocks for the future. The joint effects of high-throughput sequencing (HTS), pangenomic studies, de novo domestication, and marker technologies fasten the genomic characterization of wild relatives and landraces by developing high-density markers, large-scale studies of genotyping arrays, and enhancing the capacity to detect rare alleles and structural variations. Novel genes from wild relatives of more than hundreds of major and minor crops have now been incorporated into breeding programs for developing crop varieties with resistance to biotic and abiotic stresses, as well as high yield and nutritional qualities. Harnessing genomic tools for advanced utilization of crop genetic resources is presented in Fig. 2. A brief discussion of different genetic tools extensively used in genomic characterization, identification of genetic diversity, and novel alleles of desirable traits of genetic resources is provided under the following subheadings.

4.1 High-Throughput Sequencing (HTS) Through Genomic Studies

Genomic sequencing and re-sequencing of wild relatives through advanced genomic tools such as next-generation sequencing (NGS) and third-generation sequencing (e.g.,

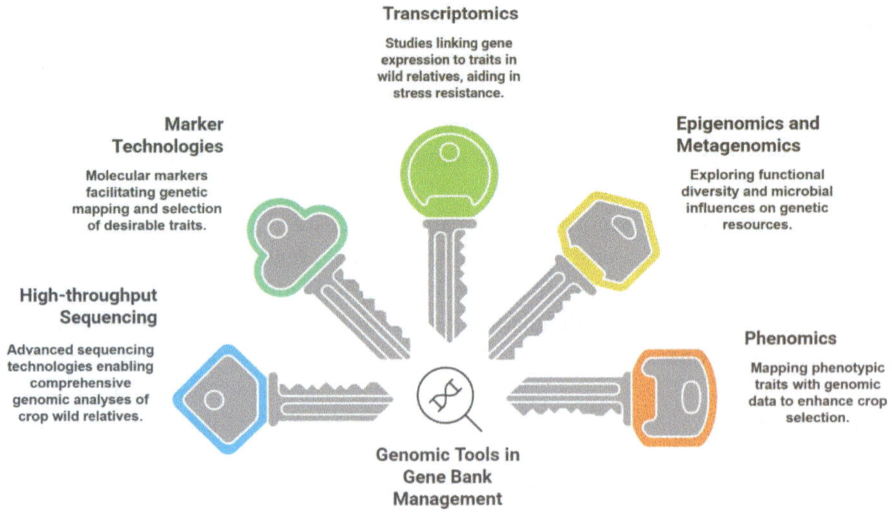

Fig. 2 Harnessing genomic tools for advanced utilization of crop genetic resources

PacBio, Oxford Nanopore) enable comprehensive genomic analyses by generating large volumes of sequence data efficiently and cost-effectively (Tripodi 2022). The whole genome sequencing of the crop wild resources and germplasm accession aids in exploiting the structural variation among and within the related species through developing the genetic maps. The chloroplast genome sequence is extensively used to manifest the genetic diversity among the accessions (Tripodi 2022). Previous studies found greater success in determining the phylogenetic relationship among the crop species through chloroplast sequencing, for example, in the identification of novel wild rice resources. Additionally, next-generation molecular markers, including single-nucleotide polymorphisms (SNPs), can also be employed in identifying the genetic diversity and gene functions for genetic resources (Tripodi 2022). For instance, greater genetic diversity among wild barley and wild sorghum accessions was identified through genetic mapping of these populations, whereas SNPs were identified for salinity and drought tolerance, respectively. Besides, re-sequencing through direct mapping has also been used in exploring the structural variation among the wild and domesticated crop species, aiding in identifying desirable alleles that can be possible to incorporate from wild to domesticated species for crop improvement. Re-sequencing of wild soya bean (soya) or wild rice (*Oryza nivara*) was found effective in the domestication of these crop species (Kim et al. 2016). The genome-wide analysis of Brassicaceae species is a revolutionary work in evolutionary studies (Kim et al. 2016). Complete sequencing and re-sequencing of the whole genome through these advanced tools provide a better understanding of the genetic architecture of the crop genetic resources and isolate key genes that control major yield and nutritional traits. These not only broaden the genetic pool of the crop species but also facilitate the domestication of new crop species that can withstand unpredictable climatic changes, including biotic and abiotic stresses. Additionally, pangenomic studies provide a comprehensive identification of structural variations present in a species, helping to exploit genetic diversity and identify many hidden genes not present in the reference genome, such as disease resistance genes in rice, rare alleles for flavor and disease resistance identified from tomato wild relatives, and drought-tolerant and pest-resistance genes in soybean (Kim et al. 2016). The uncovering of candidate genes by capturing the structural variations aids breeding efforts for better cultivars exhibiting higher adaptation under different agro-ecological niches.

4.2 Marker Technologies

Molecular marker technologies have now attained the topmost attraction in crop improvement programs with the advancement of genome sequencing protocols. These technologies provide insights into the genetic diversity, characterization, structure, and relationships among accessions and enable the feasible utilization of gene resources and pre-breeding activities. The combined utilization of molecular markers and genomic analysis facilitates the accurate taxonomic characterization of wild genetic resources and reveals the genetic distance among and within the crops and their wild relatives (Abebe et al. 2020). Besides, molecular markers are now employed in genetic mapping populations or quantitative trait loci (QTLs) linked to desirable traits, thus enabling the introgression of desirable alleles of interest into newly developed cultivated species. Over the past decades, several kinds of molecular markers have been used in molecular mapping and diversity analysis in gene bank accessions. For example, simple sequence repeats (SSRs) in maize (Prasanna 2012), rice (Quan et al. 2018), and wild tomato (Abebe et al. 2020), as well as amplified fragment length polymorphism (AFLP) in barley landraces (Baloch et al. 2017) and pea accessions (Parveen et al. 2012), were applied to get an insight into the molecular structure of these crop accessions. Compared to others, single-nucleotide polymorphism (SNP) markers are widely used to target specific breeding programs due to their high-throughput potential and affordability. However, KASP (Kompetitive Allele-Specific Polymerase chain reaction), an improved version of SNP, has now gained popularity for accurate genotyping of individuals and quick response. Genotyping-by-sequencing (GBS) is another high-throughput sequencing-based approach used for SNP discovery and genotyping in gene bank materials. SNP markers and GBS are now jointly utilized for genetic mapping and genetic diversity analysis of a large number of wild accessions and landraces of crop species that are used in marker-assisted selection (MAS) breeding to identify genes linked to resistance viz. drought resistance in maize (Dube et al. 2023), aphid resistance genes in pea (Parveen et al. 2012), genes associated with bacterial wilt resistance in tomato (Foolad and Panthee 2013), rust diseases in wheat (Kokhmetova and Atishova 2012), genes for salinity tolerance in rice (Singh et al. 2021), and genes linked to common rust in maize (Kibe et al. 2020). The use of molecular markers in wild species and landraces offers an avenue to enhance the genetic diversity of cultivated crops, thereby improving morphological, agronomic, and physiological traits—their resilience and performance in the face of changing environmental conditions.

4.3 Transcriptomics

Transcriptomics studies, which involve sequencing transcriptomes, are crucial for identifying novel alleles linked to traits in wild relatives. Next-generation sequencing technologies of RNA sequencing enable understanding of gene expression levels against biotic and abiotic stresses in crop wild resources, aiding in the development of crops resilient to changing climates. Transcriptomic investigations, facilitated by genome-wide association analysis, assist in identifying distinct transcripts and their mechanisms of action in response to biotic and abiotic stresses (Huang et al. 2024). Previously, candidate genes were identified for disease resistance in wild maize relatives, salt tolerance in wild barley, and drought tolerance in wild wheat relatives through transcriptomics studies (Niu et al. 2023). Moreover, integrating genomics and proteomics with transcriptomics reveals gene flow pathways in physiological processes, development, and stress responses of landraces and their wild relatives, helping compare them with domesticated species and thereby enhancing the understanding of the genetic basis of adaptation of genetic resources under diversified environmental conditions. Several wild crops, such as wild accessions of tomato, grapevines, and cassava, are now domesticated with the help of transcriptomics studies (Niu et al. 2023). Additionally, transcriptomics assesses the stability of gene expression in preserved accessions throughout time, maintaining the genetic integrity of conserved materials. The integration of transcriptomics with traditional germplasm bridges the gap between preserved genetic resources and their practical application in breeding by finding expressed genes linked to agronomic parameters. Furthermore, it helps to generate functional markers and identify potential genes for marker-assisted selection, which expands the importance of transcriptomics in maximizing gene bank resources for worldwide agricultural development.

4.4 Epigenomics and Metagenomics

Epigenetics and metagenomics introduce a new dimension to gene bank management by revealing functional diversity across DNA sequences. This comprises characteristics influenced by environmental factors, stress responses, and adaptability processes. Epigenomics is a useful tool for exploring the regulation of gene expression and comprehending structural changes in the genomes of wild relatives and landraces. This approach reveals structural differences that affect essential agronomic properties such as stress tolerance, disease resistance, and yield potential by examining DNA methylation, histone modification, and non-coding RNA profiles. Epigenetic changes allow plants to adapt to environmental variations by modifying gene expression, which contributes to the development of climate-resilient crops. For instance, Zhang et al. (2017) identified drought-tolerant genes in wild rice (*Oryza rufipogon*) and Liu et al. (2024) observed epigenetic variation for disease resistance in wheat landraces. Memory plasticity is one of epigenetics' most intriguing features. Secco et al. (2015) discovered that epigenetic memory of heat stress in barley influences progeny performance in similar environments. Metagenomics, on the other hand, sheds light on the microbial communities found in gene bank samples, providing new views on how microbes influence seed health, viability, and genetic stability. Gene banks may improve seed preservation practices, increase crop germplasm quality, and protect precious genetic resources for future agricultural requirements by applying metagenomic approaches to detect helpful and hazardous bacteria.

4.5 Phenomics

Phenotypic expressions of plant accessions are greatly altered by their physiological functions and environmental effects. The study of phenomics provides valuable information by mapping the phenotypic traits with genomic data through the comprehensive evaluation of molecular mechanisms of heritable traits (Wang et al. 2023). The advent of several types of high-throughput phenomics tools, such as automated imaging, sensor technologies, and robotics, helps to rapidly capture and computerize the interaction effects of environmental fluctuations on morphological, agronomic, and physiological traits (Wang et al. 2023). These approaches significantly reduce the cost of research work and assess the phenotypic variation of a large number of plant genetic resources in multiple environments. Moreover, these approaches help to identify the functional diversity and trait potential within the germplasm collections and accelerate the selection of superior accessions. Additionally, the proper documentation of phenological traits, including flowering duration, germination percentage, and different growth stages of plants, along with physiological traits, viz., stress tolerance and disease resistance, can be successfully used for developing improved cultivars in future breeding programs. For instance, phenomics tools like 2D and 3D structural models (automated imaging) were successfully employed for salt stress and yield improvement in barley (Secco et al. 2015), enhanced nutrition quality in rice (Zhang et al. 2017), drought-tolerant wheat development (Liu et al. 2024), and heat-tolerant chickpea (Pappula-Reddy et al. 2024). The integrated approaches of genome-wide association studies, transcriptomics, and phenomics reveal the genetic architecture in maize through the proper linkage between genotypic and phenotypic traits (Wang et al. 2023). In recent years, the online tool Crop-GPA has emphasized the relevance of biological processes of gene-phenotype correlations in current crop genetics, assisting in the discovery of trait-related genes for crop breeding through dynamic graphical representations (https://crop-gpa.aielab.net). In the most recent Crop-GPA, 2964 of the 10,566 genes with at least one phenotype hit demonstrated a unique gene-phenotype connection by correlating to multiple attributes. Overall, Crop-GPA provides a comprehensive collection and two computational tools for analyzing and exploring crop phenotypic data. The advancement of phenomics tools enables gene banks to go beyond genetic preservation, fostering the identification of superior germplasm for immediate and future agricultural challenges.

Integrated omics approaches, which include genomics, transcriptomics, epigenetics, and phenomics, have proven advantageous in maximizing crop genetic resources. These techniques give a complete knowledge of the molecular mechanisms underlying characteristics of interest, allowing for the effective use of genetic diversity found in gene banks.

4.6 Integration of Genomic Data into Global Gene Bank Databases

Gene banks are critical for preserving plant and animal genetic resources, and the incorporation of genomic data into these databases has increased their significance for study, breeding, and conservation. Table 1 summarizes the major components of genomic data integration, along with crop-specific examples and references.

5 Digital Gene Banking

Digital gene bank (DGB) is a groundbreaking concept in bio-conservation that combines cutting-edge technologies with genetic data preservation. It is designed to preserve and protect genetic resources in a digital format that offers an innovative solution to preserve genetic resources by ensuring the long-term sustainability of biological diversity. Unlike traditional gene banks, which primarily store physical genetic material such as seeds or DNA samples, a DGB operates in the digital realm, housing an extensive database of genomic sequences, gene expressions, and phenotypic data. This state-of-the-art facility serves as a ready-to-go repository for genetic information, enabling researchers, conservationists, and policymakers to access and utilize invaluable genetic data at any time from any place. This innovative approach is transforming traditional methods of gene conservation, ensuring that valuable genetic data is preserved for future generations.

A digital gene bank is a centralized digital repository designed to store, manage, and analyze genetic data from various sources, such as plants, animals, and microorganisms. The backbone of such a system relies on advanced technologies like cloud computing, big data analytics, blockchain, and artificial intelligence (AI). At its core, the digital gene bank serves as a digital infrastructure that aggregates genetic sequences, metadata, and related research outputs, providing researchers, policymakers, and industry professionals with seamless access to this invaluable information. Implementing a cloud-based infrastructure ensures scalability, global accessibility, and cost-effective storage of vast volumes of genetic data. Genetic sequences, associated metadata, and analytical results require terabytes or even petabytes of storage. Cloud service provides rapid data availability while enabling users to analyze genomic sequences without downloading large datasets. Blockchain technology, on the other hand, can be utilized for secure and transparent data sharing. By implementing a distributed ledger system, institutions and researchers can contribute genetic data with verifiable authenticity and maintain strict control over intellectual property

Table 1 Successful case studies of genomic data integration

Case study	Impact on crop improvement programs	References
Sugarcane genomics for conservation	Discovered genetic diversity through genomic and phenomics studies of sugarcane resources for stress resistance	Meena et al. (2022)
Cassava genome hub	Integrated genomic data from wild resources to cultivated varieties yield and pest-resistance traits	Prochnik et al. (2012)
Apple biodiversity conservation	Integrated genomic data from wild and cultivated apple species for pest resistance and improved fruit quality	Cornille et al. (2012)
SNP-seek database for rice	Integrated genomic data from 3,000 rice genomes for the identification of genes linked to stress tolerance and yield improvement	Li et al. (2014)
Lentil genomic resources	Integrated genomic sequencing data of global lentil germplasms breeding for disease resistance and climate resilience	Rajpal et al. (2023)
IRRI rice gene bank integration	Combined phenotypic, passport, and genomic data for over 100,000 rice accessions for climate resilience rice improvement programs	McCouch et al. (2016)
Peanut genomic analysis	Combined genomic and phenotypic data of peanut wild relatives for disease resistance and yield enhancement	Huang et al. (2023)
Integrated barley genome database	Integrated genomic sequencing data from wild and cultivated barley for disease resistance and agronomic traits	Li et al. (2022)
Wheat IS portal	Integrated wheat genomic, phenotypic, and environmental data for yield improvement	Alaux et al. (2018)
Banana genome hub	Consolidated genomic data from wild and cultivated banana species for disease resistance and yield improvement	Droc et al. (2022)
Chickpea meta database	Integrated resource genomic and environmental data for chickpea resources for drought tolerance and nutritional improvement	Roorkiwal et al. (2020)
Wild tomato genomic conservation	Discovered candidate genes through genomic analysis of wild tomato species for salinity and pest resistance	Wang et al. (2023)

rights. Smart contracts within blockchain systems can automate data-sharing agreements, ensuring compliance with regulations like the Nagoya Protocol for Access and Benefit Sharing. Furthermore, blockchain ensures data integrity, preventing unauthorized tampering or loss. The application of AI and machine learning (ML) enhances the functionality of a digital gene bank by enabling real-time analysis and pattern recognition. Tools such as deep learning algorithms can identify mutations, predict gene functions, and analyze relationships between genetic traits and environmental conditions. For instance, AI-powered tools can analyze plant genetic data to predict crop yields, identify disease-resistant genes, or guide precision breeding strategies. Big data analytics platforms like Apache Spark or Hadoop can process and analyze large datasets efficiently, extracting insights from petabytes of genetic data. Furthermore, Internet of Things (IoT) devices can integrate with the gene bank to enable real-time collection of genetic samples from remote locations. IoT-enabled devices, such as portable DNA sequencers or lab-on-a-chip technologies, can transmit sequencing data directly to the cloud. This enhances the speed of data accumulation and allows researchers to remotely monitor the quality of collected genetic data. Collectively, these technologies ensure that a digital gene bank is not only a static repository but a dynamic, interactive platform that fosters collaboration and drives innovation in genomics, agriculture, healthcare, and biodiversity conservation.

However, setting up a digital gene bank offers transformative benefits for preserving genetic diversity, advancing research, and fostering innovation in agriculture, medicine, and environmental conservation. Digital gene banks enable the storage and sharing of vast amounts of genetic information, safeguarding biodiversity threatened by habitat loss, climate change, and overexploitation. Unlike traditional gene banks, digital systems eliminate the need for maintaining physical samples under controlled conditions, allowing for greater accessibility and global collaboration. Advanced technologies like artificial intelligence, machine learning, and big data analytics enhance research by uncovering patterns, predicting gene functions, and streamlining data analysis. However, these benefits come with challenges, including the high costs of infrastructure, data digitization, and maintenance. Data security and intellectual property issues further complicate implementation, requiring robust systems to prevent breaches and misuse while addressing ownership and ethical concerns. Standardization and interoperability are additional hurdles, as universally accepted formats and protocols are necessary to integrate data seamlessly across platforms, and technological advancements demand continuous system upgrades.

Despite challenges, digital gene banks hold immense potential to revolutionize genetic resource management, and this feature encouraged several countries and international organizations to set up or digitize gene bank facilities. Some of the noteworthy cases are as follows. A key component of the digital gene banking system is presented in Fig. 3.

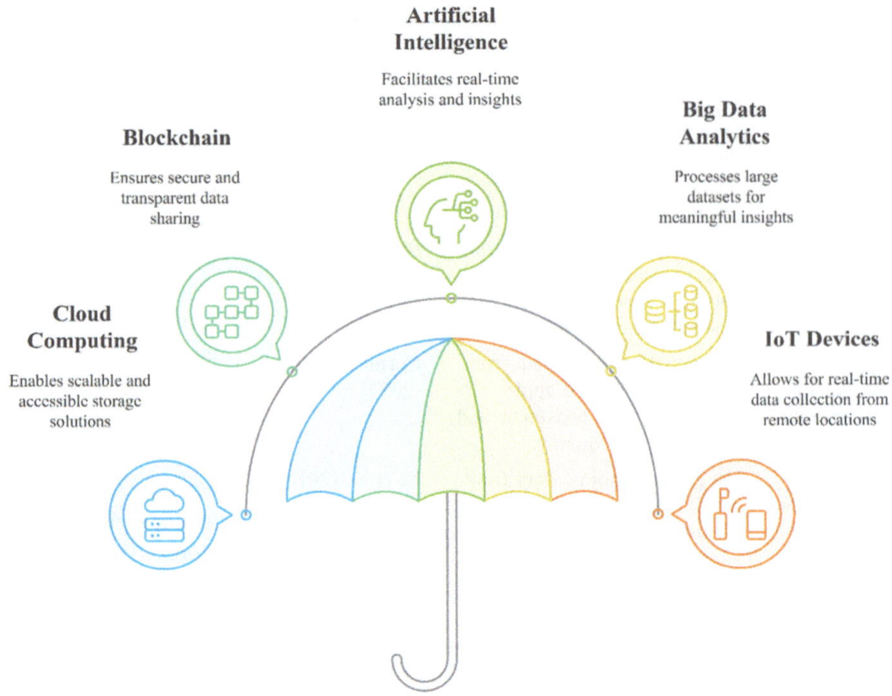

Fig. 3 Key components of digital gene banking system

5.1 CGIAR Gene Bank Platform

The CGIAR (Consultative Group on International Agricultural Research) gene bank platform manages over 750,000 crop and tree samples, making it one of the largest collections of plant genetic resources globally (Sayers et al. 2024). Its integrated advanced digitization system stores and updates data on genetic traits, metadata, and seed viability in real time through the centralized digitization tools. Moreover, its digitized inventory system allows seamless sharing of resources with global researchers. CGIAR employs bioinformatics tools and database management systems to ensure efficient data curation, access, and interoperability across multiple gene banks in its network.

5.2 Genesys

Genesys is a fully digitized global portal developed by the Crop Trust that aggregates data from gene banks worldwide. It provides a user-friendly interface to search, discover, and request genetic resources. The platform leverages cloud-based storage and sophisticated metadata tagging to categorize and manage genetic information. Genesys's digital infrastructure ensures open access to data, supporting research on crop improvement and biodiversity conservation while fostering global collaboration.

5.3 China National Gene Bank (CNGB)

CNGB incorporates cutting-edge bioinformatics and data storage technologies to digitize genetic resources at a national scale. It houses a comprehensive bioinformatics center that processes and analyzes vast datasets, ensuring accessibility and usability for researchers. The bank uses high-throughput sequencing and cloud-based databases to store genetic information, allowing for efficient data retrieval and analysis. Its digital architecture also supports public platforms for data sharing and research collaboration.

5.4 DivSeek International Thematic Hub

DivSeek's digitized approach focuses on integrating advanced genomics technologies into gene banks. By cataloging genetic diversity using next-generation sequencing and bioinformatics tools, it ensures the accessibility of data for research and breeding programs. DivSeek emphasizes the standardization of digital data formats, enabling seamless integration with global genetic repositories. Its cloud-based infrastructure allows for collaborative data analysis and resource sharing among international scientists.

5.5 NCBI GenBank

NCBI GenBank is one of the most advanced digitized nucleotide sequence databases globally, offering free public access to genetic data contributed by researchers worldwide. The database stores and organizes millions of sequences with comprehensive annotations. GenBank is integrated with tools like BLAST (Basic Local Alignment Search Tool), enabling efficient sequence comparison and analysis (Sayers et al. 2024). Its digitized systems ensure global data accessibility and interoperability with other nucleotide sequence databases, such as EMBL-EBI and DDBJ.

5.6 European Nucleotide Archive (ENA)

The ENA at EMBL-EBI exemplifies the power of digitization in genomic data management. The platform collects, processes, and archives raw sequencing data and metadata using high-performance computational tools. It ensures seamless integration with the International Nucleotide Sequence Database Collaboration (INSDC), providing researchers with consistent access to a global repository of genetic data (Sayers et al. 2024). ENA's digitized workflows allow for automated submission, validation, and curation of datasets.

5.7 DNA Data Bank of Japan (DDBJ)

The DNA Data Bank of Japan (DDBJ) is a leading nucleotide sequence database that forms part of the International Nucleotide Sequence Database Collaboration (INSDC) along with GenBank and ENA. DDBJ accepts raw and annotated genetic data from researchers worldwide, using advanced digitization tools to ensure the integrity, accuracy, and accessibility of the data. The platform's automated workflows facilitate data submission and validation, while its integration with INSDC enables global data sharing and interoperability (Sayers et al. 2024). DDBJ's user-friendly interfaces and tools support a broad range of applications, from basic research to applied genomic studies.

5.8 Svalbard Global Seed Vault

While primarily a physical seed repository, the Svalbard Global Seed Vault relies on a robust digital tracking system to catalog and monitor its inventory of millions of seed samples. Its digital system records detailed metadata, including the origin, genetic traits, and storage conditions of seeds (Sayers et al. 2024). This ensures the long-term management of genetic diversity and facilitates access for global gene banks seeking backup resources.

5.9 Nordic Genetic Resource Centre (NordGen)

NordGen employs a fully digitized system to catalog and share genetic resources across Nordic countries. Its database tracks genetic traits of plants, animals, and forests, enabling researchers and breeders to access data remotely. NordGen uses digital platforms to support sustainable agriculture and forestry by providing tools for data analysis, visualization, and resource planning (Sayers et al. 2024). Its focus on metadata standardization enhances global data sharing and interoperability with other gene banks.

6 Challenges and Their Counterstrategies Regarding the Establishment of Gene Banks

Gene banks act as vital repositories for preserving genetic resources, which are crucial for biodiversity conservation, food security, and scientific advancements. However, apart from various benefits, several challenges also hinder the establishment and management of gene banks. Below, some of these challenges are explained along with their possible solutions.

6.1 High Costs and Funding Limitations

Gene banks require substantial financial investment to build infrastructure, purchase advanced equipment, and cover operational expenses such as staff salaries and maintenance. Additionally, the cost of long-term preservation techniques, like cryopreservation, can be prohibitive for many regions, especially in developing countries. Governments and institutions often struggle to allocate sufficient funds, particularly when economic priorities are directed elsewhere. This financial burden can slow down or even halt the establishment of gene banks. To overcome this, governments and international organizations can provide grants and funding. Public–private partnerships can also help share costs. Research into cost-effective preservation methods and energy-efficient solutions, such as using renewable energy, can reduce operational expenses over time.

6.2 Technological Limitations

Effective gene banks depend on cutting-edge technologies like cryopreservation for freezing biological materials, genome sequencing for cataloging genetic diversity, and sophisticated data management systems. However, such technologies are not always accessible in limited-resource regions due to high costs, lack of expertise, or absence of infrastructure. As a result, these areas may face challenges in establishing or maintaining gene banks (Engles et al. 2021). To mitigate this, international collaborations can facilitate the transfer of technology and expertise. Additionally, local governments and research institutions can invest in developing affordable, region-specific technologies tailored to the needs and capacities of their regions.

6.3 Environmental Risks

Gene banks are vulnerable to various environmental risks, such as earthquakes, floods, fires, or other natural disasters, which can damage infrastructure and compromise stored genetic materials. Climate change adds another layer of unpredictability, with rising temperatures and extreme weather events threatening the stability of storage conditions. These risks highlight the importance of ensuring the physical safety of gene banks. To address this issue, multiple gene banks could be established in geographically diverse and unstable areas to reduce the risk of total loss. Additionally, facilities should be built with disaster-resilient designs and equipped with robust emergency response plans to protect stored materials.

6.4 Biodiversity Erosion

Genetic diversity is rapidly being lost due to habitat destruction, climate change, overexploitation, and urbanization. Many species and plant varieties become extinct before their genetic material can be collected and preserved. This poses a significant challenge for gene banks, as they may miss the opportunity to conserve valuable genetic resources. Prioritizing the collection of endangered species and threatened plant varieties is essential. Raising public awareness about the importance of biodiversity and implementing urgent conservation efforts can also help address this issue. Fieldwork to collect samples from areas at high risk of biodiversity loss should be expedited.

6.5 Legal and Ethical Issues

The ownership and access rights to genetic resources often lead to disputes between countries, organizations, or indigenous communities. Questions arise over who owns the genetic material and how the benefits derived from it (e.g., commercial profits from pharmaceuticals) are shared. These issues create legal and ethical complexities that can delay or prevent the establishment of gene banks. Clear and transparent policies based on international agreements, such as the Nagoya Protocol on Access and Benefit Sharing, can help resolve these conflicts. These policies should ensure fair distribution of benefits and respect for the rights of local communities.

6.6 Limited Expertise

Operating a gene bank requires a high level of expertise in fields such as genetics, cryobiology, data management, and conservation science. However, many regions face a shortage of trained professionals who can handle these complex operations. This lack of expertise can result in inefficient management or even the failure of a gene bank. Governments and academic institutions can address this challenge by investing in training programs and workshops to build local capacity. Partnerships with international organizations and universities can also facilitate knowledge transfer and provide the necessary technical support.

6.7 Storage and Maintenance Challenges

Maintaining genetic materials in optimal conditions requires precise environmental controls, such as maintaining ultra-low temperatures and stable humidity levels. This long-term storage can be resource-intensive, leading to high energy consumption and operational costs. In regions with unreliable energy supplies, maintaining these conditions becomes even more challenging. Adopting energy-efficient storage technologies and utilizing renewable energy sources, such as solar power, can help reduce costs and ensure reliability. Regular monitoring and maintenance of stored materials are also critical to prevent degradation and ensure their long-term viability.

6.8 Data Management Issues

Gene banks generate and manage vast amounts of data, including genetic profiles, preservation conditions, and cataloging information. Without proper data management systems, this information can become disorganized, inaccessible, or prone to errors. This can undermine the effectiveness of the gene bank and reduce its utility for researchers and conservationists. To address this, advanced digital platforms and centralized databases should be implemented to streamline data collection, storage, and sharing. Establishing international standards for data management can further enhance consistency and facilitate global collaboration.

6.9 Public Awareness and Support

Despite their critical role in conserving biodiversity and ensuring food security, gene banks often receive limited public attention. This lack of awareness can lead to insufficient funding, weak policy support, and limited community engagement. Many people do not realize the importance of preserving genetic resources for future generations. Public awareness campaigns can help educate people about the benefits of gene banks. Engaging communities in conservation initiatives, such as local seed banks or citizen science projects, can foster greater support and involvement in genetic preservation efforts (Engles et al. 2021).

6.9.1 Cross-Border Collaboration

Genetic resources often need to be shared across countries and regions to ensure effective conservation and utilization. However, political tensions, trade restrictions, and a lack of international cooperation can create barriers to the exchange of genetic materials and expertise. These challenges can limit the effectiveness of gene banks in addressing global conservation needs. Diplomatic efforts and international agreements, such as those promoted by the Svalbard Global Seed Vault, can encourage cross-border collaboration (Sayers et al. 2024). Establishing global networks and partnerships can also facilitate the sharing of resources and knowledge.

By addressing these challenges comprehensively, gene banks can be established and maintained effectively. This will play a crucial role in preserving the world's genetic heritage, ensuring biodiversity, and supporting sustainable development.

7 Future Prospects

In this dynamic era of genomics, it is obligatory to utilize the new technologies of genomics and phenomics data in gene bank management. The current genomic revolution has the potential to significantly alter gene bank operations; this potential is unlikely to materialize unless there is a fundamental change in the way gene bank observation, structure, operation, administration, and funding are conducted. Training on genomics, phenomics, bioinformatics, and population genetics is needed to improve the capacity of scientists along with other gene bank personnel. For efficient utilization of plant genetic resources of the gene bank, collaborate effectively with the stakeholders. A number of partnerships and collaborations can be developed to conserve, manage, and use Plant Genetic Resources for Food and Agriculture (PGRFA) efficiently, like Genesys, Global Information System (GLIS), and European Co-operative Programme for Plant Genetic Resources (ECPGR). Digital Object Identifiers (DOIs) can be used easily and perpetually to share information on PGRs with different organizations. These connections and partnerships will enable gene banks to make use of detailed information on PGR data, technical support, and infrastructure resources accessible to the user community, which is still a major barrier to the efficient conservation and sustainable use of PGR today. Technological advances such as next-generation sequencing (NGS), whole genome sequencing (WGS), genomic mapping, high-throughput phenotyping (HTP), DNA fingerprinting and genome editing (CRISPR Cas-9), long-read sequencing, and the application of artificial intelligence and machine learning in genomic analysis have made it possible to collect genomic and phenomic data for gene bank collections, enabling breeding programs to more easily develop climate-resilient, biotic- and abiotic-stress–tolerant, high-yielding varieties. The genomic era presents unprecedented opportunities to conserve genetic resources, which will ensure careful navigation of equitable resource sharing and of ethical, technical, and social challenges. Collaborative global efforts, equitable access to technology, and significant integration of genomics into conservation strategies will be key to safeguarding the sustainability of biodiversity for future generations.

8 Conclusion

The genomic era has revolutionized the conservation of genetic resources, offering unprecedented opportunities to preserve and utilize biodiversity. Gene banks, as critical repositories of genetic diversity, ensure food security, ecological resilience, and scientific advancement by safeguarding the genetic material of plants, animals, and microorganisms. Advanced genomic tools such as genome sequencing, gene editing, and bioinformatics have enhanced traditional conservation strategies, enabling precise characterization, efficient storage, and targeted utilization of genetic resources. Gene banks also serve as reservoirs for adaptive traits that can be harnessed to develop resilient crops, restore ecosystems, and mitigate biodiversity loss. However, their success in the genomic era depends on robust policies, international collaboration, and sustainable funding. By integrating cutting-edge genomic approaches with traditional practices, gene banks play a pivotal role in preserving the planet's genetic heritage, ensuring the survival of countless species, and supporting the sustainable development of human societies. This synergy secures a resilient future for generations to come.

References

Abebe AM, Choi J, Kim Y, Oh CS, Yeam I, Nou IS et al (2020) Development of diagnostic molecular markers for marker-assisted breeding against bacterial wilt in tomato. Breed Sci 70(4):462–473. https://doi.org/10.1270/jsbbs.20027

Alaux M, Rogers J, Letellier T, Flores R, Alfama F, Pommier C et al (2018) Linking the international wheat genome sequencing consortium bread wheat reference genome sequence to wheat genetic and phenomic data. Genome Biol 19(1):111. https://doi.org/10.1186/s13059-018-1491-4

Baloch FS, Alsaleh A, Shahid MQ, de Miera VCLES, Aasim M, Nadeem MA et al (2017) A whole genome DArTseq and SNP analysis for genetic diversity assessment in durum wheat from central fertile crescent. PLoS ONE 12(1):e0167821. https://doi.org/10.1371/journal.pone.0167821

Commission on Genetic Resources for Food (2010) The second report on the state of the world's plant genetic resources for food and agriculture, vol 2. Food & Agriculture Organization of the UN (FAO)

Cornille A, Gladieux P, Smulders JM, Roldán-Ruiz I, Laurens F, Le Cam B et al (2012) New insight into the history of domesticated apple: secondary contribution of the European wild apple to the genome of cultivated varieties. PLOS Genet. https://doi.org/10.1371/journal.pgen.100270

Droc G, Martin G, Guignon V, Summo M, Sempéré G, Durant E et al (2022) The banana genome hub: a community database for genomics in the Musaceae. Hortic Res. 9:221. https://doi.org/10.1093/hr/uhac221

Dube SP, Sibiya J, Kutu F (2023) Genetic diversity and population structure of maize inbred lines using phenotypic traits and single nucleotide polymorphism (SNP) markers. Sci Rep 13(1):17851. https://doi.org/10.1038/s41598-023-44961-3

Engels JM, Ebert AW (2021) A critical review of the current global ex situ conservation system for plant genetic resources. Plants 10(10):1904. https://doi.org/10.3390/plants10081557

Engels JMM, Ebert AW, van Hintum T (2024) Collaboration between Private and Public Genebanks in Conserving and Using Plant Genetic Resources. Plants 13(2):247. https://doi.org/10.3390/plants13020247

Foolad MR, Panthee RD (2013) Marker-assisted selection for bacterial wilt resistance in tomato using SNP markers. Plant Genome. 6(2):99–108. https://doi.org/10.3835/plantgenome2013.03.0011

Huang R, Li H, Gao C, Yu W, Zhang S (2023) Advances in omics research on peanut response to biotic stresses. Front Plant Sci 14:1101994. https://doi.org/10.3389/fpls.2023.1101994

Huang P, El-Soda M, Wolinska KW, Zhao K, Davila Olivas NH, van Loon JJA et al (2024) Genome-wide association analysis reveals genes controlling an antagonistic effect of biotic and osmotic stress on *Arabidopsis thaliana* growth. Mol Plant Pathol 25(3):e13436. https://doi.org/10.1111/mpp.13436

Kibe M, Nyaga C, Nair SK, Beyene Y, Das B, M SL et al (2020) Combination of linkage mapping, GWAS, and GP to dissect the genetic basis of common rust resistance in tropical maize germplasm. Int J Mol Sci 21(18):6518. https://doi.org/10.3390/ijms21186518

Kim H, Jung J, Singh N (2016) Population dynamics among six major groups of the oryza rufipogon species complex, wild relative of cultivated Asian rice. Rice 9:56. https://doi.org/10.1186/s12284-016-0119-0

Kokhmetova AM, Atishova MN (2012) Identification of sources of resistance to wheat stem using molecular markers. Russian J Genet Appl Res 16(1):132–141. https://doi.org/10.1134/S2079059712060081

Li T, Bian J, Tang M, Shangguan H, Zeng Y, Luo R et al (2022) BGFD: an integrated multi-omics database of barley gene families. BMC Plant Biol 22:454. https://doi.org/10.1186/s12870-022-03846-9

Li Z, Fu B, Gao Y, Wang W, Xu J, Zhang F et al (2014) The 3000 rice genomes project. The 3000 rice genomes project. Giga Sci 3:7. https://doi.org/10.1186/2047-217X-3-7

Liu Y, Fu B, Zhang Q, Cai J, Guo W, Zhai W, Wu J (2024) Genetic diversity and population structure of wheat landraces in Southern winter wheat region of China. BMC Genom 25:664. https://doi.org/10.1186/s12864-024-10564-z

Mahanayak B (2024). Ex-situ and in-situ conservation of wild life. World J Biol Pharm Health Sci 18:277–282. https://doi.org/10.30574/wjbphs.2024.18.3.0371

McCouch S, Wright M, Tung CW et al (2016) Open access resources for genome-wide association mapping in rice. Nat Commun 7:10532. https://doi.org/10.1038/ncomms10532

Meena MR, Appunu C, Kumar R, Manimekalai R, Vasantha S, Krishnappa G et al (2022) Recent advances in sugarcane genomics, physiology, and phenomics for superior agronomic traits. Front Genet 13:854936. https://doi.org/10.3389/fgene.2022.854936

Mishra H, Singh M (2023) The significance of wildlife sanctuaries for Indian agriculture. Indian Farmer 10(05):243–246

Niu Y, Li J, Sun F, Song T, Han B et al (2023) Comparative transcriptome analysis reveals the key genes and pathways involved in drought stress response of two wheat (*Triticum aestivum* L) varieties. Genomics 115(5):110688. https://doi.org/10.1016/j.ygeno.2023.110688

Pappula-Reddy S, Kumar S, Pang J, Chellapilla B, Pal M, et al (2024) High-throughput phenotyping for terminal drought stress in chickpea (*Cicer arietinum* L.). Plant Stress 11:100386. https://doi.org/10.1016/j.stress.2024.100386

Parveen I, Singh M, Pratap A (2012) Molecular marker-based strategies for genetic improvement in peas. Plant Breed 131(5):565–573. https://doi.org/10.1111/j.1439-0523.2012.01999.x

Prasanna BM (2012) Diversity in global maize germplasm: characterization and utilization. J Biosci 37:843–855). https://doi.org/10.1007/s12038-012-9227-1

Prochnik S, Marri PR, Desany B, Rabinowicz PD, Kodira C, Mohiuddin M et al (2012) The cassava genome: current progress, future directions. Trop Plant Biol 5(1):88–94. https://doi.org/10.1007/s12042-011-9088-z

Quan R, Wang J, Hui J, Bai H, Lyu X, Zhu Y et al (2018) Improvement of salt tolerance using wild rice genes. Front Plant Sci 8:2269. https://doi.org/10.3389/fpls.2017.02269

Rajpal VR, Singh A, Kathpalia R, Thakur R, Khan MK et al (2023) The prospects of gene introgression from crop wild relatives into cultivated lentil for climate change mitigation. Front Plant Sci 14:1127239. https://doi.org/10.3389/fpls.2023.1127239

Roorkiwal M, Bharadwaj C, Barmukh R, Dixit GP, Thudi M, Gaur PM et al (2020) Integrating genomics for chickpea improvement: achievements and opportunities. Theor Appl Genet 133:5. https://doi.org/10.1007/s00122-020-03584-2

Sayers EW, Cavanaugh M, Clark K, Pruitt KD, Sherry ST, Yankie L, Karsch-Mizrachi I (2024) GenBank 2024 Update. Nucleic Acids Res 52(D1):D134–D137. https://doi.org/10.1093/nar/gkad903

Secco D, Wang C, Shou H et al (2015) Stress induced gene expression drives transient DNA methylation changes at adjacent repetitive elements. elife 4:1–26. https://doi.org/10.7554/eLife.09343.001

Singh RK, Kota S, Flowers TJ (2021) Salt tolerance in rice: seedling and reproductive stage QTL mapping come of age. Theor Appl Genet 134(11):3495–3533. https://doi.org/10.1007/s00122-021-03890-3

Tripodi P (2022) Next generation sequencing technologies to explore the diversity of germplasm resources: achievements and trends in tomato. Comput Struct Biotechnol J 20:6250–6258. https://doi.org/10.1016/j.csbj.2022.11.028

Walters C, Pence VC (2021) The unique role of seed banks in the conservation of plant genetic resources. Plant Sci 311:110984. https://doi.org/10.1016/j.plantsci.2021.110984

Wang C, Yang X, Zhang Y, Shen C, Shi J, Xia C et al (2023) Barley FASCIATED EAR genes determine inflorescence meristem size and yield traits. Crop J 11:679–691. https://doi.org/10.1016/j.cj.2022.10.001

Zhang J, Long Y, Xue M, Xiao X, Pei X (2017) Identification of microRNAs in response to drought in common wild rice (*Oryza rufipogon* Griff.) shoots and root. PLoS ONE 12(1):e0170330. https://doi.org/10.1371/journal.pone.0170330

Plant Biodiversity Conservation and Biotechnology

Mehrana Koohi-Dehkordi, Mozhdeh Shafaei, Sara Ghasemzadeh, and Mehran E. Shariatpanahi

Abstract

Biotechnology greatly enhances the efficacy of plant conservation strategies. Techniques including tissue culture, cryopreservation, and molecular markers hold the options in regeneration, conservation, and genetic assessment of the plant genetic assets. Through cryopreservation, long-ternetic diversity in naturalm conservation programs can store plant germplasm, and this does not lead to the loss of sensitivity, while through tissue culture, large quantities of endangered species of plants can be regenerated. Furthermore, molecular methods allow for the evaluation of population/genotypic variability, which is crucial when it comes to auspices for creating improved adaptability of species. As the rate of biological diversity declines so rapidly, these applications represent essential biotools in establishing and supporting sustainable systems of agriculture and ecosystems. The combination of biotechnology in conserving the natural plant diversity and other conventional approaches is important in the conservation of ecosystems and the management of plant genetic resources for sustainable development.

Keywords

Biodiversity conservation · Plant propagation · Cryopreservation · Endangered species · DNA barcoding · Genetic engineering

M. Koohi-Dehkordi · M. Shafaei · S. Ghasemzadeh · M. E. Shariatpanahi (✉)
Agricultural Biotechnology Research Institute of Iran (ABRII), Agricultural Research, Education and Extension Organization (AREEO), Karaj, Iran
e-mail: mehran.shariatpanahi@abrii.ac.ir

1 Introduction

There are several practical biotechnological tools for conservation. One of the essential in vitro methods for the cloning of plant materials and creating disease-free stocks is plant tissue culture. Plant tissue culture allows for the quick and sterile multiplication of plants and valuable germplasm. The long-term genetic resource storage is possible due to cryopreservation, where the plant genetic materials are stored at very low temperatures, more preferably at the temperature of liquid nitrogen, that is, − 196 °C (Fonseka et al. 2020). Plant species have been recovered using gene engineering methods, including genetically modified organisms (GMOs), to make plants withstand environmental pressure, diseases, and pests. Some plant species have been preserved by employing genetic engineering methods, including GMOs, to reverse the impacts of environmental pressures, diseases, and pests (Coelho et al. 2020). Plant genetic resource conservation involves in vitro storage technology, where medium-term storage involves slow growth techniques and long-term storage is achieved through cryopreservation for species with resistant seeds or vegetatively propagated ones. The use of biotechnology in the management or conservation and sustainable use of medicinal plants is also eased by the fact that biotechnological methods ensure the conservation of therapeutic species without altering their core characteristics (Kumar et al. 2024).

It was established that the use of more developed techniques in sequencing genomes is crucial to the identification and preservation of endangered plant species through the creation of large genetic databases. Biotechnological tools have been incorporated into plant conservation efforts, and more effective and efficient means of preserving the plant diversity, certifying the survival of threatened species, and maintaining the balance of the ecosystem have been achieved. Through ecosystem restoration and the conservation of plant species, these biotechnological activities are beneficial for environmental recovery. The conservation of

N. Kumar (ed.), *Biotechnology in Plant Conservation*, Sustainable Landscape Planning and Natural Resources Management,
https://doi.org/10.1007/978-3-032-02605-7_9

plants as a subject matter is an essential and interesting discipline in the management of global biodiversity, and the continuous advancement of biotechnological tools and techniques offers perspectives in the handling of issues in this sub-area. Overall, biotechnology has multi-faceted roles within plant conservation, which involve instruments and strategies for sustainable plant genetic resource management (Fonseka et al. 2020).

In the hope of achieving these objectives, seed banks and gene banks have been established to ensure that future generations will also benefit from existing plant genetic resources. Therefore, there is a high possibility of new conservation opportunities given the recent advances in genome editing technologies that can modify plant genomes, including the CRISPR/Cas9 system. These developments with the tools and techniques that enhance the management and utilization of plant genetic resources demonstrate why biotechnology is essential in the conservation of genetic plant diversity (Fonseka et al. 2020). The transition from early theories to the present-day biotechnological use shows that plant conservation continues to evolve to address more challenges in species conservation (Vasil 2008). The adaptation of various biotechnological developments into conservation plans enhances the reliability of plant species conservation; simultaneously, by extension, it enhances the chances of rehabilitating and restoring deteriorating ecosystems. Indeed, it is expected that biotechnology will feature a more significant role in plant conservation as it unleashes further sophisticated techniques to guard the earth's botanical wealth (Fonseka et al. 2020). Looking at the historical development of biotechnology for plant conservation, the true "pathway of progress" is well marked, and each important step has made the prospect of plant conservation and support of the world's plant resources stronger (Vasil 2008). This continuous evolution underlines the need for further research and funding in the biotechnological processes to combat the worsening issues of global biodiversity conservation. In conclusion, the progressive development of biotechnology techniques within plant conservation shows a general improvement of methods and approaches, which together collectively make the nature of plant conservation stronger and more effective in protecting its species (Fonseka et al. 2020).

2 Tissue Culture and Micropropagation Techniques

This section describes the fundamentals of tissue culture for biodiversity conservation, highlighting the technique's application in propagating plant species, especially those considered endangered. It also covers the disadvantages and shortcomings associated with tissue culture, owing to the high costs, contamination, and instability of genes. Nonetheless, tissue culture continues to be a valuable technology for plant germplasm preservation and the promotion of the sustainable use of plant resources.

2.1 Principles of Tissue Culture in Biodiversity Conservation

The idea of using tissue culture strategy in the management of biochemical diversity in plants is anchored on the in vitro processing for the prescription of reproductive procedures for plant species that are often difficult to propagate or do not have functional seeds. By employing such a biotechnological method of plant tissue culture, there is the possibility of multiplying plants at high rates, which helps in the management of plant genetic resources and the storage of germplasm in different environments. Methods including slow growth culture and cryopreservation remain crucial for short- and long-term conservation to store plant genetic resources for an effective length of time (Rai 2022). Moreover, tissue culture helps improve medicinal plant growth as the stress due to overharvesting and habitat destruction results in genetic depletion and enables the production of secondary metabolites relevant to pharmaceutical industries (Icro 2022). Moreover, the application of molecular biology with tissue culture technology has paved the way for genetic variability and biodiversity improvement to be one of the components that have been successful in raising crop improvement. Therefore, tissue culture plays an important role in the protection of genetic resources, the provision of opportunities for endangered species, and rational plant usage.

2.1.1 Methods and Protocols for Plant Tissue Culture

Tissue culture, a remarkable method in plant biotechnology, has now become essential for the conservation of species diversity. It covers the process of cultivating plant cells, tissues, or organs under aseptic conditions and in a controlled system (Fig. 1). This technology gives several advantages over traditional conservation techniques, such as the ability to propagate germplasm quickly, eliminate diseases, and store material for long-term purposes.

Micropropagation, somatic embryogenesis, and cryopreservation are some of the tissue culture techniques of major importance in the conservation of biodiversity. One of the most important tissue culture techniques described is micropropagation, which accelerates the production of plant material with identical genetic markers required for the conservation of endangered species and guarantees the usage of high-quality planting material, such as in the case of banana plants (Sandeep 2022). This method is also important in preserving medicinal plants because it is a creation of a large stock of uniform plant materials that are important

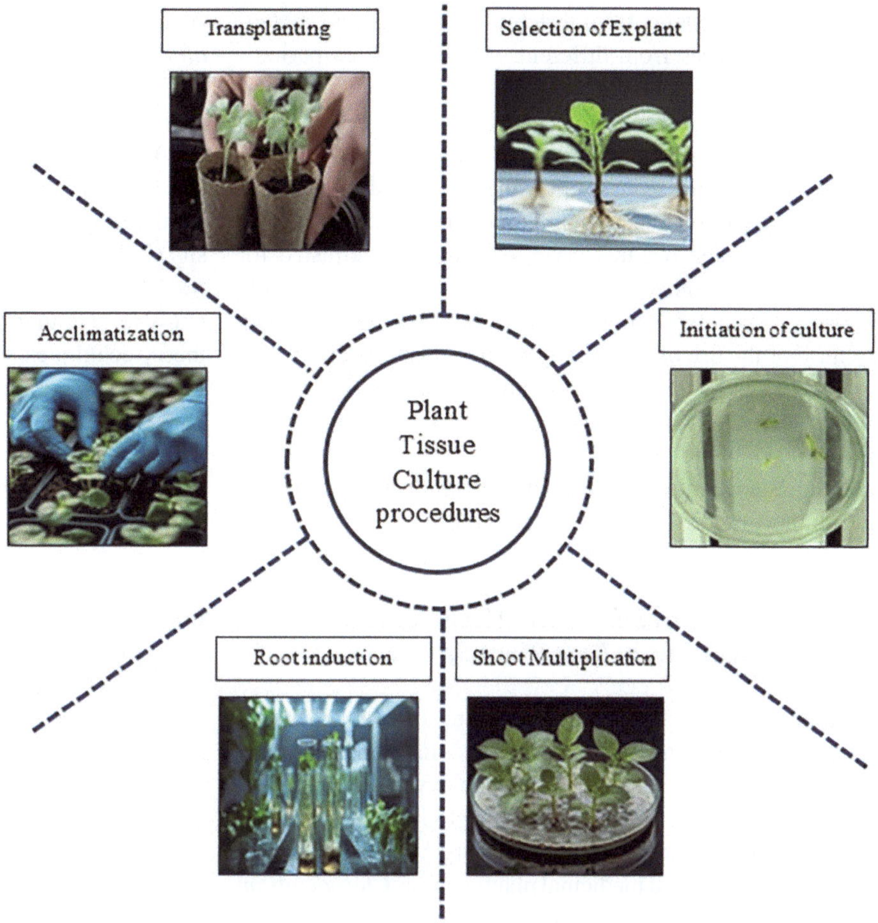

Fig. 1 The schematic diagram of plant tissue culture procedures

in the formulation of standardized phytopharmaceuticals. The other tissue culture technique is somatic embryogenesis, which enables the regeneration of plants from somatic cells that are able to overcome the hurdles of sexual reproduction and act as a good technique for conservation, particularly for species that produce seeds with low vigor. In addition, cryopreservation offers a long-term conservation regime by preserving plant tissues at extremely low temperatures to promote the conservation of the plant material for future recovery. This method, combined with slow-growth storage, is often employed for the banking of various plant species, many of which have recalcitrant seeds or are endangered species (Rai 2022). Globally, these tissue culture techniques provide synergistic strategies for plant genetic resource management and utilization for plant biodiversity conservation (Rai 2022).

2.1.2 Complete Protocol for Micropropagation

The standard process of micropropagation comprises a set of key procedures, the major stages of which are as follows: explant preparation, sterilization, culture medium establishment, and acclimatization. Primary explants can be leafing sectors, nodal sectors, or shoot tips, and the result of the studies conducted using explants from *Mucuna pruriens* and *Zamioculcas zamiifolia* on G0 indicated that the plant growth regulators benzyl amino purine (BAP) and naphthaleneacetic acid (NAA) are vital for shoot and root formation (Thakur et al. 2024). The type of culture medium, which can be Murashige and Skoog (MS) or Woody Plant Medium (WPM), determines the level of shoot multiplication and elongation; highly effective PGR combinations boost these processes (Özkul et al. 2024). For example, in black mulberry and pomegranate micropropagation, regeneration discovered that the appropriate amount and types of plant growth regulator (PGR) achieved different levels of shoot and root formation. Lastly, there is a need to acclimatize the plantlets to the ex vitro environment. Recovery of the plantlets ranges from 85.8 to 93% when the right media and environmental conditions are provided (Özkul et al. 2024).

2.1.3 Additional Considerations for Biodiversity Conservation

Genetic variation in the tissue culture collections of plants needs to be preserved because it increases their resistance to diseases and unfavorable conditions. This can be done by harvesting explants from different organisms and replenishing the cultures from different samples at least periodically, as genetic problems are identified in in vivo populations, like that in *Hancornia speciosa* (Fajardo et al. 2018). The main challenge that must be avoided is contamination by employing strict aseptic measures and continually screening for infections to maintain cultural wellness. It has also been pointed out that cryopreservation is adjusted for each species to enhance the chances for subsequent viable recovery. Successful reintroduction has also been known because of the use of genetically diverse plant material; such approaches have demonstrated enhanced survival in experiments, especially when the genetic mix is brought from different populations (Sandeep 2022). Hence, tissue culture techniques are central to preserving the biodiversity of plants and supporting the cause of conservation.

2.2 Examples of Successful Tissue Culture Applications in Rare Plant Species

It is possible to show that worthy techniques of tissue culture corresponding to rare plant species have been involved through some strategies related to conservation and reproduction. For instance, the micropropagation technology has been effectively used to produce a large number of short plantlets of threatened medicinal plants such as *Clerodendrum serratum* and *Uraria picta* through 100% success in rooting and constant, exponential shoot multiplication in the extended culture period (Sharma et al. 2016). Moreover, the culturing of plant species and cultures on the Murashige and Skoog (MS) media supplemented with growth regulatory agents such as benzyl aminopurine (BAP) has made it possible to clone some scarce species ex vivo, as well as maintain genome stability and conserve optimum levels of valuable secondary metabolites (Sharma et al. 2016). New techniques using some optimal environmental conditions have been used in managing risk species such as *Lamium glaberrimum*, proving the capability of tissue culture in maintaining plant viability during the slow-growing phase (Ivanova et al. 2022). These examples call for the postnatal actions of tissue culture in the ex situ conservation of rare and endangered plant species, especially amid habitat destruction and climate change (Sharma et al. 2016).

Micropropagation through tissue culture has subsequently proved important for both plant breeding and genetic conservation, and good results have been reported across most taxonomic ranges. For example, the propagation of cycads, *Encephalartos woodii*, has been done through micropropagation, greatly boosting genetic variation and conservation (Thakur et al. 2024). The Lady's Slipper Orchid (*Cypripedium calceolus*) is another plant that originated in tissue culture, where genetic multiplication enabled the formation of a population vulnerable to habitat destruction (Thakur et al. 2024). A few others to mention include the making of a *Santalum album* to support sustainable cutting and the making of the Himalayan Blue Poppy (*Meconopsis betonicifolia*), which is essential for making conservation programs (Thakur et al. 2024). Moreover, tissue culture has enabled the conservation of endemic ferns and Hoya plants and hence an appropriate approach to deal with the problem of decreasing species diversity (Pithiya et al. 2022). In general, these techniques not only fund the propagation of endangered species at very fast rates but also guarantee that genetic purity will be upheld; thus, these techniques serve a central role in the formulation of global conservation policies.

2.3 Challenges and Limitations of Tissue Culture

Although the tissue culture technology has many benefits for plant propagation and conservation, some technical and economic obstacles prevent its wide adoption. The first is the cost factor because tissue culture requires sophisticated equipment and expert staff, which increases the cost (Kumari et al. 2023). The second factor is time, which makes tissue culture costly because it takes time to provide the necessary facilities and skilled human resources (Pithiya et al. 2022). The third is contamination, further increasing the costs. Genetic instability is another problem because somaclonal variations are possible, and this is reflected in the uniformity and quality of the propagated plants (Pithiya et al. 2022). However, tissue culture is still crucial in biotechnology and agriculture, where solutions are, among others, conservation of endangered species and aseptic propagation of plants (Thakur et al. 2024). Consequently, there is a need to address these technical and economic factors to enhance the abilities of tissue culture technologies toward sustainable development and improvements in the cultivation of crops (Pithiya et al. 2022). Several restraints have the potential to limit the efficiency of tissue culture, but if these are solved, then tissue culture can become an even more powerful weapon in the fight for biodiversity conservation and ecosystem restoration. This is why scientists and conservationists are working toward the development of better and cheaper tissue culture methods.

3 Genetic Engineering for the Preservation of Species

Genomics and transgenic technologies are very helpful for the conservation of endangered species, as the techniques serve as stronger facilities in studying and preserving the species. Such technology aids in the discovery of genes that induce adaptation and fitness to construct tracking tools for biodiversity. Several strategies that involve genomics have been presented concerning the management of small populations, genetic rescue, and coping with new environments. Reference genomes of model organisms, public genomic data, and population resequencing help with biodiversity evaluation, the management, and restoration of populations, as well as support their decisions. Although these technologies are useful, one has to assess the costs, time, and skills that are required to introduce the technologies (Theissinger et al. 2023).

3.1 Applications of Genetic Engineering

Genetic engineering is a relatively young science used to make changes to the DNA sequences of organisms to bring about the desired traits. This includes cell culture, protoplast fusion, and recombinant DNA technology. Although various methodologies have been used, including homologous recombination and random DNA integration, the CRISPR/Cas9 system has emerged as the gold standard and the most widely used technique for gene modification. This technology has greatly transformed the area of genetics and has several applications in gene mapping. This chapter discusses CRISPR, gene cloning, and the establishment of transgenic plants as the most vital techniques of genetic engineering in the field of biodiversity conservation.

3.1.1 CRISPR as a Gene-Editing Tool

Over the recent past, CRISPR/Cas9 has attracted interest as a great gene-editing tool that is precise, efficient, and versatile in the modification of plant genomes (Fig. 2). This technology has been used in different crop species, thus allowing for the modification of the desired agroecological characteristics (Pandey et al. 2022). At this time, there are at least six Cas proteins available for directing targeted mutagenesis. Of these, Cas9 is the most commonly used CRISPR system adopted for genetic engineering in plants. This property of high capacity and low complexity comes from the need for only Cas9 and one guide RNA for construction, and subsequently, Cas9 excises two genomic sites upon identifying specific DNA sequences with the help of protospacer adjacent motif (PAM). To allow Cas9 to work on a larger set of target sites, proteins have been either evolved or designed to accommodate various PAMs. It has been found that CRISPR/Cas9 could be used to cope with issues in agriculture, such as climate change and food shortage, by designing climate change-resistant and sustainable produce (Haque et al. 2018). This will help fight both abiotic and biotic stresses, hence improving crop production in a dynamic environment. The current technology enjoys benefits over conventional plant breeding practices, including time efficiency in the development of the required traits and production of non-transgenic gene edited (GE) plants (Haque et al. 2018). However, the application of CRISPR/Cas9 has some problems with efficiency and delivery, especially in the difficult target species of crops. Further research and development continue to enhance this technology for use in producing foods that are safe for human consumption. It can be concluded that CRISPR-Cas9, as a tool, can be used to stop invasions of certain species, enhance the key adaptive traits, and overall protect endangered species. However, the experience of its application in conservation produces ethical concerns and may bring some risks to ecosystems. To further optimize their appropriate enhancement and utilization, researchers, regulators, conservationists, and the public at large are needed. Furthermore, accepting the convention of biodiversity might help in the establishment of rules governing gene editing in the conservation processes for wildlife (Pandey et al. 2022).

3.1.2 The Role of Gene Cloning

The application of the gene cloning technique holds great prospects in the conservation of biodiversity, especially for endangered species. Conservation genomics is useful in understanding population structures and adaptive evolution to inform management (Rode et al. 2019). However, the technology is associated with some issues and possible ecological risks when used for species conservation (Rode et al. 2019). Some other areas, which still require a lot of focus in developing an ethical perspective, relate to the manipulation of natural processes. Thus, although technologies of cloning are effective in the conservation of species on the verge of extinction, their usage should not be linear and non-critical, weighing the positive consequences for a particular species

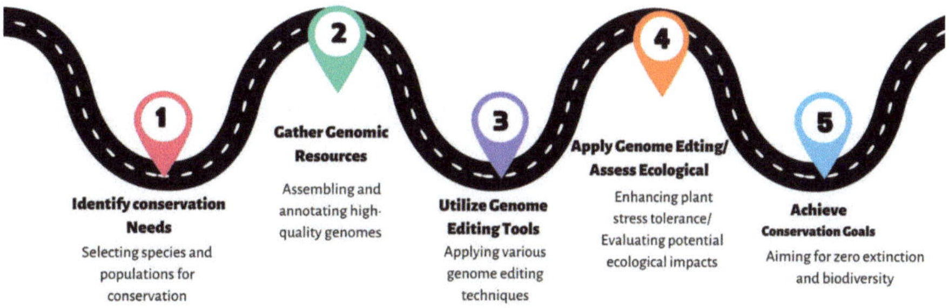

Fig. 2 The roadmap describes the steps involved in creating viable GE variants

reintroduction against the possible negative effects for the ecosystems and general biodiversity.

3.1.3 Transgenic Plants for Environmental Conservation

It is highly possible that transgenic plants would be useful in the area of conservation of biological diversity in that they decrease the use of agrochemicals and improve the restoration of ecosystems. Such transgenic products may be developed to have traits that include pest and disease tolerance and resistance to abiotic stresses that foster the growth of diminished biophysical production of ecosystem services in degraded ecosystems. The overexpression in transgenic plants can also be used to increase their tolerance to astonishing environmental factors such as drought or temperature stress that would increase their survival chance within a species rapidly altering environment. Larger African figs, for example, could be used to recover disturbed ecosystems by reestablishing self-sustaining plant species that bolster regional species diversity (Pathak and Abido 2014).

3.2 Legal Ethical Considerations and Social Impact

The use of GMOs in agricultural practices has raised controversies regarding whether or not to include their effects (Fig. 3). Although some advantages are claimed, like better changes in genotype and greater yield, controversy and skepticism are still related to their effects on the environment and health. These disadvantages entail horizontal gene flow, pest resistance, loss of biodiversity, and possible allergenicity effects (Pretty et al. 2001). Recent developments, particularly regarding the cultivation of genetically modified (GM) crops, have doubly fueled such debates. People's perceptions of GMOs are as diverse: GMOs can be seen by some people as technologies addressing global sustainability questions, by others as approaches threatening traditional agriculture (Pretty et al. 2001). Concerning all these multiple concerns, risk assessment processes should consider social, economic, and other comprehensive processes, such as the strategic environmental assessment. In this respect, the process of weighing the advantages against the disadvantages of GMOs still poses several concerns to policymakers and scientists (Meyer 2011). Biosafety regulatory measures and biochemical policies differ among countries that try to achieve innovative progress as well as minimize risks. While the penetration of GM crops in Africa has remained bleak, this is owing to cultural, social, economic, and environmental differences on the continent. The reason behind this is the need to respond to the public's concern, standardization, and ethical issues. This makes the question of whether the benefits of GMOs or the preservation of the biological variety to be among the most basic and tough questions in the advancement of the farming industry.

4 Molecular Markers and Their Applications

Molecular markers are significantly useful in genetic diversity studies, species identification, and the development of conservation strategies (Li et al. 2020). Different methods of molecular markers exist, including hybridization-based (restriction fragment length polymorphism, RFLP), PCR-based (Random amplified polymorphic DNA (RAPD), Amplified fragment length polymorphism (AFLP), Simple sequence repeat (SSR), Inter simple sequence repeat (ISSR)), and sequence-based (SNP) markers. Microsatellites, commonly known as SSRs, are preferred due to the high resolution in population and individual differentiations (Wenne 2023). SNPs have also been identified as another important marker type underused together with SSR for quantitative trait loci (QTL) mapping, as pointed out by Wenne (2023). These markers are used

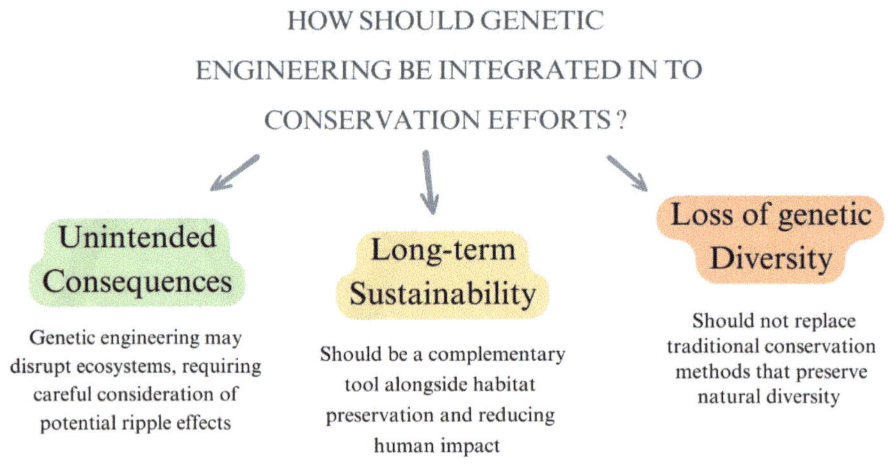

Fig. 3 The ripple effects surrounding genetic engineering

in many conservation processes involving genetic studies, germplasm identification, and control of the genetic purity of conserved seeds in gene pools. Through the use of molecular markers, the estimation of the definite number of conserved species and the formation of appropriate conservation plans and strategies have been enhanced (Li et al. 2020).

4.1 Microsatellites, Single-Nucleotide Polymorphisms, and Their Applications in Conservation Biology

Among the molecular markers, SSRs and SNPs are becoming more important when assessing the genetic structure and differentiation in plant populations. Microsatellites, or SSRs, have become very popular because of their high polymorphism, flexibility, and ability to detect differences within and between populations. They are very important in linkage mapping, association analysis, and plant breeding applications. Given this ability, many conservation applications of SSRs and SNPs include estimating population genetic diversity, detecting the existence of population bottlenecks, and inferring phylogeographic relationships that are fundamental for developing effective conservation strategies. SSR and SNPs have been inconclusive, and the two perform similarly in genetic analysis. SSRs generally show higher polymorphism and efficiency in germplasm clustering and the estimation of genetic distance. However, SNPs can provide more accurate estimations of genetic variation and divergence if applied in greater numbers (Prunier et al. 2023). SNPs and microsatellite markers have proved to be equally reliable in defining population structure and genetic diversity. Thus, the decision regarding the use of SSRs and SNPs is based on research objectives and available sources. Importantly, the combination of both marker types may help to improve the resolution of genetic analyses, allowing for a better assessment of the genetic structure and relationships that are so important for the development of appropriate conservation programs (Prunier et al. 2023). However, it should be noted that the use of these markers has tremendously enhanced the practice of conservation genetics, but caution must be taken to incorporate the theory in population genetics to avoid misinterpretation of data and information acquired while making important decisions in the course of conservation. The microsatellite development techniques have been enhanced by the next-generation sequencing breaks that have made it more cost-effective. Future developments in genomic technologies will invariably prefer both microsatellites and SNPs for conservation genetics and parentage analyses.

4.2 DNA Barcoding as a Tool for Plant Identification

DNA barcoding has been established as a potent identification tool and an effective means to safeguard plant biodiversity (Fig. 4). In this method, DNA sequences that are normally about 500–700 base pairs long are used to differentiate species at the molecular level. The widely used plant DNA barcodes such as *rbcL*, *matK*, *psbA-trnH*, and *ITS* have proven to be efficient in species differentiation (Dhivya Selvaraj et al. 2013). This technique has been useful in discovering endangered wild-collected plant material for medicines, tracking movements of CITES species (Convention on International Trade in Endangered Species) across borders, and discovering adulteration in herbal products or cases of food fraud. DNA barcoding is helpful in the management of species as it returns information on the genetic variation, evolutionary history, and recognized, concealed, and invasive species (Farooq et al. 2020). With the increasing number of species in the DNA barcode reference library, the method remains critical in identifying the evolutionary history of species, interactions between species, and species conservation (Gostel and Kress 2022). Its integration with other disciplines, such as ecology and taxonomy, is suggested for improved general biodiversity conservation approaches.

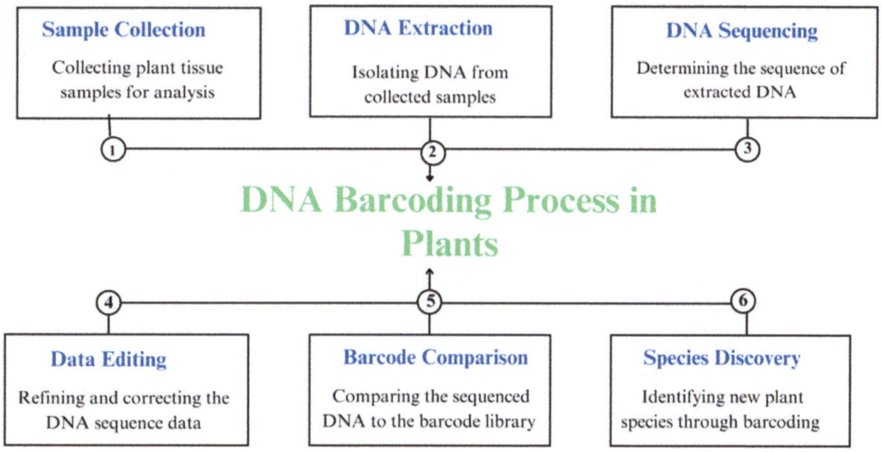

Fig. 4 Application of DNA barcoding in the discovery of plant species

4.2.1 Importance of DNA Barcoding in Assessing Plant Diversity

Species identification through DNA sequences is an important part of DNA barcoding that helps to check multiple species rapidly and accurately. This approach will help in surmounting the problems associated with the conventional morphological identification of taxonomic groups that may often be time-consuming as well as subjective, particularly in places with high species densities. Other factors that impede the use of morphological keys for identification purposes include the emergence of phenotypical traits that result from convergent evolution or exist in cryptic species. In such cases, DNA barcoding can be used to deliver accurate identifications. Integrating DNA barcoding has been successful in offering species resolution that is fast and accurate and therefore assists researchers in establishing phylogenetic structure and assessing habitat protection possibilities effectively (Farooq et al. 2020). It has revolutionized ecological research by offering a detailed view of biological diversity, thus making it easier to preserve species (Gostel and Kress 2022). In addition, barcoding is useful in the development of DNA barcode reference collections (such as the Barcode of Life Database or BOLD) required for accurate plant taxa identification in different ecosystems, which supports improved inventory of species diversity (Farooq et al. 2020).

4.3 Genomic Resources and Conservation Management

Whole-genome sequencing and genome-wide association studies (GWAS) have become especially essential instruments to extend conservation genetics. These resources can also be used to perform complete diagnostics of the genetic polymorphisms in populations to enable management plans to be adopted conveniently. For instance, knowledge of the genetic variation for the traits concerning climate can be used to improve breeding programs for different endangered plant species. With the advancement in whole-genome sequencing and genome-wide association studies, adaptive alleles and inbreeding patterns important for conservation have become of interest. Reference genomes are now seen as the basic infrastructure in conservation genomics, providing support for biological analyses in diverse taxa (Theissinger et al. 2023).

4.4 Successful Applications of Molecular Markers in Conservation Efforts

4.4.1 Use of Molecular Markers in Species Identification

Microsatellite markers, or SSRs, are highly efficient for species differentiation and genetic diversity evaluation in date palms (*Phoenix dactylifera* L.). Several research works have established and described SSR markers in date palms and used these to distinguish between related genotypes and estimate genetic diversity (Akkak et al. 2009). These markers have exhibited significant polymorphisms between different dissected samples that make them excellent for cultivar identification, pedigree analysis, and, most importantly, genetic mapping studies. The application of SSR markers has helped to evaluate the genetic closeness between the accessions of date palms from different regions for conservation and breeding purposes (Khierallah et al. 2017).

4.4.2 Endangered Species Assessment

The development of genomics has recently emerged as a major tool for the conservation of endangered species. Thus, nuclear microsatellites and the DNA barcoding system allow for efficient and detailed genetic assessment, which provides fundamental information based on which conservation efforts can be applied. These tools are essential for measuring genetic divergence, the influences of population structure, and phylogenetic trees, which are important in the identification of conservation units and priority setting. The recent technology of genotyping on a large scale enables the management of endangered plant species, protection against gene pollution, and assessment of genetic characteristics in different generations. Furthermore, the genomic information offers clues to adaptive genetic changes, levels of inbreeding, and disease susceptibilities to improve the knowledge regarding the level of resilience and the ability of the species to adapt to the changing environmental conditions (Li et al. 2020).

4.4.3 Invasive Species Management

Molecular genetics has emerged as essential in determining the invasiveness of the plant species as well as in managing invasive plants. Thus, genetic studies have shown that inviolate populations may vary in the level of differentiation, and therefore, local adaptation may be an insignificant factor in invasion biology, but genetic constitution, such as phenotypic plasticity, may be inherent to invaders (Ward et al. 2008). Hybridization has been identified as a relevant factor in some of the plant invasions, the effect of which can be seen by augmenting the genetic stock that produces new genotypes (Ward et al. 2008). Molecular markers are particularly important when the taxonomy is ambiguous, when it is necessary to determine the source of invasions, to assess the ability of the invader to spread, and to determine whether genetic variability is related to invasion success. Such genetic instruments may be applicable in the control of existing biotic invasions and as a means of avoiding new ones.

4.4.4 Climate Change Effect Evaluation

Global climate change influences plant distribution and genetic variation; therefore, plants must evolve to counter any effects caused by climate change. Importantly, plants can hypo- or hypertrophy, migrate, adapt locally in situ, or face local extirpation with numerous potential constraints on these states and processes (Christmas et al. 2016). Molecular data provide essential information regarding species' evolutionary dynamics and the ability to adapt to abiotic and biotic stimuli. Case studies have used novel methods to assess biotic responses to climatic change, such as spatial niche modeling, cost-benefit evaluation of adaptation strategies, and exploration of relevant traits in various habitats (Scholes et al. 2008). For instance, studies on Proteaceae in the Cape Floristic Region predicted that whereas most of the species would be stable, others would need to be moved or that their habitats in the present sites would disappear (Scholes et al. 2008). In view of these pieces of evidence, genetic diversity evaluation should form part of conservation undertakings and climate change effects analysis (Christmas et al. 2016).

4.4.5 Application of Herbarium Specimens

Due to advancements in the extraction and sequencing of DNA, it has become more possible to conduct research on the plant materials that have been collected in the past to identify the previous evolutionary processes and the population structure that occurred during the process (Bieker and Martin 2018). These genomic approaches enable the examination of plant responses to environmental shifts, document temporal changes in genetic variation, and detect genes affected by recent selection (Burbano and Gutaker 2023). Research involving herbarium samples has discovered relationships between human actions and shifts in haplotype frequency and plant pathogen evolution (Bieker and Martin 2018). Incorporating herbarium genomics with data from other sources provides significant information on the distribution and diversification of plant species, which makes the herbarium invaluable for addressing conservation goals that are exacerbated by current environmental changes (Burbano and Gutaker 2023).

4.4.6 Construction of the DNA Barcode Reference Collections

Conventional DNA barcoding involves the use of informative gene regions, namely rbcL, matK, trnH-psbA, and ITS2, to generate reference databases for swift species identification (Mosa et al. 2019). These libraries are especially valuable for conservation, especially in regions with many plant species, such as the Arabian region, according to Mosa et al. (2019). DNA barcoding helps to identify the species that are hidden and invasive, estimate and compare the phylogenetic richness, and help in establishing the conservation measures (Farooq et al. 2020). However, there are still problems related to creating global barcode libraries and incorporating genomic sequencing technologies. This implies that for better results in the studies and conservation of biodiversity, DNA barcoding should be combined with other traditional taxonomical and ecological methods.

5 Conclusion

This chapter discusses the importance of biotechnology in conservation efforts for endangered species. Techniques like somatic embryogenesis and micropropagation are crucial for the propagation and storage of endangered species. Advanced tools like CRISPR can genetically modify population genetics, supporting population monitoring and genetic recuperation of threatened species. Biotechnological approaches can revolutionize ecosystems by enhancing and degrading species, increasing genetic stock, and producing artificial species to replace degraded species. However, biotechnological enhancements may introduce target genes into wild-type populations and disrupt ecological balance. Sustainable biotechnological enhancements can help achieve agricultural productivity, medicine, and environmental management. Biotechnology can also be used for bioremediation, addressing nutritional deficiencies in susceptible populations. These technologies can develop robust agricultural systems, protect species diversity, and adapt to climate change, fostering future Earth sustainability.

References

Akkak A, Scariot V, Torello Marinoni D, Boccacci P, Beltramo C, Botta R (2009) Development and evaluation of microsatellite markers in Phoenix dactylifera L. and their transferability to other Phoenix species. Biol Plant 53:164–166

Bieker VC, Martin MD (2018) Implications and future prospects for evolutionary analyses of DNA in historical herbarium collections. Bot Lett 165(3–4):409–418

Burbano HA, Gutaker RM (2023) Ancient DNA genomics and the renaissance of herbaria. Science 382(6666):59–63

Christmas MJ, Breed MF, Lowe AJ (2016) Constraints to and conservation implications for climate change adaptation in plants. Conserv Genet 17:305–320

Coelho E, Ballesteros LF, Domingues L, Vilanova M, Teixeira JA (2020) Production of a distilled spirit using cassava flour as raw material: chemical characterization and sensory profile. Molecules 25(14):3228

Fajardo CG, Costa DF d, Chagas KPT d, Vieira F d A (2018) Genetic diversity in natural populations of Hancornia speciosa Gomes: implications for conservation of genetic resources. Ciência e Agrotecnologia 42(6):623–630

Farooq Q, Shakir M, Ejaz F, Zafar T, Durrani K, Ullah A (2020) Role of DNA barcoding in plant biodiversity conservation. Sch Int J Biochem 3(3):48–52

Fonseka R, Fonseka H, Abhyapala K (2020) Crop wild relatives: an underutilized genetic resource for improving agricultural productivity and food security. In: Agricultural research for sustainable food systems in Sri Lanka: volume 2: A pursuit for advancements, pp 11–38

Gostel MR, Kress WJ (2022) The expanding role of DNA barcodes: indispensable tools for ecology, evolution, and conservation. Diversity 14(3):213

Haque E, Taniguchi H, Hassan MM, Bhowmik P, Karim MR, Śmiech M, Zhao K, Rahman M, Islam T (2018) Application of CRISPR/Cas9 genome editing technology for the improvement of crops cultivated in tropical climates: recent progress, prospects, and challenges. Front Plant Sci 9:617

Icro M, Pasamonik D (2022) Tissue culture: A perpetual source for the conservation of medicinally important endangered plant species. In: Advances in plant tissue culture, pp 373–393

Ivanova NN, Tsiupka VA, Mitrofanova OV, Mitrofanova IV (2022) Rare and endemic plant species of mountain Crimea flora: a morphogenic response to conservation in the gene bank in vitro. Turczaninowia 25(1):24–30

Khierallah HS, Bader SM, Hamwieh A, Baum M (2017) Date palm genetic diversity analysis using microsatellite polymorphism. In: Date palm biotechnology protocols volume II: germplasm conservation and molecular breeding. Springer, pp 113–124

Kumar R, Sharma V, Kumar S (2024) Biotechnological tools for conservation and sustainable utilization of medicinal plants. In: Role of science and technology for sustainable future: volume 1: sustainable development: a primary goal. Springer, pp 303–319

Kumari N, Gupta A, Pandey BC, Kushwaha R, Yaseen M (2023) In vitro cultures: challenges and limitations. In: Plants for immunity and conservation strategies. Springer, pp 371–383

Li Y, Liu C, Wang R, Luo S, Nong S, Wang J, Chen X (2020) Applications of molecular markers in conserving endangered species. Biodivers Sci 28(3):367

Meyer H (2011) Systemic risks of genetically modified crops: the need for new approaches to risk assessment. Environ Sci Eur 23:1–11

Mosa KA, Gairola S, Jamdade R, El-Keblawy A, Al Shaer KI, Al Harthi EK, Shabana HA, Mahmoud T (2019) The promise of molecular and genomic techniques for biodiversity research and DNA barcoding of the Arabian peninsula flora. Front Plant Sci 9:1929

Özkul M, Sevgin N, Özkul M, Negi NP, Eyduran SP, Yilmaz B, Akin M (2024) Developing micropropagation protocol for black mulberry. Turk J Agric For 48(4):557–566

Pandey VC, Gajić G, Sharma P, Roy M (2022) 6-Designer plants for climate-resilient phytoremediation. In: Pandey VC, Gajić G, Sharma P, Roy M (eds) Adaptive phytoremediation practices, pp 227–274

Pathak MR, Abido MS (2014) The role of biotechnology in the conservation of biodiversity

Pithiya M, Sharma SK, Sharma M, Sharma M (2022) Advancements and challenges in plant tissue culture: a comprehensive overview. J Plant Biota 1(1):12–16

Pretty J, Brett C, Gee D, Hine R, Mason C, Morison J, Rayment M, Van Der Bijl G, Dobbs T (2001) Policy challenges and priorities for internalizing the externalities of modern agriculture. J Environ Plan Manag 44(2):263–283

Prunier JG, Veyssière C, Loot G, Blanchet S (2023) Comparing the utility of microsatellites and single nucleotide polymorphisms in conservation genetics: insights from a study on two freshwater fish species in France. Diversity 15(5):681

Rai MK (2022) Plant tissue culture targeting germplasm conservation. In: Advances in plant tissue culture. Elsevier, pp 205–221

Rode NO, Estoup A, Bourguet D, Courtier-Orgogozo V, Débarre F (2019) Population management using gene drive: molecular design, models of spread dynamics and assessment of ecological risks. Conserv Genet 20:671–690

Sandeep R, Bharti G, Kanaka S, Rao AS, Naganjaneyulu P (2022) Plant tissue culture: advancing quality material production through micro propagation, with emphasis on Banana cultivation. J Plant Biota 1(1):23–26

Scholes RJ, Mace GM, Turner W, Geller GN, Jürgens N, Larigauderie A, Muchoney D, Walther BA, Mooney H (2008). Toward a global biodiversity observing system. vol 321. American Association for the Advancement of Science, pp 1044–1045

Selvaraj D, Park JI, Chung MY, Cho YG, Ramalingam S, Nou IS (2013) Utility of DNA barcoding for plant biodiversity conservation. Plant Breed Biotechnol 1(4):320–332

Sharma A, Sharma M, Jain M, Arora K, Rai S, Purshottam D (2016) Plant tissue culture approach for cloning and conservation of some important RET medicinal plants. In: Plant tissue culture: propagation, conservation and crop improvement. Springer, pp 199–217

Thakur S, Shruti H, Mishra S, Ekka SK, Kushwaha A, Kujur R (2024) A review on plant tissue culture. Asian J Biol 20(2):14–18

Theissinger K, Fernandes C, Formenti G, Bista I, Berg PR, Bleidorn C, Bombarely A, Crottini A, Gallo GR, Godoy JA (2023) How genomics can help biodiversity conservation. Trends Genet 39(7):545–559

Vasil IK (2008) A history of plant biotechnology: from the cell theory of Schleiden and Schwann to biotech crops. Plant Cell Rep 27:1423–1440

Ward SM, Gaskin JF, Wilson LM (2008) Ecological genetics of plant invasion: what do we know? Invasive Plant Sci Manag 1(1):98–109

Wenne R (2023) Microsatellites as molecular markers with applications in exploitation and conservation of aquatic animal populations. Genes 14(4):808

Plant Genetic Resource Conservation Using Molecular Tools: An Overview

Muhammad Mumtaz Khan, Muhammad Tahir Akram,
Feroz Ahmed Tipu, Rashad Qadri, and Rashid Al-Yahyai

Abstract

Plant genetic resource conservation is essential for biodiversity, food security, and ecosystem sustainability. Seed banks and field germplasm collections serve as a foundation for conserving plant genetic resources. However, the use of molecular tools in plant conservation has improved the understanding and management of genetic resources. Molecular techniques such as DNA barcoding, marker-assisted selection (MAS), and next-generation sequencing (NGS) have enabled accurate identification and monitoring of genetic diversity across plant species. These molecular approaches facilitate the evaluation of genetic variability and the detection of rare or endangered genetic trait genotypes. Molecular markers such as amplified fragment length polymorphism (AFLP), SSRs, and SNPs are particularly valuable for assessing plant populations and guiding breeding programs for improving crops. Additionally, molecular data availability bolsters ex situ and in situ conservation efforts by mitigating genetic erosion and supporting the restoration of plant species and ecosystems. Integrating molecular tools with conventional methods offers a comprehensive framework for preserving PGRs. This combined approach is crucial for addressing challenges posed by habitat loss, climate change, and the overexploitation of plant species.

M. M. Khan (✉) · R. Al-Yahyai
Department of Plant Sciences, College of Agricultural and Marine Science, Sultan Qaboos University, Al-Khoud 123, P.O. Box 34, Muscat, Oman
e-mail: mumtaz@squ.edu.om

M. T. Akram · F. A. Tipu
Department of Horticulture, PMAS-Arid Agriculture University, Rawalpindi 46300, Pakistan

R. Qadri
Institute of Horticultural Sciences, University of Agriculture, Faisalabad 38000, Pakistan

Keywords

Biodiversity · Diversity · Genetic markers · Genetic erosion · Sustainability

1 Introduction

Plant genetic resources (PGRs) are the essential genetic materials containing the alleles of various genes found in different plant species. Any form of vegetative or reproductive plant material capable of propagating is known as a "Plant Genetic Resource." It is the basic elementary unit involved in crops' genetic enhancement, particularly horticultural and agronomic crops (Hasan and Abdullah 2015). Worldwide, the demand for PGRs is increasing due to climate change, which has increased pest infestation, disease incidence, and abiotic stresses. Furthermore, population increase has resulted in the erosion of essential germplasm from different regions.

Food security is a global issue due to the increasing population, soil erosion, climate change, and urbanization that deplete natural plant genetic resources. In addition, adopting high-yielding selected crop varieties is a significant cause of genetic diversity distinction in plant genotypes. It is essential to protect available crop genetic resources from extinction and to utilize them for crop improvement. Therefore, conserving and exploiting are necessary to cope with global food challenges.

The recent advancements in biotechnology, such as genetic engineering, molecular markers, and sequencing, have improved the conservation process. These techniques helped in exploring the native germplasm and underutilized plant species. It maximized the use of traditional local landraces with active involvement from local communities. Within the framework of the United Nations Conference on Environment and Development, global initiatives such

N. Kumar (ed.), *Biotechnology in Plant Conservation*, Sustainable Landscape Planning and Natural Resources Management,
https://doi.org/10.1007/978-3-032-02605-7_10

as the Convention on Biological Diversity and international undertakings on plant genetic resources are collaborating to support the conservation and sustainable use of plant genetic resources.

1.1 Importance of Genetic Diversity in PGRs

The information regarding genetic diversity serves as a foundation for crop improvement. A diverse range of PGRs allows breeders to develop new varieties with desirable characteristics. To improve crop efficiency, it is important to consider the farmer's preference. Plant breeders emphasize the development of climate-resilient crops that integrate desirable traits. Furthermore, genetic diversity manifested by wild species and breeding stocks is vital in enhancing and developing modern crop varieties. In breeding programs, it enables the utilization of genetically different parents to enhance the efficiency of agricultural and horticultural crops (Raffard et al. 2019).

1.2 Plant Genetic Resources Conservation

Selection, cultivation, utilization, and preservation of seeds from native well-adapted crops, called "landraces," have been practiced since the beginning of agriculture (Panis et al. 2020). These practices continued until the end of the nineteenth century. Afterward, Mendel's discovery led to breeding programs focused on developing high-yielding and stress-resistant crop varieties. In the mid-twentieth century, the Green Revolution was introduced, resulting in a drastic increase in production, and landraces were substituted with improved cultivars that promoted the monoculture cropping system. As a result, over 90% of crop varieties and 75% of genetic diversity in plant genetic resources were lost from farmers' fields.

With time, it was realized that there is a dire need to conserve endangered PGRs for the sustainability of agricultural systems against climate change, environmental challenges, threats of pests and diseases, and rapid population growth. Therefore, over 80,000 crop species worldwide were collected and conserved in approximately 3400 different botanical gardens. The basic aim was to conserve the diversity of PGRs for different breeding programs. Likewise, in the twentieth century, ex situ conservation of plant genetic resources for food and agriculture began in specialized repositories known as "gene banks." Its primary focus is the conservation of inter- and intra-specific diversity of crops. Currently, more than 17,000 international institutes are involved in the conservation and sustainable utilization of PGRs for food and agriculture (Bélanger and Pilling 2019). Moreover, over 711 gene banks and 16 international institutes across 90 countries conserve 5.4 million accessions of more than 7051 genera, focusing on crops, wild relatives, modern cultivars, and breeding material. Different international treaties, aligned with the Convention on Biological Diversity (CBD), have been implemented to promote preservation, sustainable utilization, and equitable benefit-sharing and handling of genetic resources. The International Treaty on PGRs for Food and Agriculture (ITPGRFA), effective since 2004, works alongside the CBD to ensure agricultural sustainability and food security. The treaty aims to preserve and sustainably use Plant Genetic Resources for Food and Agriculture (PGRFA). Conserving and sustainably using PGRFA is vital for achieving food security, adapting to changing climates, and helping to overcome human needs. Likewise, since 2014, the Nagoya Protocol has facilitated access to genetic resources. This incentivizes the preservation and sustainable use of genetic resources, contributing to biodiversity. Gene banks developed in different countries for crop conservation are shown in Table 1.

In situ conservation maintains genetic resources and enables species to grow naturally with the least human intervention. It is particularly effective in conserving wild species, forests, and fruit trees. In situ conservation involves two methods to enhance genetic diversity: (i) farm conservation and (ii) genetic reserve conservation. In farm conservation, the breeders' focus is on conventional crop varieties, while in genetic reserve, conservation maintains wild species without or with less human intervention in their natural habitats.

Genetic conservation involves areas, like forest reserves, that are specially designated to preserve genetic diversity for long-term sustainability. Whether on-farm conservation involves sustainable management of locally developed landraces and their wild relatives within farming systems, this process allows for farmers' continuous evolution and selection of desirable plants. In situ conservation also supports open pollination among different genotypes, resulting in populations with diverse alleles.

In ex situ conservation, genetic resources are conserved in artificial habitats instead of their natural habitats. It involves methods like preservation in seed gene banks, plant tissue cultures, cryopreservation, and greenhouse cultivation. This method plays a vital role in protecting endangered and overexploited resources from habitat destruction, climate change, and human interference, preventing their extinction. Ex situ conservation offers an alternative for safeguarding valuable genetic resources that might otherwise be lost. The collected genetic resources should be assessed, characterized, and documented to prevent repetition and conserve them in controlled environments to protect them from external threats (Malhotra et al. 2019). Hence, to mitigate the risks of natural disasters and climate change, both

Table 1 List of international gene banks for different crop conservation

Gene bank	Country	City	Crop
Svalbard Global Seed Vault	Norway	Longyearbyen	Maize, rice, lettuce, cowpea, potato, etc.
CIAT, Colombia	Colombia	Cali	Cassava
AVRDC, Taiwan	Taiwan	Taipei	Tomato, onion, peppers, Chinese cabbage
NBPGR, New Delhi, India	India	New Delhi	Fruits, tubers, spices, bulbous crops, etc.
ICRISAT, India	India	Hyderabad	Sorghum, pearl millet, finger millet, groundnut (peanut), pigeonpea, chickpea
CIP, Peru	Peru	Lima	Potato, Andean root, tubers, etc.
CIAT, Colombia	Colombia	Palmira	Grain legumes, roots, tubers, cassava, banana, yam, etc.
CIMMYT, Mexico	Mexico	El baton	Maize, wheat, triticale, barley, sorghum, etc.
IRRI, Los Banos	Philippines	Los Banos	Rice
Croatian Plant Genetic Resources Database, Croatia	Croatia	Zagreb	Cereals, maize, vitis, etc.
National Inventory on Plant Genetic Resources in Germany	Germany	Bonn	Peas
The Australian Seed Bank	Australia	Canberra	Endemic and threatened plants from southeastern Australia
Plant Gene Resources of Canada (GRIN-CA)	Canada	Saskatoon	Cereals, pulses, oilseed, tomato, forages, etc.
National Bureau of Plant Genetic Resources	India	New Delhi	Cowpea, sorghum, pearl millet, ginger, etc.

in situ and ex situ conservation methods should be integrated and complement each other. Various research institutes have been established worldwide to preserve and improve crops. The details of these institutes have been provided in Table 2.

1.3 Significance of Conserving PGR for Food Security, Sustainability, and Biodiversity

The conservation of PGRs is vital to ensure food security. The sudden loss of valuable plant genetic resources has prompted global communities to conserve PGRs to maintain genetic diversity. It was emphasized in Rio de Janeiro in 1992 during the Convention on Biological Diversity (CBD) signing, where the conservation of PGRs and biodiversity became a key international topic (De Oliveira and Martins 2002). The primary objective of CBD is to protect biodiversity and ensure future food security by preventing the loss of essential genes. To ensure these, effective policies are implemented.

These strategies, including in situ conservation, ex situ conservation, and biotechnological conservation, work synergistically to avoid genetic resource erosion. Plant genetic diversity is vital to developing and releasing new and well-adapted climate-resilient cultivars. The main aim of PGR preservation is to maintain a broad range of genetic diversity within a species, holding specific or potential value, to ensure that these resources can be utilized and explored by current and future generations. Under changing environmental conditions, genetic variation allows crop species to adapt and survive. However, cultivars with low genetic diversity are at risk of biotic and abiotic stress. Preserving genetic diversity allows breeders to develop climate-resilient crops with desirable characteristics.

1.4 Challenges for PGR Conservation

The conservation and sustainable utilization of PGRs are globally important in addressing food security challenges. In past decades, the loss of biodiversity and genetic resources has been tremendously depleted by natural disasters like forest fires, floods, volcanic eruptions, and anthropogenic activities, which are the primary reasons for plant genetic resource losses (Fig. 1). The primary issues in PGR conservation include proper collection and maintenance of plants in artificial environments, the risk of genetic multiplicity loss, inbreeding, and adaptation issues. Additionally, there are challenges such as high costs for conservation and maintenance, the need for skilled personnel, a lack of proper infrastructure, and reliance on external resources like electricity and water.

Since the Green Revolution, industrial agriculture and the global expansion of markets, food preferences, tastes, and cultures have severely affected genetic diversity on farms and in gene banks. The integrity of these resources is further compromised by genetically modified organisms (GMOs) and other challenges, including population growth, climate change, and social and economic instability. These modifications have resulted in habitat destruction and the loss

Table 2 List of different research institutes dedicated to crop improvement and conservation

Research institute	City	Country	Establishment year	Crops
International Rice Research Institute (IRRI)	Los Banos	Philippines	1960	Rice, maize, wheat, and vegetables
International Maize And Wheat Improvement Center (CIMMYT)	Mexico city	Mexico	1971	Maize, wheat, and other crops such as barley and oats
International Potato Center (CIP)	Lima	Peru	1972	Potato and other root and tuber crops
International Center for Tropical Agriculture (CIAT)	Cali	Colombia	1967	Cassava, rice, maize, beans, and other tropical crops
International Center for Agriculture Research in the Dry Areas (ICARDA)	Beirut	Lebanon	1977	Wheat, chickpea, barley, lentil, and other dry area crops
International Crops Research Institute for the Semi-Arid Tropics (ICRISAT)	Hyderabad	India	1972	Sorghum, chickpea, pearl millet, pigeonpea, groundnut, and other semi-arid region crops
International Plant Genetic Research Institute (IPGRI)	Rome	Italy	1974 (as IBPGR), renamed IPGRI in 1991, and later Bioversity International in 2006	Maize, banana, wheat, cassava, potato, rice, and other fruits and vegetables
West Africa Rice Development Association (WARDA)	Abidjan	Ivory coast	1971	Rice

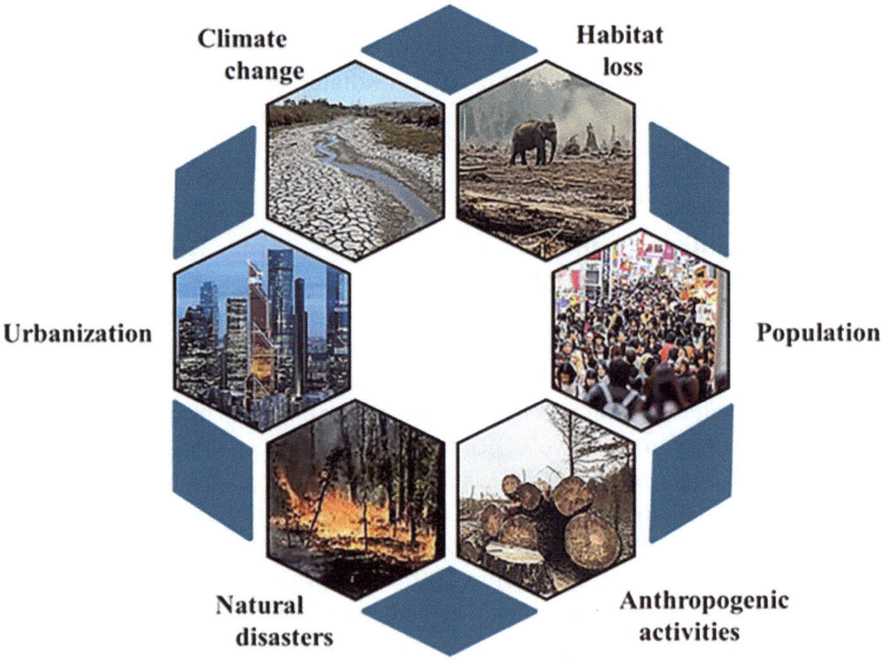

Fig. 1 Reasons for the depletion of plant genetic resources

of plant genetic resources. Furthermore, climatic disruptions, including droughts, floods, diseases, irregular rainfall patterns, and extreme events, affect the environment and plant genetic resources (PGRs). Moreover, rising sea levels have resulted in the loss of coastal areas and the intrusion of salty water, contributing to plant genetic diversity. Undoubtedly, advancement and mechanization in agriculture have increased productivity manyfold. However, it has reduced plant diversity due to selected varieties. Other factors contributing to genetic erosion are irregular seed dissemination, lack of knowledge, and the homogenization of market products.

1.5 Strategies for PGR Conservation

Successful PGR conservation can be obtained by a clear scientific and technical understanding of the species, such as factors affecting plant growth, adaptation, and efficient and economical methods for conservation. Further conservation strategies vary with the nature of the material and the time duration. Life cycle length, reproduction methods, individual size, and ecological status influence the nature of the conserved material. The objectives of conservation may be research, introduction, and breeding for genetic integrity. In addition, the scope of conservation refers to the time scale and geographical area involved, whether local, regional, or global. However, practical limitations to the scale and scope of conservation exist, and it is not feasible to maintain plant material with complete genetic integrity over time.

Without proper and rigorous conservation practices and policies, these resources are at high risk in the future (Hammer and Teklu 2008). This issue can be tackled by sharing information, raising awareness about PGR diversity, creating new technologies, and implementing systematic, scientific conservation methods to relevant stakeholders.

1.6 PGR for Ecosystem Stability and Food Security

Plant genetic resources (PGRs) are pivotal in ecosystem stability and food safety. To prevent further loss of these valuable resources, it is essential to prevent PGR erosion and promote their sustainable usage. To address global food challenges, there is a need to investigate, collect, and conserve these PGR resources in a sustainable way. PGRs consist of cultivated crops, landraces, wild species, and accessions that serve as a foundation for breeding and highly resilient crops. PGRs also contribute to enhancing biodiversity and maintaining balance among species, which is essential for sustainable agriculture. Preserving and sustainably utilizing PGRs can ensure a sustainable food supply and conserve natural resources for future generations (Singh et al. 2020).

2 Molecular Tools Employed for Conservation

Molecular markers play a vital role in biodiversity conservation by assessing genetic diversity, analyzing population structure, and determining crop evolution. These tools facilitate the identification of genotypes, management of species, and mitigation of the threat of crop extinction. Modern molecular techniques, such as DNA sequencing methods including RAPD, simple sequence repeats (SSRs), SNPs, and AFLP and genomic tools such as genome-wide association study (GWAS) and whole genome sequencing, help identify a species' genetic trait. New molecular tools, such as Clustered Regularly Interspaced Short Palindromic Repeats (CRISPR), are new crop identification and conservation interventions. These new molecular tools support conservation strategies to protect endangered species.

2.1 DNA Barcoding

DNA barcoding is a technique used for species identification through a short section of DNA from specific genes. Each barcode is distinctive and provides clear-cut identification of closely related species. This technique has been extensively adopted and has increased the availability of plant DNA sequence data. DNA barcodes consist of standardized, short DNA sequences from either the organelle or nuclear genome, allowing for distinguishing one species from another. In DNA sequencing, the internal transcribed spacer (ITS) rRNA region is commonly used for plants, while the 16S rRNA gene is used for prokaryotes. DNA barcoding also plays a pivotal role in the identification of species and environmental monitoring, determining the source species of natural medicines and detecting contamination, especially in powdered or small particle forms (Antil et al. 2023). Furthermore, it is important for biodiversity studies because it enables scientists to quickly and accurately identify unknown species by comparing barcode sequences to a complete reference database. This method increases our understanding of biodiversity, supports conservation initiatives, and notifies evidence-based decision-making for sustainable ecosystem management (Vine et al. 2009). The procedure of DNA barcoding is shown in Fig. 2.

2.2 Polymerase Chain Reaction (PCR)

Polymerase chain reaction (PCR) is a technique used to amplify a specific segment of DNA and produce billions of copies from a small initial sample. It was developed by Kary Mullis in 1983 and became a vital tool in agriculture, especially for breeding purposes; it has a vital role in molecular biology, forensic science, medical diagnostics, and evolutionary biology. The PCR technique is based on the enzymatic replication of DNA. During PCR, a specific short segment of DNA is amplified using primers and enzymes. DNA polymerase synthesizes new DNA strands that are complementary to the template DNA. Since DNA polymerase can only add nucleotides to an existing 3′–OH group, a primer is necessary to start the process. This allows additional nucleotides to be added to the 3′ end of the DNA strand (Ehtisham et al. 2016).

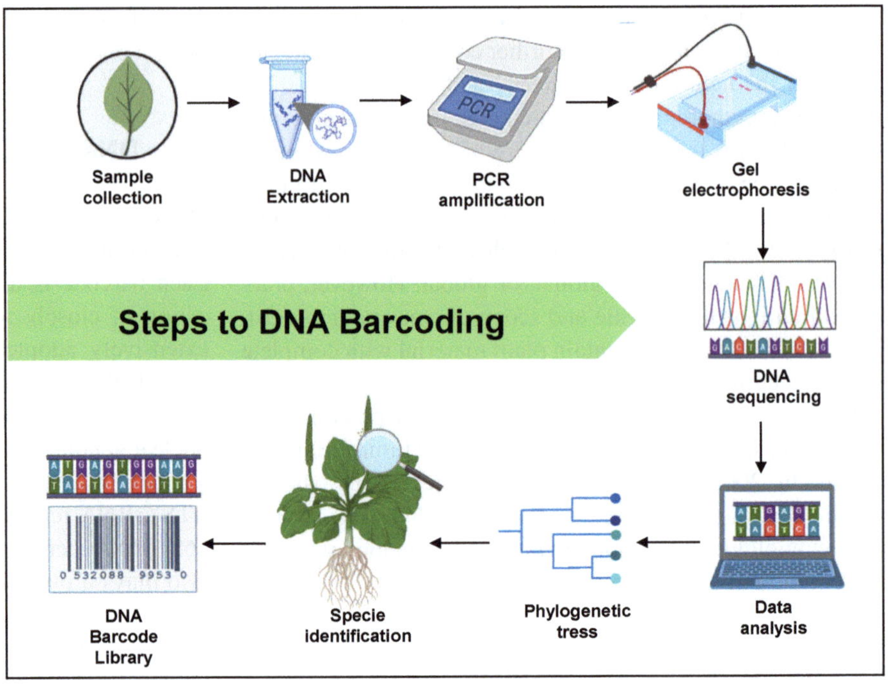

Fig. 2 Different steps related to DNA barcoding

Major Steps Involved in PCR

The PCR technique involves three main steps, which are repeated for 25–35 cycles. These steps include (i) denaturation, (ii) annealing, and (iii) extension. In denaturation, the double-stranded DNA is heated to about 94–98 °C to disperse it into two single strands. In annealing, the reaction is cooled to 50–65 °C to allow the primers to bind to their opposite sequences on the single-stranded DNA, while in the extension phase, the temperature is again increased to 72 °C, which is the optimal temperature for DNA polymerase to make new DNA strands by adding nucleotides to the primers.

2.3 Next-Generation Sequencing

Next-generation sequencing (NGS) is also known as high-throughput sequencing. It is used for sequencing millions of DNA molecules simultaneously. This powerful technology has revolutionized genomics and molecular biology by reducing the time and cost required for sequencing. NGS technologies are widely used to explore genetic resources and are contributing to eradicating food hunger across the world by improving crop yield for self-sufficiency in food resources. This technology enhances the precise and efficient design of targeted ESTs, SSRs, and single-nucleotide polymorphism (SNP) markers. It also assists in developing genome editing and gene engineering strategies in plants. Furthermore, it helps achieve sustainable production by accurately identifying specific traits or phenotypes under challenging environmental conditions.

Next-generation sequencing (NGS) technologies have significantly enhanced the ability to characterize genetic and molecular variations in gene bank collections, including wild species and landraces. Whole genome sequencing of diverse crop accessions provides researchers with the ability to reveal genetic variations that are necessary for enhancing agronomic characters. Over 3000 accessions of chickpeas and rice have been recently re-sequenced and produced genome data to associate genetic variations with fundamental traits like stress resistance (Varshney et al. 2021). Similar work is being conducted for crops like maize, soybean, brassica species, tomato, and potato. Extending these studies to include other crops and their wild relatives (CWRs) could categorize valuable genes and alleles for stress resistance and other important traits. This knowledge will help genome editing introduce novel genes and alleles from CWRs and landraces into cultivated gene pools, addressing challenges such as climate adaptation, resource scarcity, and improved crop efficiency. Moreover, PGRs offer opportunities to investigate epigenomic variations, which involve chemical modifications that regulate gene expression and offer considerable potential for crop enhancement.

Major Steps Involved in Next-Generation Sequencing (NGS)

The major steps involved in next-generation sequencing include library preparation, cluster generation, sequencing, and data analysis. In library preparation, DNA is extracted from the sample and fragmented into smaller pieces. Adapters are then added to the fragments to facilitate their binding for sequencing. In cluster generation, the DNA fragments are amplified to create clusters of identical fragments on a solid surface, such as a flow cell. After cluster generation, the next step is sequencing. In this step, DNA fragments are sequenced by sequencing by synthesis (SBS), where fluorescently labeled nucleotides are added, detected, and incorporated into the growing DNA strand. After sequencing, the data is processed and analyzed using bioinformatics tools to assemble the sequences and identify genetic variants.

2.4 Marker-Assisted Selection

Marker-assisted selection (MAS) is a modern breeding method used to achieve desirable characteristics in plants using molecular markers. This method offers greater accuracy and efficiency than traditional selection techniques. It enables breeders to select individuals with the desired characteristics through genetic profiling instead of relying on physical characteristics. MAS is based on molecular markers, genetic linkage, and the selection process. It is commonly used in plant breeding with various objectives, like selecting individuals with genetic resistance to specific diseases and environmental stresses, such as drought, salinity, and extreme temperature. It further assists in developing hybrid varieties by ensuring the presence of desirable traits in parent lines. The effectiveness of MAS depends on the strength of the connection between the marker and the gene of interest. However, traits controlled by quantitative traits are more challenging to manipulate using MAS (Collard and Mackill 2008).

2.5 Genotyping Technologies

Genotyping technologies in plants determine the genetic makeup (genotype) by examining their DNA sequence. These technologies are essential for understanding genetic variation, identifying disease-associated genetic markers, and studying evolutionary relationships. The major genotyping technologies involved in the determination of genetic makeup are random amplified polymorphic DNA (RAPD), restriction fragment length polymorphism (RFLP), single-nucleotide polymorphisms (SNPs), amplified fragment length polymorphism (AFLP), next-generation sequencing (NGS), and massively parallel sequencing (MPS).

2.6 Advantages of Molecular Tools over Traditional Methods in Assessment and Conservation of Plant Genetic Resources (PGRs)

Conventional approaches, such as morphological characterization and seed banking, were extensively used globally for PGR evaluation and conservation. However, new molecular tools have offered distinct advantages over traditional methods. Molecular tools deliver better resolution and precision in measuring genetic diversity than conventional morphological approaches. Markers such as SSRs, SNPs, and AFLPs enable recognition of genetic changes at the DNA level, which may not be detectable through morphological characters. These permit more exact recognition of genetic relations and changes between crop accessions (Akçura et al. 2005). Furthermore, these techniques are rapid and more effective than conventional methods, such as PCR and NGS, examining large numbers of trials at once, leading to a significant decrease in the time required for genetic analysis.

Molecular tools can detect genetic changes that are not indicated phenotypically and disclose hidden diversity within crops. This identification of genetic change can serve as an important resource for breeding and conservation. The genetic information is used in population identification, seed storage procedures, and habitat rehabilitation. Furthermore, molecular markers such as marker-assisted selection (MAS) make breeding precise, choosing plants with desirable traits. It is helpful in the development of quick and better varieties with high yields, disease resistance, and other necessary characteristics. However, conventional techniques are laborious, require more time, and are less precise. The other features of molecular and conventional techniques are highlighted in Table 3.

3 Evaluating the Plant Genetic Diversity Through Molecular Markers

Evaluating crop genetic diversity is significant for preservation, breeding, and understanding evolutionary processes. Various molecular markers, including amplified fragment length polymorphism (AFLP), simple sequence repeats (SSRs), and single-nucleotide polymorphisms (SNPs), are used in plant diversity evaluation.

3.1 Amplified Fragment Length Polymorphism

Amplified fragment length polymorphism (AFLP) is a PCR-based technique that selectively amplifies a subset of

Table 3 Comparison of molecular and conventional techniques used for assessment and conservation of plant genetic resources (PGRs)

Feature	Molecular methods	Conventional methods
Accuracy	High accuracy in identifying genetic resources, including minor genetic variations	Limited accuracy; unable to detect minor genetic variations between populations or individuals
Speed	Faster identification and analysis using techniques like PCR and DNA extraction	Time consuming due to manual data collection, sequential processing, and manual record-keeping
Resolution	High resolution through DNA-based markers to evaluate genetic diversity	Low resolution; genetic variations at the DNA level cannot be detected
Cost-effectiveness	Cost-effective for large-scale analysis over time despite high initial investment	Requires significant and ongoing investment in infrastructure and maintenance
Plant material requirement	Requires minimal plant material for analysis	Requires entire plants or large parts (roots, leaves, seeds) for morphological characterization
Data management	Enhanced data storage and sharing for better global conservation efforts	Limited data storage and sharing capacity, reducing conservation efficiency
Conservation and verification	Improved seed bank management and genetic uniformity verification	Limited conservation capacity and difficulties in verifying seed certification and genetic accuracy
Labor demand	Utilizes automation and high-throughput technologies, reducing labor requirements	Labor-intensive due to manual morphological characterization and low-throughput processes
Genetic insight	Capable of detecting genetic changes at the DNA level	Unable to detect genetic changes; relies solely on morphological characterization
Breeding efficiency	Enables quicker development of high-yield and disease-resistant varieties	Phenotypic selection for hybrid varieties is slow and time consuming
Examples	RAPD, SSR, NGS, SNP, and other molecular markers	Gene banking, cryopreservation, and ex situ conservation methods, like botanical gardens

digested DNA fragments to produce unique fingerprints of interest. AFLP is a multipurpose and powerful molecular marker that measures genetic diversity, population building, and evolutionary relations among plant species. It combines restriction enzymes and the polymerase chain reaction (PCR) to generate many DNA fragments, separated by gel electrophoresis.

In AFLP, DNA is extracted from plant tissues and then digested with two restriction enzymes to produce DNA fragments. After the production of fragments, adapters are ligated to the sticky ends of the restriction fragments. After ligation, a subset of fragments is amplified using primers that match the adapter sequences and then extended into the restriction fragments. Finally, gel electrophoresis separates PCR amplified fragments, resulting in unique band patterns.

3.2 Simple Sequence Repeats (SSRs)

Simple sequence repeats (SSRs), or microsatellites, are short, repetitive DNA sequences dispersed throughout the genome. These repeats vary in length among individuals, making them admirable markers for genetic studies. The steps involved in SSR include DNA extraction, PCR amplification, and electrophoresis. In simple sequence repeats (SSRs), plant tissues are used to extract DNA, and then regions of SSR are amplified using specific primers. After amplification, gel electrophoresis is used to separate the fragments.

3.3 Single-Nucleotide Polymorphisms (SNPs)

Single-nucleotide polymorphisms (SNPs) are single base-pair variations in the DNA sequence, representing the most common type of genetic variation. They are also dispersed throughout the genome, making them valuable markers for genetic studies. The steps involved in single-nucleotide polymorphisms are DNA extraction, genotyping, and data analysis. In single-nucleotide polymorphisms (SNPs), plant tissues are used to extract DNA. Afterward, SNPs are identified using next-generation sequencing and allele-specific PCR. The presence or absence of SNPs is analyzed to determine genetic diversity and their relationships. The comparison of molecular markers AFLP, SSR, and SNP is shown in Table 4.

4 CRISPR-Based Approaches for Crop Improvement

The CRISPR-Cas system generates double-strand DNA breaks that are repaired through the cell's DNA repair mechanisms. The non-homologous end joining (NHEJ) pathway may alter gene function by inducing deletions or insertions. CRISPR technology also includes epigenome editing, which allows for the targeted manipulation of epigenetic marks. This method combines sequence-specific

Table 4 Comparison of different molecular markers used in genetic diversity

Characteristics	SNPs	AFLPs	SSRs
Genome coverage	Whole genome	Whole genome	Whole genome
Polymorphism level	Variable	Moderate to high	High
PCR requirement	Yes	Yes	Yes
Inheritance/expression	Co-dominant	Dominant/co-dominant	Co-dominant
Initial cost	High	Moderate	Low
Locus specificity	Yes	No	Yes
DNA quantity required	Low	Moderate	Low
Major application	Genetic diversity	Gene tagging	Genetic diversity
Technical requirements	Moderate	Moderate	Moderate
Marker index	High	Moderate	Moderate
Genome abundance	Very high	High	Moderate to high
Status	Present	Past	Present
Reproducibility	High	Moderate to high	High

DNA-binding modules with effectors, offering precise control over gene expression. For example, catalytically inactive Cas9 (dCas9) is fused with DNA methyltransferase domains to add repressive methylation marks, silencing specific genes.

Other Techniques for Molecular Identification and Conservation
The other molecular techniques for molecular identifications include 16S rRNA gene sequencing, fluorescence in situ hybridization (FISH), matrix-assisted laser desorption/ionization time-of-flight mass spectrometry (MALDI-TOF MS), and whole genome sequencing (WGS). In 16S rRNA, the gene sequencing technique targets the 16S ribosomal RNA gene, which is conserved between bacteria and archaea. Sequencing through this gene aids in recognizing and classifying microorganisms. Fluorescence in situ hybridization (FISH) uses fluorescently labeled DNA probes that bind to specific DNA sequences within the cells. This method allows for the visualization and recognition of microorganisms directly in their natural condition (Ashfaq et al. 2022).

Matrix-assisted laser desorption/ionization time-of-flight mass spectrometry (MALDI-TOF MS) identifies microorganisms by analyzing the mass and charge of their proteins. It is a fast and precise method used in clinical, food, and environmental microbiology. Whole genome sequencing (WGS) includes sequencing the entire genome of an organism, providing complete genetic data. It helps recognize species, study genetic diversity, and notice genes encoding antibiotic resistance (Shokralla et al. 2014).

Another advancement is single-cell transcriptomics, which enables in-depth analysis of gene expression at the level of individual cells. This technology provides insights into plant development and cellular processes, enabling targeted genetic modifications to improve crop characteristics (Levengood et al. 2024). Moreover, improvements in plant transformation techniques, such as nanoparticle-mediated transformation and non-tissue culture regeneration systems, are overcoming past limitations and expanding the range of species that can be genetically modified. These advancements are essential in addressing global challenges related to food security and environmental sustainability.

Precise characterization of crop species is essential for maintenance. The recognition of new species aids in protecting the Earth's evolutionary heritage. Phylogenetic tools are important for considerable community assembly and finding centers of diversification (Gohil et al. 2019). The advent of molecular biology methods has enabled identification and detection procedures by strengthening specific conserved DNA sequences. With time, great modifications have been employed in search of a fast and inexpensive microbial recognition technique, and now, a whole genome to be sequenced is possible (Huber et al. 2018).

5 Biorepositories and Seed Bank

Biorepositories and seed banks are important in conserving and utilizing biological multiplicity. Biorepositories are essential for research and conservation as they are facilities for collecting, storing, and distributing biological materials. Cell lines, microorganisms, tissues, and DNA samples are present in such material (Bhattacharya and Mummenhoff 2024). Apart from providing material biorepositories are vital for numerous fields, like pharmaceutical research, personalized medication, and genetic studies.

Seed banks are specialized facilities that store seeds to preserve genetics. It plays a vital role in habitat restoration and biodiversity preservation and is significant in supporting the ecosystem. Seed banks have a key role in seed collection, storage, and distribution. It maintains seed viability and genetic integrity for an extended period. Preserving various biological materials, biorepositories, and seed banks conserves genetic diversity for breeding programs and for restoring and conserving degraded ecosystems (Mastretta-Yanes et al. 2024).

Optimal storage conditions are necessary to preserve the genetic material for the long term. Therefore, access to genetic resources is essential for researchers to achieve maximum benefits. Another challenge is developing a sustainable technique for collecting, storing, and distributing long-term genetic material conservation. Biorepositories and seed banks are fundamental to preserving biodiversity and ensuring food security. They are essential to address global challenges, like climate change and habitat degradation, and are vital for sustainable agricultural practices.

5.1 Importance of Biorepositories in the Preservation of Genetic Resources

Biorepositories are amenities that play a pivotal role in managing and conserving hereditary resources. They ensure the long-term preservation, storage, and convenience of biological materials, playing a dynamic role in the conservation, research, and sustainable use of biodiversity. Biorepositories preserve genetic material, such as seeds, tissues, and DNA. The preservation of these genetic resources has a significant role in preserving biodiversity and is vital for the flexibility and adaptability of species. Genetic diversity is important for breeding programs and environmental research and aids in comprehending evolutionary methods.

The fundamental function of biorepositories is to provide secure and long-term storage of genetic materials by maintaining optimal storage conditions to preserve the viability and integrity of genetic material for an extended period. Biorepositories play an essential role in ensuring equitable access to genetic resources. This conservatory improves the effective utilization of genetic resources to maximize stakeholder benefits. It also facilitates reintroduction and restoration efforts to maintain biodiversity and support the ecosystem. For example, seeds stored in different seed gene banks restore degraded ecosystems and help maintain biodiversity.

5.2 Methods for Evaluation of Seed Viability and Vigor

Different methods are used to check seed viability and vigor. It includes seed germination tests, tetrazolium tests, electrical conductivity tests, seedling vigor tests, and DNA-based tests. Each method is individual and has its own features.

Seed Germination Test In this method, seeds are placed in a controlled environment to observe germination percentages and rates.

Tetrazolium Test This method is used to check the seed viability and vigor, in which seeds are stained with a dye to identify viable tissues.

Electrical Conductivity Test This test is performed to measure the leakage of electrolytes from seeds to check the damage.

Seedling Vigor Test It is important to assess the growth and development of seedlings under different stress conditions.

DNA-Based Test In this method, seed viability and vigor evaluation tests are performed through molecular techniques to assess seed viability.

5.3 Practices for Long-Term Storage of Genetic Materials

The long-term storage of genetic material relies on best storage practices. Therefore, before storage, it should be ensured that genetic materials are pure, clean, and properly dried. Furthermore, it is necessary to maintain temperature and humidity storage conditions as they significantly affect seed preservation and viability. For long-term and efficient storage, moisture and air exposure should be avoided. Furthermore, it is necessary to check the stored materials regularly to ensure seed viability and safety.

6 Genomics and Plant Resources Preservation

Genomics has played an essential role in preserving plant genetic properties, offering innovative tools and methods to improve conservation efforts. It permits the inclusive evaluation of genetic diversity through NGS and molecular markers that help recognize genetic changes in emerging resilient plant varieties. It supports both in situ and ex situ conservation policies. MAS techniques have helped to recognize plants with required traits, like disease resistance or drought tolerance. It helped in the fast breeding of climate-resilient cultivars.

Cryopreservation includes freezing plant tissues at ultra-low temperatures to preserve their genetic material for the long term. Genomic practices help detect the best candidates for cryopreservation and have confirmed the viability of deposited genetic resources (Zhang et al. 2024). Genomics enables global collaboration and information sharing among researchers, conservationists, and policymakers. International agreements, like the Convention on Biological Diversity (CBD) and the International Treaty on Plant Genetic Resources for Food and Agriculture (ITPGRFA), encourage the exchange of genetic resources and information (Li et al. 2024). Incorporating genomics into plant resource preservation improves the efficacy and effectiveness of conservation attempts. By exploiting innovative genomic tools, we can understand, defend, and use plant genetic diversity for sustainable agriculture and food security.

6.1 Application of Genomics in Understanding Plant Adjustments and Resilience

Genomics has changed our understanding of plant adaptation and flexibility, particularly under climate change and ecological stressors. It has enabled the recognition of genes stimulated in response to various stress circumstances, like drought, heat, salinity, and pests. By considering stress-responsive genes, researchers are making efforts to develop strategies that can cope with environmental stresses (McCoy et al. 2023). A genome-wide association study (GWAS) is used to find genetic variants related to specific traits, like drought tolerance or disease resistance. This data can help the breeders to choose plants with prominent traits, leading to substantial crop diversity. Furthermore, techniques like CRISPR/Cas9 enable exact alterations to plant genomes, permitting scientists to familiarize or enhance traits related to resilience. Consumption of crop wild relatives (CWRs) holds genetic diversity to recover cultivated plants. Genomics also helps recognize valuable traits in domesticated species through breeding or genetic engineering (Supple and Shapiro 2018).

6.2 Role of Genomic Selection in Enhancing Conservation Strategies

Genomic selection has developed as a potent tool in increasing conservation plans by providing thorough genetic information that updates better management and preservation practices for biodiversity. Genomic selection offers the complete analysis of genetic variation within and among species. By recognizing genetic variants that contribute to resilience, adaptability, and other critical traits, conservationists can prioritize these alternatives in their conservation efforts. Preserving a comprehensive genetic base is important for a species' long-term existence.

In conservation breeding programs, genomic selection aids in identifying individuals with the most necessary genetic characteristics, such as disease resistance, climate adaptability, and reproductive fitness. By integrating these characteristics into breeding decisions, the health and capability of captive or managed populations can be considerably improved. Genomic selection helps select the most valuable species for reintroduction into the indigenous or ecosystem restoration programs. By confirming that reintroduced species have the necessary genetic characteristics for survival and variation, the success rate of these programs is increased. Continued genomic monitoring of indigenous and captive species allows conservationists to assess genetic health and identify potential issues, like inbreeding or loss of genetic diversity. Early detection permits timely interventions to keep genetic robustness and population viability. Genomic selection delivers detailed insights into the genetic composition of species, allowing more informed decision-making regarding habitat protection, resource allocation, and preservation priorities.

6.3 Genomic Selection and its Potential for Crop Improvement

Genomic selection is an essential breeding method that predicts the genetic value of individuals in a breeding population by using various genome-wide markers. This method shows significant potential for crop enhancement and offers numerous benefits over conventional breeding methods. Genomic selection permits the efficient selection of superior genotypes, reducing the time involved in developing new crop cultivars. These interventions make it easy for breeders to make timely decisions (Funk et al. 2019). It leads to the accurate and precise selection of plant species with desirable traits, higher yield, stress tolerance, and disease resistance. However, conventional breeding requires extensive phenotyping, which is expensive and time consuming. Genomic selection reduces the need for phenotyping by using various genetic markers to predict qualities that make the breeding process cost-effective. Combining genomic techniques with new technologies, such as speed, enhances its efficiency. This combination permits rapid development of improved crop cultivars (Shaw et al. 2024).

6.4 Integrating Genomic Information into Conservation Organizations

Integrating genomic information into conservation organizations offers an important approach to enhancing biodiversity protection and management. Collaborations among conservation organizations, academic institutions, governmental and non-governmental organizations, and private industry allow the sharing of genomic data and resources. This association helps collect and generate data (Latch et al. 2020). Integrating genomic data into preservation policies also helps in making timely decisions. The conservationists may utilize this data to make decisions on protection, resource allocation, and management practices (Lau et al. 2021). Integrating genomic information into conservation organizations offers several benefits, including improving genetic diversity, improving the effectiveness of conservation initiatives, and promoting global collaboration. By implementing these strategies, conservationists efficiently address the challenges of environmental change and biodiversity loss.

7 International Agreements

The aim of the Convention on Biological Diversity (CBD) is to preserve biodiversity and ensure future food security. In 1992, an agreement was signed at the Earth Summit in Rio de Janeiro to conserve genetic resources for human welfare and secure food supply in the future. Later, CBD made additional agreements, "Cartagena Protocol on Biosafety" and the "Nagoya Protocol on Access to Genetic Resources and the Fair and Equitable Sharing of Benefits Arising from their Utilization."

The Nagoya Protocol is a supplementary contract of the CBD to provide an authentic organization for the fair and reasonable profit sharing arising from the valuable use of genetic resources. It was adopted in Nagoya, Japan, in 2010 and was executed on 12 October 2014. It provides a framework to evaluate the genetic resources and equitable sharing of benefits derived from their use.

Likewise, the Plant Treaty's primary objective is to preserve and promote sustainable utilization of plant genetic resources for food and agriculture. It assured the fair and equitable sharing of profit arising from their use. The treaty was signed in 2001 and came into effect on 29 June 2004. The treaty recognizes farmers' rights and sets up a Multilateral System of Access and Benefit-Sharing to facilitate the exchange of plant genetic resources. These international agreements are crucial for enhancing the sustainable utilization and preservation of genetic resources and supporting worldwide food safety and biodiversity preservation.

7.1 Role of Policy in Promoting Molecular Conservation Efforts

Policies play a pivotal role in promoting molecular conservation efforts by providing a structure for the sustainable usage and protection of genetic resources. Policy is essential in promoting molecular conservation by providing a legal and regulatory structure, supporting research and development, encouraging collaboration, and ensuring compliance with preservation goals. Initiatives like the Nagoya Protocol under the Convention on Biological Diversity (CBD) ensure fair sharing of benefits from genetic resources and help prevent biopiracy. The International Treaty on Plant Genetic Resources for Food and Agriculture (ITPGRFA) encourages collaboration and data sharing between countries. Integrating policy with molecular tools enhances the effectiveness of conservation strategies and biodiversity protection for future generations.

The preservation of PGRs is of global concern due to the disappearance of valued genetic resources. Addressing this challenge requires the collective efforts of various shareholders, farmers, ethnobotanists, crop breeders, government and non-governmental organizations, seed groups, strategy-makers, and international collaborators. Significant progress can be achieved by fostering collaboration, sharing knowledge, raising awareness about PGR diversity, advancing technology, and employing systematic and scientific conservation strategies.

8 Conclusion

Plant genetic resources (PGRs) are essential for conserving plant diversity and food security. The introduction of molecular tools offers more accurate and efficient methodologies for assessing, detecting, and preserving genetic diversity than traditional practices. DNA barcoding, polymerase chain reaction (PCR), and next-generation sequencing (NGS) are new molecular techniques for verifying plant genetic variation. AFLPs, SSRs, and SNPs are the essential molecular markers to study plant populations and help improve plants. In addition, it also assists in the preservation of endangered species. Combining traditional conservation practices with molecular tools builds a holistic and integrative framework for PGR conservation. This approach is necessary to counteract the threats to plant biodiversity, agriculture, and environmental sustainability imposed by habitat loss, climate change, and overexploitation.

Acknowledgments We greatly acknowledge SQU, PMAS-AAUR, and UAF for their support in writing the manuscript.

References

Akçura M, Kaya Y, Taner S (2005) Genotype-environment interaction and phenotypic stability analysis for grain yield of durum wheat in the central Anatolian region. Turk J Agric For 29(5):369–375

Antil S, Abraham JS, Sripoorna S, Maurya S, Dagar J, Makhija S, Bhagat P, Gupta R, Sood U, Lal R, Toteja R (2023) DNA barcoding, an effective tool for species identification: a review. Mol Biol Rep 50(1):761–775

Ashfaq MY, Da'na DA, Al-Ghouti MA (2022) Application of MALDI-TOF MS for identification of environmental bacteria. a review. J Environ Manag 305:114359

Bélanger J, Pilling D (2019) The state of the world's biodiversity for food and agriculture. FAO

Bhattacharya S, Mummenhoff K (2024) Effective seedbank management to ensure food security and preserve biodiversity. Pl Syst Evol 310(3):15

Collard BC, Mackill DJ (2008) Marker-assisted selection: an approach for precision plant breeding in the twenty-first century. Philos Trans R Soc B: Biol Sci 363(1491):557–572

de Oliveira L, Martins ER (2002) A quantitative assessment of genetic erosion in ipecac (Psychotria ipecacuanha). Genet Resour Crop Evol 49:607–617

Ehtisham M, Wani F, Wani I, Kaur P, Nissar S (2016) Polymerase chain reaction (PCR): back to basics. Indian J Contemp Dent 4(2):30

Funk WC, Forester BR, Converse SJ, Darst C, Morey S (2019) Improving conservation policy with genomics: a guide to integrating adaptive potential into US endangered species act decisions for conservation practitioners and geneticists. Conserv Genet 20:115–134

Gohil N, Panchasara H, Patel S, Singh V, 2019 Molecular biology techniques for the identification and genotyping of microorganisms. In: Microbial genomics in sustainable agroecosystems. vol 1, Springer, Singapore, pp 203–226

Hammer K, Teklu Y (2008) Plant genetic resources: selected issues from genetic erosion to genetic engineering. J Agric Rural Dev Trop Subtrop 109(1):15–50

Hasan M, Abdullah HM, 2015 Plant genetic resources and traditional knowledge: emerging needs for conservation. In: Plant genetic resources and traditional knowledge for food security, pp 105–120

Huber D, Voith von Voithenberg L, Kaigala GV (2018) Fluorescence in situ hybridization (FISH): history, limitations and what to expect from micro-scale FISH? Micro Nano Eng 1:15–24

Latch EK (2020) Integrating genomics into conservation management. Mol Ecol Res 20(6):1455–1457

Lau JD, Song AM, Morrison T, Fabinyi M, Brown K, Blythe J, Allison EH, Adger WN (2021) Morals and climate decision-making: insights from social and behavioural sciences. Curr Opin Environ Sustain 52:27–35

Levengood H, Zhou Y, Zhang C (2024) Advancements in plant transformation: from traditional methods to cutting-edge techniques and emerging model species. Plant Cell Rep 43(11):1–23

Li X, Li N, Wen D, Yu J, Hong J, Wu M, Cheng L, Meng S (2024) Identification of stress responsive NAC genes in *Casuarina equisetifolia* L. and its expression analysis under abiotic stresses. Agronomy 14(3):535

Malhotra N, Panatu S, Singh B, Negi N, Singh D, Singh M, Chandora R (2019) Genetic resources: collection, conservation, characterization and maintenance. In: Lentils. Academic, pp 21–41

Mastretta-Yanes A, da Silva JM, Grueber CE, Castillo-Reina L, Köppä V, Forester BR et al (2024) Multinational evaluation of genetic diversity indicators for the Kunming-Montreal global biodiversity framework. Ecol Lett 27(7):e14461

McCoy JC, Spicer JI, Ibbini Z, Tills O (2023) Phenomics as an approach to comparative developmental physiology. Front Physiol 14:1229500

Panis B, Nagel M, van Den Houwe I (2020) Challenges and prospects for the conservation of crop genetic resources in field genebanks, in in vitro collections and/or in liquid nitrogen. Plants 9(12):1634

Raffard A, Santoul F, Cucherousset J, Blanchet S (2019) The community and ecosystem consequences of intraspecific diversity. Meta-Anal Biol Rev 94:648–661

Shaw RE, Brockett B, Pierson JC, Sarre SD, Doyle P, Cliff HB, Eldridge MDB, Miller KA, Ottewell K, Parrott ML, Rossetto M, Sunnucks P, Banks SC, Byrne M, Campbell BL, Chong C, Ens EJ, Ferraro PA, Grueber CE, Harrison PA, Hogg C, Hutton M, Roycroft E, Sowersby W, Neaves LE (2024) Building meaningful collaboration in conservation genetics and genomics. Conserv Genet 25(6):1127–1145

Shokralla S, Gibson JF, Nikbakht H, Janzen DH, Hallwachs W, Hajibabaei M (2014) Next-generation DNA barcoding: using next-generation sequencing to enhance and accelerate DNA barcode capture from single specimens. Mol Ecol Res 14(5):892–901

Singh K, Gupta K, Tyagi V, Rajkumar S (2020) Plant genetic resources in India: management and utilization. Vavilov J Genet Breed 24(3):306

Supple MA, Shapiro B (2018) Conservation of biodiversity in the genomics era. Genome Biol 19:1–12

Varshney RK, Roorkiwal M, Sun S, Bajaj P, Chitikineni A, Thudi M, Singh NP, Du X, Upadhyaya HD, Khan AW, Wang Y, Garg V, Fan G, Cowling WA, Crossa J, Gentzbittel L, Voss-Fels KP, Valluri VK, Sinha P, Singh VK, Ben C, Rathore A, Punna R, Singh MK, Liu X (2021) A chickpea genetic variation map based on the sequencing of 3,366 genomes. Nature 599(7886):622–627

Vine SJ, Crowther MS, Lapidge SJ, Dickman CR, Mooney N, Piggott MP, English AW (2009) Comparison of methods to detect rare and cryptic species: a case study using the red fox (*Vulpes vulpes*). Wildl Res 36(5):436–446

Zhang F, Long R, Ma Z, Xiao H, Xu X, Liu Z, Wei C, Wang Y, Peng Y, Yang X, Shi X, Cao S, Li M, Xu M, He F, Jiang X, Zhang T, Wang Z, Li X, Yu L-X, Kang J, Zhang Z, Zhou Y, Yang Q (2024) Evolutionary genomics of climatic adaptation and resilience to climate change in alfalfa. Mol Plant 17(6):867–883

Plant Genetic Resources Conservation: Genetic Erosion to Genetic Engineering

Susmita Shukla, Himanshu, Tushar Joshi, Rajbeen Mazumder, Ankita Sharma, and Shiv Kant Shukla

Abstract

Plant genetic resources serve as the basis for biodiversity conservation, agricultural productivity, and food security. Genetic erosion is an important threat, driven by exploitative farming practices, habitat destruction, and climate change, that has led to the reduction in genetic diversity. This chapter explores the shift from genetic erosion to advancements in genetic engineering that provide effective solutions for conservation. Traditional methods of conservation, including in situ and ex situ approaches, play a significant role in maintaining plant genetic resources, whereas genetic engineering introduces innovative approaches such as the development of nutrient-enriched, stress-resilient crops, enhanced productivity, enhanced genetic diversity, and the restoration of endangered plant species. The present study also emphasizes successful conservation efforts highlighting the success stories of the International Rice Research Institute (IRRI) Gene Bank, Peru Potato Conservation, and the Svalbard Global Seed Vault. Ethical concerns and policy frameworks, such as the International Treaty on Plant Genetic Resources for Food and Agriculture (ITPGRFA) and the Convention on Biological Diversity (CBD), draw attention to the need for balanced strategies. The sustainable use of plant genetic resources for food and agriculture can be achieved by combining traditional and modern methods, which are crucial for future agricultural resilience and global food security.

S. Shukla (✉) · Himanshu · T. Joshi · R. Mazumder · A. Sharma
Applied Plant Biotechnology Research Lab, Centre for Plant and Environment Biotechnology, Amity Institute of Biotechnology, Amity University, Noida, Uttar Pradesh, India
e-mail: sshukla3@amity.edu

S. K. Shukla
Kaushalaya-The Skill University, Ahmedabad, Gujarat, India

Keywords

Plant genetic resources · Biodiversity · Agriculture · Genetic erosion · Genetic engineering

1 Introduction

Plant genetic resources (PGR) are the genes of plant species worth breeding, conserving, and cultivating; they include landraces, wild relatives of cultivated crops, and improved cultivars. Their role does not stop at providing food, fiber, or fuel, but extends beyond that to incorporate services to ecosystems, cultural traditions, and even economic development (Pathirana and Carimi 2022). According to the Food and Agriculture Organization (FAO), plant genetic resources are "the diversity of genetic material contained in traditional varieties, modern cultivars, crop wild relatives, and other plant species of present or potential value to food and agriculture." In addition, the Convention on Biological Diversity endorses the use of plant genetic resources for conserving genetic diversity, sustainable utilization, and upholding the intrinsic value of genetic materials for future generations. Plant genetic resources sustain agricultural productivity using breeding technologies that allow breeding to develop resilient and high-yielding crops. The genetic resource serves as a reservoir of important traits for overcoming the biotic and abiotic stress factors. Resistance to stem rust in wheat, for instance, or salinity tolerance in rice has once in a while been derived from wild family or conventional types. Breeding applications depend upon PGR to improve crops' resistance to biotic stresses, such as those imposed by insects and pathogens, and to abiotic stresses, consisting of drought, salinity, and immoderate temperatures. Climate change's exacerbation of these stresses similarly makes the state of affairs noticeably essential. Genes introgression from wild relatives has made vegetations like tomatoes

N. Kumar (ed.), *Biotechnology in Plant Conservation*, Sustainable Landscape Planning and Natural Resources Management,
https://doi.org/10.1007/978-3-032-02605-7_11

and bananas resistant to such powerful pathogens. Plant genetic resources uphold international biodiversity, as a consequence making sure of the integrity of ecosystems and the couple of services they offer. Agricultural landscapes that are genetically diverse better withstand environmental shocks and ailment outbreaks, as a consequence representing stable food production. For instance, conventional polyculture promotes genetic heterogeneity, which reduces the probability of crop failure. Plant genetic resources embody the continuum among conventional agricultural structures that are relevant to genetic diversity and sustainability and cutting-edge systems that leverage superior breeding techniques. Indigenous knowledge structures have long preserved PGR, such as the Andes' substantial potato landraces. Today, those resources are integrated into excessive-end genetic engineering, which quickens the selection and adaptation process for desirable traits.

1.1 Historical Milestones in Plant Conservation

The conservation of plant genetic resources (PGRs) is crucial for maintaining biodiversity, ensuring agricultural resilience, and addressing global challenges such as climate change and food security. The history of plant conservation is shaped by both traditional knowledge and scientific advancements. From ancient seed-saving practices to modern genetic engineering, efforts to safeguard plant diversity have evolved significantly over time.

1.1.1 Traditional Agricultural Practices

Early conservation methods began as farming traditions. Seed saving involves selecting seeds from the best-yielding crops and storing them, which permitted ancient farmers to hold genetic variance and make crops more adaptive to local environments. For instance, farmers in the Fertile Crescent have long preserved the wild relatives of barley and wheat, guaranteeing the success of these crops. Even the indigenous communities of the Andes region have cultivated diverse varieties of potatoes, making an enormous reservoir of genetics for future adaptation.

Sacred Groves and Cultural Traditions

Sacred groves are those areas, however small, that have been declared as sacred by humans within their community. In the case of India, the sacred groves of the Khasi Hills in Meghalaya have much of their biodiversity because the collection of resources is prohibited in these areas, which shows the level of spirituality attached to conservation.

1.1.2 Modern Agricultural Practices: Establishment of Gene Banks and Botanical Gardens

Gene banks have created a new paradigm in the field of ex situ conservation by introducing systematic collection and storage of seeds. The Philippines has one such International Rice Research Institute (IRRI), with an extensive collection of rice germplasm. The National Center for Genetic Resources Preservation (NCGRP), in the United States, conserves seeds under controlled conditions for long-term storage. Botanical gardens have expanded their focus on conservation and research for institutions. For example, the Royal Botanic Gardens, Kew, London, documents and preserves rare and endangered species (Meyer and Barton 2019).

1.2 Global Milestones in Plant Conservation

The Convention on Biological Diversity (CBD), adopted at the Earth Summit in Rio de Janeiro in 1992, has since provided an umbrella framework for biodiversity conservation, sustainable use, and equitable benefit sharing. Some of the signatures or results include in situ conservation (protected areas, community-based initiatives) and ex situ conservation (gene banks, seed vaults).

Importance of the CBD in Biodiversity Governance

The CBD has been instrumental in strengthening national and global governance systems for biodiversity. It has established frameworks for protected areas and promoted the idea of ecosystem-based management. It led to the introduction of National Biodiversity Strategies and Action Plans (NBSAPs), which are required for all countries signatory to the conference. The CBD paved the way for frameworks on access and benefit sharing (ABS) of genetic resources, main to the Nagoya Protocol in 2010, which strengthens the truthful sharing of benefits from the usage of genetic resources. Additionally, the CBD's Aichi Targets (2011–2020) provided specific goals related to biodiversity conservation, including targets to protect ecosystems, reduce pollution, and halt biodiversity loss.

Scientific Breakthroughs

Marking the times, genetic markers were developed in the 1980s. By identifying the specific genetic diversity within populations, possible conservation priorities could be inferred. Molecular biology is now being integrated with the use of traditional plant breeding as evidenced by the identification of genes enhancing drought or pest tolerance as new PGR in breeding programs (Henry 2014; Shruti et al. 2023; Tripathi et al. 2024).

2 Genetic Erosion

Genetic erosion refers to the gradual decline or loss of genetic variation within plant populations, often led by human activities such as industrial agriculture, monocropping, and the replacement of traditional crop varieties with modern, high-yielding cultivars. This loss of genetic diversity reduces the ability of plant species to withstand diseases, pests, and environmental stresses, thereby jeopardizing agricultural productivity and ecosystem stability.

2.1 Causes of Genetic Erosion

Genetic erosion has become an important area of concern, as it affects the sustainability of world agriculture, biodiversity, and food security. Understanding the key causes of genetic erosion is an important step in formulating appropriate conservation strategies for plant genetic resources (PGR) protection. Some of the causes are listed below.

Agricultural Modernization

Monocultural high crop-growing high-yielding varieties have been one major cause to contribute most significantly to genetic erosion. Monoculture farming can be defined as growing only one variety of a crop on a very large scale, meaning there is no variation at all in an agricultural landscape. The introduction of high-yielding varieties promoted during the "Green Revolution" displaced many traditional varieties that thrived and adapted to the local and indigenous farming systems. Most of these varieties offer a narrow genetic base and make themselves susceptible to many diseases, pests, and changing conditions of the climate. Loss of genetic diversity has reached greater heights because of heavy applications of agrochemicals like synthetic pesticides and fertilizers. The chemicals ensure homogeneity in crops by suppressing natural variation and discouraging the cultivation of existing traditional and diverse play varieties.

Deforestation and Habitat Loss

Deforestation is the process by which trees in forests are cleared along with small areas where they remain. It mainly serves agriculture and settlement as well as the construction of roads and other infrastructures. These are particular genetic contributors as far as the erosion of wild plant species is concerned. Natural vegetation, whether forests, wetlands, or other such habitats, shelters scores of plant species, as well as their wild relatives, and provides ecosystems for more than millions of genetic resources ranging from wild ancestors of cultivated crops to plants with prospective medicinal as well as agricultural value. Habitat loss also directly takes away the ability of wild relatives of crops to adapt to different environmental stresses. Wild relatives are extremely crucial because they serve as genetic banks from where genetic traits can improve disease resistance, drought tolerance, and other important features of cultivated crops. Propagation of aromatic and medicinal plants, as well as threatened and endangered species through plant tissue culture, is one of the prominent areas for conservation and crop improvement (Singh et al. 2023; Shukla et al. 2025).

Climate Change

Global biodiversity is affected, and the most serious cause of genetic erosion is climate change, which has an impact on plant genetic diversity. Changes in humidity, temperature, and the increasing number of severe climatic events critically change the conditions under which plants survive and distribute. This creates the necessity to "mobilize" different plant species or to adapt them to new conditions under changing climate zones. For many of them, especially those with restricted ecological niches, such changes lead to decreased genetic diversity, because populations become either fragmented or extinct. The continual increases in temperature and the changing precipitation patterns are changing habitats for both cultivated crops and wild plants. Species that have adapted to specific climates may not thrive effectively under new conditions, resulting in reduced genetic diversity and availability.

Overexploitation of Plant Resources

Overexploitation of plant species, whether for commercial purposes, medicine, or food, has caused overharvesting in populations of wild plants and the loss of important genetic material. Overexploitation of timber, medicinal plants, or other natural products contributes to unsustainable harvesting practices that reduce the genetic diversity in populations and exhaust them. Overharvesting reduces not only the abundance of a species but also genetic variability in populations, which makes them more vulnerable to environmental changes and diseases. The effects of such overexploitation on genetic diversity also affect the possibilities of the development of agriculture and medicine that are sustainable. Before plants with rare or unique genetic characteristics could be fully studied or exploited, most of them became extinct, depriving the future generation of valuable resources (Miladinovic et al. 2021).

2.2 Impact of Genetic Erosion

Genetic erosion has significantly impacted biodiversity, agriculture, and food security. It is important to address these challenges for the green and healthy ecosystem.

Impact on Biodiversity

Genetic erosion means various reductions in biodiversity, especially via the loss of wild relatives and indigenous crop forms: species-related subspecies losses also will count.

Loss of wild relatives and indigenous crops: These wild relatives of crops—the wild ancestors of modern wheat or rice—contain precious genetic materials for constructing resistance to pests and diseases and tolerance to some environmental stresses such as drought or salinity.

Species resilience to environmental stressors: When populations are genetically homogeneous, the different individuals are less likely to survive the changed condition, which is particularly significant when looking at climate change because variations in temperature and precipitation may exceed the capacity of species to adapt without enough genetic variation.

Role of genetic diversity in ecosystem services: All of this means genetic diversity is of prime importance for the many ecosystem functions that humankind uses, such as pollination, soil fertility, or pest control. It depends on the genetic diversity of the plants in terms of the number of pollinator species, for instance, bees, butterflies, birds, etc. The more an erosion reduces plant diversity, the more it cascades into a decline within pollinators and finally threatens food production.

2.2.1 Impact on Agriculture

Genetic erosion poses adverse effects on crop and livestock productivity and resilience. The less diversified the genetic base of agricultural systems, the higher the likelihood of the failure of crops to pests, diseases, and environmental stress. Genetic erosion is highly detrimental to agricultural sustainability—certainly where monocultures replace traditional agricultural systems relying on diverse crop varieties. Erosion of genetic diversity in crops and livestock: The present agricultural system has been increasingly concentrated on a very few crop species and breeds, and through such concentration, there has been a loss of genetic diversity. Risks of narrow genetic bases in major crops: Erosion of genetic diversity in crops increases risks of environmental stress factors. These crops have been bred for uniformity to the irreversible loss of genetic variation on which identification may occur for developing responses to new challenges, Maintaining traditional agricultural practices: The diversity of traditional cropping practices of landrace cultivation and diverse crop rotations are important for better access to available genotypes for farmers preparing themselves to face change. The increase in genetic diversity and resilience in agriculture also comes from traditional practices like growing landraces and crop rotation diversity (Engels and Ebert 2021).

Table 1 Impact of genetic erosion across biodiversity, agriculture, and food security

Aspect	Impact of genetic erosion
Biodiversity	Decrease in the number of species
	Absence of unique genes and specific traits
	Loss of genetic resistance leading to susceptibility to diseases and pest attacks
	Disturbance of ecosystem services with extinction of endemic species
Agriculture	Poor resilience of crops to environmental stresses (drought, salinity, climate change)
	Decreased genetic resources for breeding improved varieties
	Increasing dependence on fewer crops predisposes the system to monoculture vulnerabilities
	Declining productivity and lower income accrued to farmers
Food security	Diminished availability of diverse and nutritious food sources
	More dependence on a few staple crops makes crises more famine prone
	Fading of traditional and indigenous crop varieties that are better adapted to local conditions
	Lesser ability to feed the increasing number of humans over the globe

2.2.2 Impact on Food Security

Losses in genetic resources in crops and livestock would diminish adaptation to changing environmental conditions, whether due to droughts, floods, or changes in pest pressures. In areas where agricultural systems already suffer from the direct effects of climate change, further loss of genetic resources leads to the development of a situation where food security will become threatened (Das et al. 2024). Genetic erosion and the ability to adapt is about the genetic diversity that connects crops and livestock concerning their ability to adapt to new conditions. The lower the genetic erosion has hit the pool of genetic material available, the lower the response to change under challenges such as climate change, pest proliferation, or diseases. Table 1 depicts the impacts of genetic erosion in different aspects.

2.3 Strategies for Mitigating Genetic Erosion

It is widely agreed that the problem of genetic erosion will continue to erode the gene pool, particularly that of plant species, especially food crops. The issues of global biodiversity, agricultural resilience, and food security are then jeopardized. Such dispossession may warrant conservation measures that are extremely intensive and rich to mitigate genetic erosion and save what appear to be inexhaustible gene banks, which are imperative for maintaining ecosystems and agricultural systems.

2.3.1 In Situ Conservation Strategies

In situ, conservation means protecting genetic diversity by preserving it within its natural environment or the area where it is traditionally cultivated. It is an extremely important method of conserving the genetic resources of plant species in the wild and on farms. This in situ conservation also precludes maintaining evolutionary processes acting on genetic marks, and hence, it is essential for conservation to ensure the continuity and stability of ecosystems and agricultural systems in the future.

Ex Situ Conservation Strategies

Genetic resources are kept in secure environments away from their natural settings as part of ex situ conservation. In situ conservation would be the ideal, but ex situ conservation is often coupled with conserving endangered or otherwise inaccessible species and critters when the ideal is not an option. Ex situ conservation involves preserving genetic resources in safe conditions distant from their natural habitats. Ex situ conservation is frequently used in conjunction with the preservation of endangered or unreachable species and animals when in situ conservation is not feasible (Ebert and Engels 2020).

2.3.2 Sustainable Agricultural Practices

In place of maintaining the ancestral genetic stock of each food crop, agroecological and sustainable farm practices are hallmarks of reducing genetic erosion, primarily because of their contribution to the increased crop diversity and reduced dependence on monocultures. They build further resilience within farming systems because they also have the potential to conserve the genetic diversity of crops and their ecosystems.

2.3.3 Breeding for Genetic Diversity

Most breeding programs are playing a critical factor in creating a genetic diversity storehouse in crops. The breeding programs develop new varieties with enough genetic diversity and also can withstand environmental stressors by putting in germplasm within the wild relatives and landraces and even through heterosis (hybrid vigor).

3 Plant Genetic Resources Conservation

Plant genetic resources are the building blocks of our food systems and ecosystems. They represent the genetic material of plants that have value for agriculture, food security, and biodiversity. As the world faces challenges like climate change, habitat loss, and intensive farming, conserving these resources is more important than ever. Plant genetic resources are essential for maintaining healthy ecosystems and ensuring agriculture remains productive and resilient. Conservation of these plants can take place in two main ways: in situ (on-site) and ex situ (off-site) methods (Fig. 1).

3.1 In Situ Conservation

In situ conservation method means shielding plant genetic sources in their natural habitats, where they naturally develop and evolve. By preserving plant life within their ecosystems, we make certain that their genetic range is preserved in a way that lets them to evolve and thrive in their environment.

Protected Areas and Biosphere Reserves

One of the simplest ways to preserve plants within the wild is via establishing protected regions, including national parks, reserves, and biosphere reserves. These regions are set apart

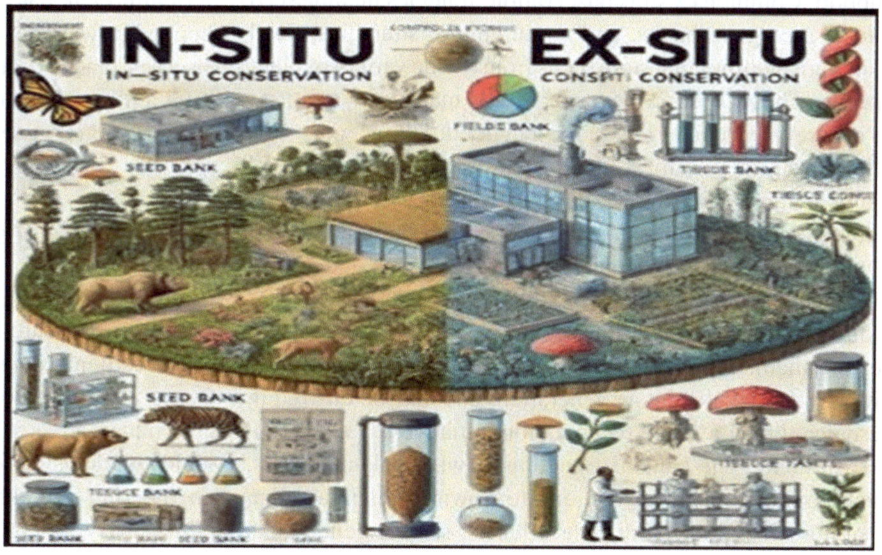

Fig. 1 In situ and ex situ conservation

to safeguard both the vegetation and the ecosystems in which they exist. Biosphere reserves, for instance, are designated by UNESCO to stabilize biodiversity conservation with sustainable development. These reserves guard no longer just rare or endangered plant species, but complete ecosystems, like the tropical rainforests of Southeast Asia or the Andes Mountains in South America. In these protected areas, human activities are cautiously managed to restrict damage to the environment. The gain of in situ conservation is that it preserves plants of their herbal surroundings, allowing them to continue evolving and interacting with other species, which facilitates preserving ecological balance (Schneider 2023).

On-Farm Conservation

Another essential method of in situ conservation is on-farm conservation, where farmers preserve traditional plant varieties alongside modern-day plants. This technique helps preserve the genetic diversity of flora used in agriculture, making plants extra resilient to pests, illnesses, and environmental stresses. For instance, many farmers in Asia preserve conventional rice varieties, while farmers in Africa do the same with maize varieties. On-farm conservation also supports food security by retaining these culturally great and nutritious crops alive, ensuring they are available for future generations. By integrating conservation into farming practices, we now not only defend plant range but also make sure that vegetation continue to adapt to neighborhood conditions (Olorode 2004).

3.2 Ex Situ Conservation

Ex situ conservation entails keeping plant genetic sources outside their natural habitats. This method is essential for species at threat because of habitat destruction or for plants with small populations that cannot survive in the wild without support. Common ex situ conservation strategies include seed banks, area gene banks, and tissue culture strategies. Seed banks are facilities that store seeds from a wide variety of plant species. These seeds are kept in carefully managed conditions to ensure they remain viable for lengthy durations, once in a while even for centuries. One of the most well-known seed banks is the Svalbard Global Seed Vault in Norway, which houses over a million samples from around the arena. Seed banks provide a protection net for plant species that are probably endangered or prone to extinction. If plants disappear from the wild, seeds stored in seed banks may be used to help restore them. This technique guarantees that genetic diversity is preserved for future generations and may be utilized in research, breeding, and recuperation initiatives (Yadav et al. 2024). Field gene banks are regions in which plant species, such as wild family of vegetation, are grown and maintained in residing collections. Unlike seed banks, subject gene banks allow vegetation to develop in soil, which is closer to their natural surroundings. These gene banks are important for retaining species that could not be stored as seeds or for plants that require particular developing conditions. Field gene banks not only hold genetic range but also function as research centers for plant breeding and genetic studies. They are particularly important for species that need specific conditions to thrive, which includes tropical or perennial vegetation. Examples consist of the International Rice Research Institute's (IRRI) gene financial institution and the Kew Gardens' Millennium Seed Bank within the United Kingdom (Abiha et al. 2021).

Tissue Culture and Cryopreservation

Tissue subculture and cryopreservation are contemporary techniques for maintaining plant genetic material. Tissue culture involves growing plant cells or tissues in a controlled, sterile environment, which can produce clones or genetically identical plants. This technique is regularly used to preserve uncommon or endangered species that are difficult to propagate via traditional methods.

Cryopreservation takes matters similarly through freezing plant tissues, along with seeds or embryos, at extraordinarily low temperatures, regularly in liquid nitrogen. This stops cellular activity, successfully "pausing" time and preserving the genetic material for the long term. Cryopreservation is in particular beneficial for conserving species which cannot be stored as seeds or grown through traditional propagation methods (Louwaars 2020).

4 Genetic Engineering in Plant Conservation

Genetic engineering, the process of enhancing an organism's genetic makeup using biotechnology, has revolutionized agriculture. It has allowed scientists to create crops with stepped forward developments, along with resistance to pests, diseases, and extreme weather conditions. But it also has giant ability for plant conservation, in particular in protective treasured plant genetic resources.

4.1 Genetic Engineering in Agriculture

Genetic engineering has significantly impacted agriculture by creating genetically modified organisms (GMOs) which are more resilient to environmental stresses, pests, and diseases. Crops such as drought-resistant maize and pest-resistant cotton have already been developed to assist farmers increase yields while using fewer chemicals. In conservation, genetic engineering can assist plants to live in a rapidly changing climate. By enhancing plants to resist drought, excessive temperatures, or flooding, it makes sure that these plants continue to grow in locations wherein conventional varieties might not survive (Tripathi et al. 2024).

4.2 Genetic Engineering's Role in Plant Conservation

Genetic engineering can play several key roles in conserving plant genetic resources.

Developing stress-resilient crops: Climate change is one of the greatest threats to plant biodiversity, with severe climate events like droughts and floods becoming more common. Genetic engineering can assist create crops which are more resilient to these stresses. For instance, genetically modified rice and maize varieties that are drought-tolerant may help sustain food production in areas wherein water is scarce, thereby safeguarding crop range.

Enhancing genetic diversity: Genetic engineering can also help to enhance the genetic variety of crops. By introducing new tendencies into the gene pool, we can improve the resilience of crops to pests, diseases, and changing environmental conditions. This ensures that crops maintain to adapt and thrive even as their developing situations trade. In addition, genetic engineering can assist reintroduce beneficial genes which had been misplaced through selective breeding or genetic erosion. For example, genes that offer resistance to certain pests or diseases may be reintroduced into crops, enhancing their long-term viability (Ritambhara et al. 2021).

4.3 Restoring Endangered Plant Species

Some plant species are on the verge of extinction due to habitat loss, overexploitation, and climate change. Genetic engineering gives a promising way to restore those species. Tools like CRISPR gene editing can help introduce genetic tendencies that make endangered plants more resilient to their changing environments, improving their chances of survival. An example of this is the recuperation of the American chestnut tree, which was devastated by a fungal disease. By introducing ailment-resistant genes into the tree, scientists preference is to supply this iconic species again to life. Conserving plant genetic resources is critical for maintaining biodiversity, food security, and the health of our ecosystems. Both in situ conservation—through protected areas and on-farm conservation—and ex situ methods, such as seed banks and field gene banks, are essential for preserving plant diversity. Genetic engineering, with its ability to create stress-resilient crops, enhances genetic diversity, restores endangered species, and offers exciting new possibilities for plant conservation. While traditional methods will remain crucial, the integration of genetic engineering provides an important tool for ensuring that our plant genetic resources are protected and available for future generations (Priyanka et al. 2021).

5 Linking Genetic Erosion and Genetic Engineering

Genetic drift refers to the loss of genetic variety in a species, which could have vital effects on ecological balance, biodiversity, and species survival. These losses of range may be because of a range of factors in outcomes, together with habitat degradation, climate alternate, overuse and great use of monoculture in agriculture Genetic variety is lost and so is the capacity to adapt to changing environments, giving them accelerated vulnerability to sicknesses, parasites and weather trade In this case, genetically technology has emerged as a ability device to fight genetic harm. However, the use of genetics in this field is fraught with complications and ethical dilemmas that must be carefully considered. Genetic engineering, which involves the application of biotechnology to the genome of an organism, holds significant promise for addressing the challenges posed by genetic degradation. They may be used to conserve, restore, and enhance the genetic diversity of both wild and domesticated species. However, the application of genetic engineering technologies in conservation and agriculture raises many scientific and ethical questions. To understand the role of genetic modification in combating genetic extinction, it is important to examine applicability, challenges, and ethical considerations. Figure 2, which depicts link between genetic erosion and genetic engineering.

5.1 Genetic Engineering as a Tool to Combat Genetic Erosion

Genetic engineering plays a crucial role in countering genetic erosion by reintroducing lost traits and enhancing crop diversity. By integrating genes from wild relatives or other resilient sources, scientists can restore resistance to pests, diseases, and environmental stresses that may have been lost through selective breeding. This technology helps maintain biodiversity, ensuring food security and sustainability while protecting valuable plant genetic resources from extinction.

5.1.1 Conservation of Endangered Species

One of the maximum pressing problems in conservation biology is the lack of genetic variety in endangered species. Small populations are especially susceptible to genetic erosion due to the fact that they are frequently isolated and experience inbreeding, which results in a reduction in genetic variation. This loss of genetic variety could make

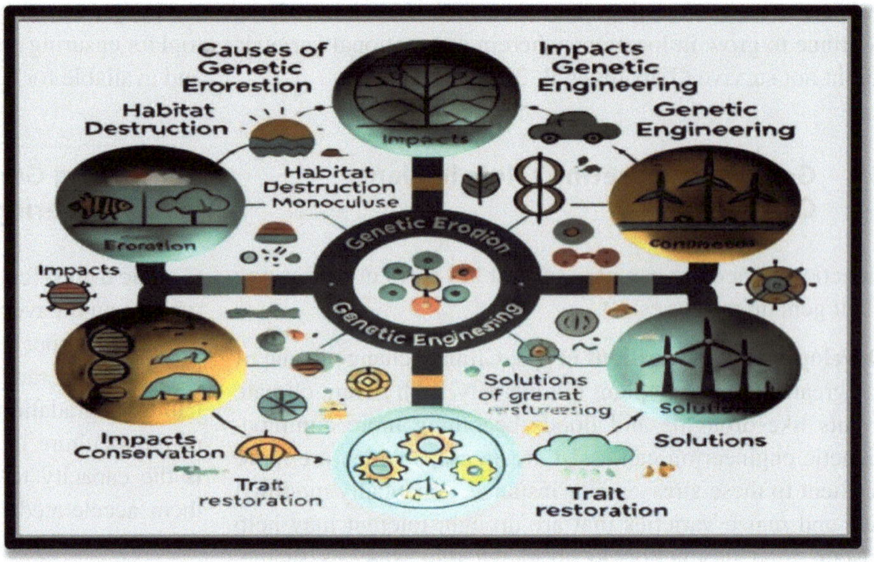

Fig. 2 Link between genetic erosion and genetic engineering

it greater difficult for species to conform to environmental modifications or face up to sicknesses and pests. Genetic engineering gives a potential way to this hassle by facilitating the restoration of lost genetic developments. For instance, scientists can use genetic engineering to introduce genes from different populations or closely related species to improve the genetic pool of endangered species. In a few cases, it can be possible to reintroduce genes that have been misplaced due to preceding bottlenecks within the population. This form of intervention could help to prevent extinction and promote the long-term survival of endangered species. An instance of such a method is the potential use of genetic engineering to restore the genetic variety of the northern white rhinoceros, a species this is now functionally extinct in the wild. With only two females left and no viable male, scientists are exploring advanced breeding technologies including cloning and genetic manipulation to restore the species While still in the experimental stage, these efforts emphasize categories of how genetic modification can help reduce the effects of genetic damage in endangered species.

5.1.2 Enhancing Agricultural Diversity

Farming systems are another area where genetic decline poses a serious threat to biodiversity. The widespread adoption of monocropping, where a single variety is grown over large areas, reduces genetic diversity and produces crops more susceptible to diseases, pests, and environmental changes. This is a particular problem in the context of global challenges such as climate change and the emergence of new pathogens. Genetic engineering can help solve these problems by introducing genetic traits that make crops more resistant to disease, drought, or changing temperatures. For example, genetically modified (GM) crops such as Bt cotton are engineered to resist insect pests, reduce the need for pesticides and produce fighting GM crops such as corn and rice against the drought to cope with periods of water scarcity, which can help protect global food in the face of a changing climate In crops in addition to creating a stronger crop, genetic engineering can help increase diversity agricultural production by enabling new crop varieties that were previously not breedable through conventional breeding methods has been done. This may want to assist revitalize agricultural diversification and ensure crop adaptability to changing environmental conditions. In some instances genetic modification can be used to introduce traits from wild spouse and children of domesticated crops, boom genetic diversity and growth resistance to pests, diseases and the environment have advanced (Yadesa 2022).

5.1.3 Gene Banks and Biotechnological Solutions

Another promising software of genetic engineering in preventing genetic erosion is using gene banks to save genetic cloth from endangered species and vegetation. Gene banks keep genetic range via gathering and storing seeds, tissues, and DNA from a huge variety of species. These genetic assets can then be used to regenerate populations or restore genetic tendencies which can were misplaced because of genetic erosion. Biotechnological advances, particularly in gene editing technology like CRISPR, offer thrilling opportunities for the use of gene banks to combat genetic erosion. With CRISPR, scientists can selectively edit genes inside stored genetic material, probable bringing decrease returned lost developments or enhancing the genetic range of a populace. This could be specially beneficial for reviving genetic range in plants and species which might be liable to extinction due to environmental or human-delivered about factors. Moreover, gene banks might be used to maintain and restore genetic cloth from wild populations which might be underneath hazard, preserving unique traits that could be vital for the survival of these species. For instance, genetic material from the remaining final human beings of a threatened species may be stored and edited to reintroduce misplaced tendencies or boom genetic range, supplying a critical aid for destiny conservation efforts (Rao and Hodgkin 2002).

5.2 Challenges and Ethical Considerations

While genetic engineering offers enormous potential for combating genetic erosion, its use in conservation and agriculture is not without challenges and ethical concerns. The deployment of genetic engineering technology in the herbal international and agricultural systems increases crucial questions on the potential dangers, unintended effects, and broader implications for biodiversity (Salgotra and Chauhan 2023) Loss of Traditional Breeding Practices Genetic engineering poses risks of accidental effects for organisms and ecosystems. Modified genes from genetically modified organisms (GMOs) can spread to wild populations, disrupting herbal genetic range and ecological balance. For instance, genetically changed crops may additionally crossbreed with wild loved ones, transferring engineered trends and reducing biodiversity. Additionally, unintended gene transfer have to result in resistant pests or new illnesses, similarly threatening genetic variety. While advanced gene-modifying equipment like CRISPR permit particular modifications, their lengthy-term outcomes continue to be unsure, potentially inflicting unpredicted interactions amongst trends and lowering the health of organisms in their environments (Aubry 2023) Social and Economic Equity The improvement and manage of genetic engineering technologies are dominated by a few companies and wealthy nations, growing disparities in get entry to. Small-scale farmers, mainly in developing global locations, may additionally lack the belongings to adopt those generation, while huge industrial farms advantage disproportionately. This should exacerbate present day inequalities in worldwide agriculture and biodiversity conservation. Furthermore, the monopolization of genetic resources via organizations increases issues of intellectual property and fairness, limiting equitable get right of entry to crucial genetic substances. Gene Editing and Genetic Ownership The big use of gene-editing equipment like CRISPR intensifies debates approximately ownership and manipulate of genetic fabric. Questions about who owns changed genes and who has the proper to govern them spotlight worries approximately commodifying life. Patenting genetically modified organisms concentrates strength among some businesses, proscribing get right of entry to genetic sources and elevating the risk of exploitation, including biopiracy using genetic substances from indigenous communities without consent or repayment.

6 Policy and Governance

Policy and governance in biodiversity conservation: Biodiversity is crucial for the health of ecosystems and human well-being. However, global biodiversity faces substantial threats, starting from habitat loss, climate change, overexploitation of natural resources, and invasive species. Effective coverage frameworks and governance systems are therefore necessary to conserve and sustainably use biodiversity. International frameworks, inclusive of the Convention on Biological Diversity (CBD) and the International Treaty on Plant Genetic Resources for Food and Agriculture (ITPGRFA), offer a foundation for global efforts to conserve biodiversity and manage genetic resources. In parallel, national governments play an essential role in imposing conservation strategies, aligning their domestic rules with global agreements while addressing local biodiversity challenges. This essay will discover the global frameworks, agreements, and the position of national governments in biodiversity conservation, with an emphasis on CBD and ITPGRFA.

6.1 International Frameworks and Agreements

The Convention on Biological Diversity (CBD), adopted at the 1992 Earth Summit in Rio de Janeiro, is one of the most complete global agreements aimed toward preserving biodiversity and ensuring the sustainable use of its additives. The CBD has three targets: conservation of biodiversity, sustainable use of biodiversity, and fair and equitable sharing of the benefits arising from the use of genetic resources (Rajpal et al. 2023).

The International Treaty on Plant Genetic Resources for Food and Agriculture (ITPGRFA), adopted in 2001 under the Food and Agriculture Organization (FAO) of the United Nations, is another critical framework associated with biodiversity, focusing on the conservation and sustainable use of plant genetic resources for food protection and agriculture.

Main Objectives of the ITPGRFA

i. Conservation and sustainable use of plant genetic resources: It promotes the conservation of a wide range of plant species, many of which are important for food protection and nutrition.
ii. Access and Benefit-sharing: The treaty ensures that nations proportion the advantages springing up from the use of plant genetic sources. For example, it created a

Multilateral System of Access and Benefit-sharing (MLS) that provides farmers, breeders, and researchers with smooth get entry to to plant genetic material, whilst also making sure that advantages are shared with the carriers of such sources.

iii. Farmer's rights: It recognizes the contributions of farmers in maintaining and growing agricultural biodiversity and ensures certain rights, including the right to store, use, trade, and promote farm-saved seeds.

The ITPGRFA is essential because it directly addresses the genetic resources necessary for global food production, which are under increasing threat from weather change, land degradation, and genetic erosion (Schmitz et al. 2023).

6.1.1 Interlinkages Between CBD and ITPGRFA

While the CBD focuses of the conservation of all biodiversity, the ITPGRFA is a specialized agreement dealing with the conservation and sustainable use of plant genetic resources for food and agriculture. These frameworks are closely associated, as both agreements emphasize the equitable sharing of advantages arising from genetic resources. However, the CBD mostly deals with genetic sources in a general context, while the ITPGRFA focuses particularly on agricultural vegetation. While the CBD offers with the broader conservation of biodiversity, the ITPGRFA tackles specific issues related to agriculture and meals protection, consisting of ensuring get admission to to the genetic diversity that helps crop breeding and agricultural development. Together, those two frameworks work in the direction of included governance of biodiversity and the assets that underpin agricultural productiveness, food security, and sustainable improvement.

6.2 Role of Central Governments in Conservation

While global agreements which include the CBD and ITPGRFA provide an international framework for biodiversity conservation, national governments play an important role in imposing these frameworks and ensuring that biodiversity is conserved and sustainably used at the national level.

6.2.1 Implementation of International Agreements

Central governments are responsible for ratifying global agreements and adapting them to local contexts. For instance, signatories of the CBD are required to: develop and implement National Biodiversity Strategies and Action Plans (NBSAPs); establish protected areas and integrate biodiversity conservation into numerous sectors which include agriculture, forestry, fisheries, and tourism; and implement regulation on access and benefit-sharing (ABS) to make certain that genetic resources are accessed in compliance with CBD's pointers. Similarly, for the ITPGRFA, national governments should make sure that their domestic guidelines and practices align with the provisions of the treaty, particularly promoting sustainable agricultural practices and protecting plant genetic resources.

6.2.2 Policy Development for Biodiversity Conservation

Central governments integrate biodiversity conservation into broader policy frameworks. This may include:

Environmental Laws: Enacting legislation to shield critical ecosystems, regulate looking, manipulate invasive species, and manipulate natural resources sustainably,

Sectoral policies: Ensuring that sectors including agriculture, forestry, fisheries, and power contribute to biodiversity conservation rather than causing further degradation.

Sustainable development plans: Incorporating biodiversity conservation into national sustainable development plans to ensure that financial growth does not come at the cost of biodiversity. For instance, policies such as sustainable agriculture and forestry management can lessen pressures on natural habitats and make sure that biodiversity is maintained while permitting monetary improvement (Hufford et al. 2019).

6.3 Success Stories of Genetic Resource Conservation

The conservation of plant genetic resources is critical for ensuring global food security, protecting biodiversity, and mitigating the impacts of climate change. Various institutions, governments, and local communities worldwide have implemented innovative conservation strategies that serve as models for safeguarding genetic diversity. These success stories demonstrate the effective application of ex situ and in situ conservation methods, often integrating traditional knowledge with modern scientific advancements. Some of the success stories are mentioned below.

6.3.1 The Svalbard Global Seed Vault, Norway

The Svalbard Global Seed Vault, popularly known as the "Doomsday Vault," is an important ex situ conservation project meant to preserve the genetic diversity of the world's crops. Nestled within the remote Svalbard archipelago, the facility in Norway is around 1300 km from the North Pole and was opened in 2008 through a collaboration among the Norwegian government, the Global Crop Diversity Trust (or Crop Trust), and the Nordic Genetic Resource Center (NordGen).

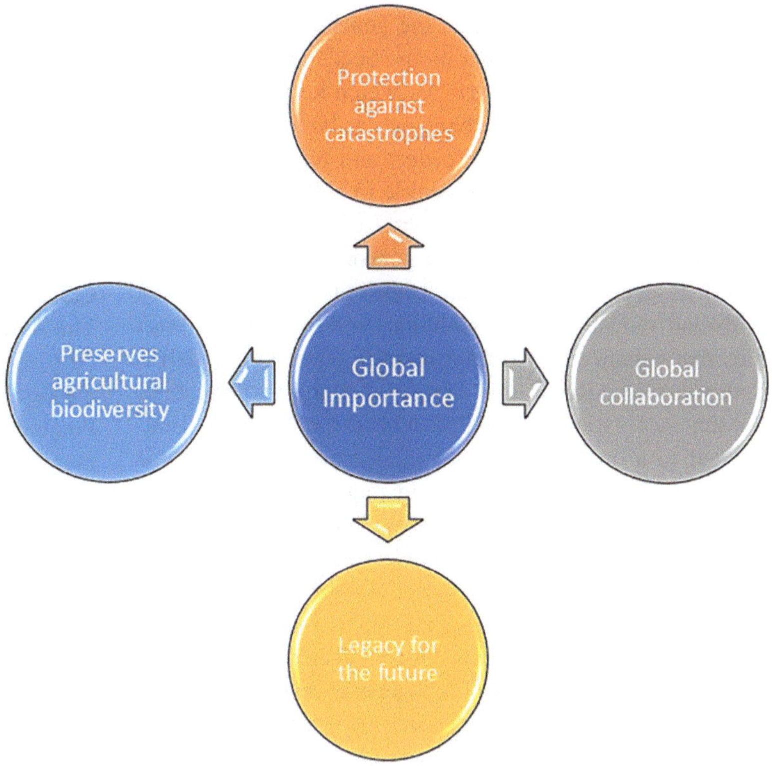

Fig. 3 Global importance of Svalbard Global Seed Vault

Structure and Design

Because the Seed Vault is built deep within a mountain, it is safe from both natural and human-made disasters. It was designed to be resilient against extreme weather events, earthquakes, and sea-level rise associated with climate change. Its location within the permafrost guarantees that the seeds will stay frozen without power, even in worst-case scenarios. Inside the vault, the temperature is a crisp − 18 °C, the perfect environment for preserving seeds long-term, as it slows down seed metabolism and extends their viability. The facility consists of three vault rooms extending 100 m into the mountain, an airlock entry system, multiple layers of security to ensure protection, and an advanced monitoring system that regulates humidity, temperature, and safety.

Seed Collections and Storage

The Svalbard Global Seed Vault already contains seed samples from crop varieties from almost every country, numbering more than 1 million in total. The samples include staples, such as wheat, rice, maize, and barley, as well as indigenous crops and wild plant species. The seeds stored in the vault serve as backup repository for existing gene banks, which could be threatened by events such as wars, natural disasters, or lack of funds. The seeds are provided by individual countries, agricultural research organizations, and gene banks. All depositors are the rights holders of their seeds and can retrieve them, if needed (Fig. 3). Key contributors are the International Rice Research Institute (IRRI), the United States Department of Agriculture (USDA), and national seed banks in the regions of Africa, Asia, and Latin America.

6.3.2 Rice Biodiversity, Philippines—International Rice Research Institute (IRRI) Gene Bank

IRRI, based in Los Baños, Philippines, is an internationally recognized center for rice conservation and research. Established in 1960, IRRI aims to reduce hunger, malnutrition, and poverty through the creation of better rice varieties and sustainable agricultural practices. Its efforts focus on enhancing food security, especially in Asia, where rice is a staple food for millions. IRRI Gene Bank is also known as "the treasure trove of rice diversity." This gene bank is one of the largest and most prominent repositories of rice genetic resources in the world. It houses more than 130,000 rice accessions gathered from around the world, including:

(a) Landraces: traditional rice varieties cultivated by farmers for generations and tailored to local conditions.
(b) Wild relatives: wild species of rice that provide valuable genetic traits such as drought resistance, pest tolerance, and disease immunity.
(c) Improved cultivars: modern rice varieties developed through breeding programs to achieve high yields and resilience to stresses.

The gene bank ensures that the genetic diversity of rice is preserved for future generations. Seeds are stored under controlled conditions at − 18 °C, which keeps them viable for decades. The IRRI gene bank acts as both a conservation facility and a resource for breeders and researchers worldwide. IRRI has played a crucial role in helping rice farmers solve some of their biggest problems, climate change, pests, and diseases. IRRI harnesses the genetic diversity contained within its gene bank especially the traits of landraces and wild relatives to create high-yielding, stress-resistant rice varieties that enhance global food security (Ogwu et al. 2014). Under the Green Revolution, IRRI created short-stature, high-yielding rice varieties like IR8, dubbed the "Miracle Rice." IR8 led to a dramatic increase in rice production in Asia, alleviating hunger and poverty in many places. IRRI has developed rice varieties like Sahbhagi Dhan that can thrive in water-scarce regions, ensuring stable yields in drought-prone areas, and flood-tolerant rice varieties such as Swarna-Sub1 have been engineered to survive submergence for up to two weeks, helping farmers in flood-affected regions maintain production. IRRI has bred genetic resistance against significant pests and diseases, such as bacterial leaf blight and the brown plant hopper, as well as the blast fungus, by transferring traits from wild relatives into existing rice varieties. This minimizes the use of chemical pesticides, thus encouraging environmentally sustainable farming practices. One of the most prominent biofortified rice varieties is Golden Rice, which has been genetically engineered to produce beta-carotene, a precursor of vitamin A, to combat vitamin A deficiency in regions where this malnutrition is common.

The IRRI Gene Bank's large collection of rice accessions is a vital resource in overcoming modern agricultural obstacles, including water and drought scarcity (wild relatives of rice have genes for drought tolerance, allowing the development of resilient rice varieties that need less water), diseases and pests (genetic traits from landraces and wild species improve resistance to evolving pests and diseases, decreasing crop losses and the need for chemical inputs), and climate Change (IRRI's work ensures that rice farming remains sustainable in the face of extreme weather events, rising temperatures, and changing rainfall patterns). IRRI's contributions to rice research have had a transformative impact on global food security. Rice varieties developed by IRRI are cultivated across millions of hectares, particularly in Asia and Africa, feeding billions of people. By improving yields and resilience, IRRI's work has empowered smallholder farmers to achieve economic stability and adapt to changing environmental conditions. IRRI's focus on sustainable practices, such as reducing water usage and minimizing pesticide reliance, promotes environmental conservation. IRRI works with national governments, research organizations, and farmer communities around the world to conserve and use rice genetic resources. It also provides seeds and expertise to breeders and researchers around the world through collaborations to help solve food security problems on a global scale.

6.3.3 Potato Conservation in Andes, Peru

The Andes region in South America, including countries like Peru, Bolivia, Ecuador, and Colombia, is the birthplace of the potato, one of the world's most important food crops. Potatoes were domesticated in the Andean highlands more than 8000 years ago by indigenous communities, who have cultivated and preserved an extraordinary diversity of potato landraces adapted to various climates and altitudes. This region remains a global hub for in situ conservation, where genetic diversity is preserved directly in farmers' fields and natural habitats. The Andes region is home to over 4,000 varieties of native potatoes, representing a rich reservoir of genetic diversity. These varieties differ in shape, size, color, adaptation, and nutritional content. This genetic wealth is not only culturally significant but also critical for global food security. As climate change intensifies, the genes found in these traditional landraces are being used to develop stress-tolerant potato varieties that can adapt to drought, heat, and emerging pests and diseases. For millennia, indigenous farmers in the Andes have acted as custodians of potato diversity. Their extensive understanding of local ecosystems, soils, and climates has allowed them to breed dozens of different varieties of potatoes for varying conditions (Guzzon et al. 2022). Farmers employ traditional techniques, such as:

(a) Selective breeding: identifying and cultivating the best-performing plants for resilience, taste, and yield.
(b) Crop rotation and mixed cropping: enhancing soil fertility and reducing pest pressures by growing potatoes alongside other native crops like quinoa and maize.
(c) Altitude-based farming: cultivating different potato varieties at elevations ranging from 2500 to 4500 m, each suited to specific microclimates.

By maintaining these traditional practices, Andean farmers ensure that potato landraces remain genetically diverse, resilient, and capable of adapting to environmental changes. The Potato Park in Peru, located in the Sacred Valley near Cusco, is a pioneering initiative in community-led conservation. Established by indigenous communities with support from organizations like the International Potato Centre (CIP), the Potato Park covers over 10,000 hectares of high-altitude land. Figure 4, which depicts the important objectives of the Potato Park.

Fig. 4 Potato Park's objectives

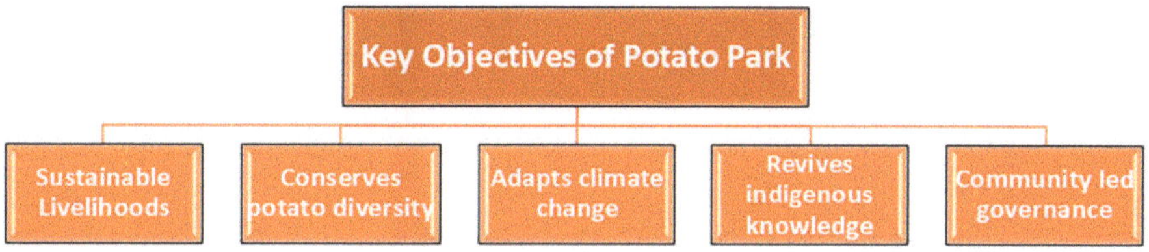

7 Role of Genetic Engineering in Addressing Conservation Challenges

The use of genetic engineering has become a revolutionary and progressive approach to the pressing issues of conservation of plant genetic resources. While there are traditional approaches like in situ and ex situ conservation, the hurdles that the world is facing in terms of climate change, genetic erosion, and food insecurity are unprecedented, and traditional solutions will not suffice alone to address these issues. Genetic engineering provides powerful tools to protect and restore plant genetic diversity through precise and target-specific interventions, contributing to sustainable agriculture and ecosystem resilience.

7.1 Addressing Genetic Erosion

Addressing genetic erosion involves executing conservation methods such as in situ preservation, seed banks, and sustainable farming practices to preserve plant genetic diversity. Figure 5, describes genetic engineering as a mitigation for genetic erosion.

7.2 Enhancing Stress Resilience

Climate change has intensified environmental stresses such as droughts, floods, extreme temperatures, and soil salinity, posing significant threats to global agriculture. These stresses not only jeopardize food security but also accelerate genetic erosion by pushing vulnerable plants and their wild relatives toward extinction. Traditional crop varieties, though productive under normal conditions, often lack the resilience to withstand such harsh and unpredictable environments. Genetic engineering offers innovative solutions for developing stress-resilient crops by introducing specific genes that enhance their ability to thrive under adverse conditions. This approach plays a crucial role in preserving plant genetic resources while ensuring sustainable food production, especially in regions most severely affected by climate change.

7.2.1 Drought-Tolerant Crops

Drought is one of the major consequences of climate change, particularly in arid and semiarid regions where water availability is already scarce. Prolonged drought can lead to severe crop failures, threatening food security and the livelihoods of millions of farmers. Genetic engineering has enabled scientists to develop plants that can tolerate water scarcity by identifying and transferring drought-resistant genes from wild species or drought-adapted plants. These genes enhance the plant's capability to help regulate water-use efficiency, maintain cellular integrity, and trigger stress-responsive pathways, enabling plants to better cope with adverse environmental conditions. For example, drought-tolerant maize varieties have been developed by incorporating dehydration-responsive element binding (DREB) genes, ensuring stable yields in water-scarce regions like sub-Saharan Africa and

Fig. 5 Genetic engineering as a solution for genetic erosion

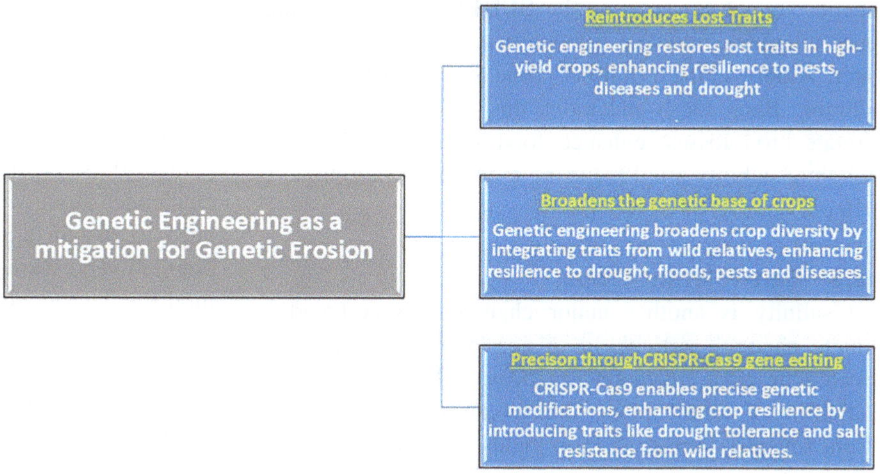

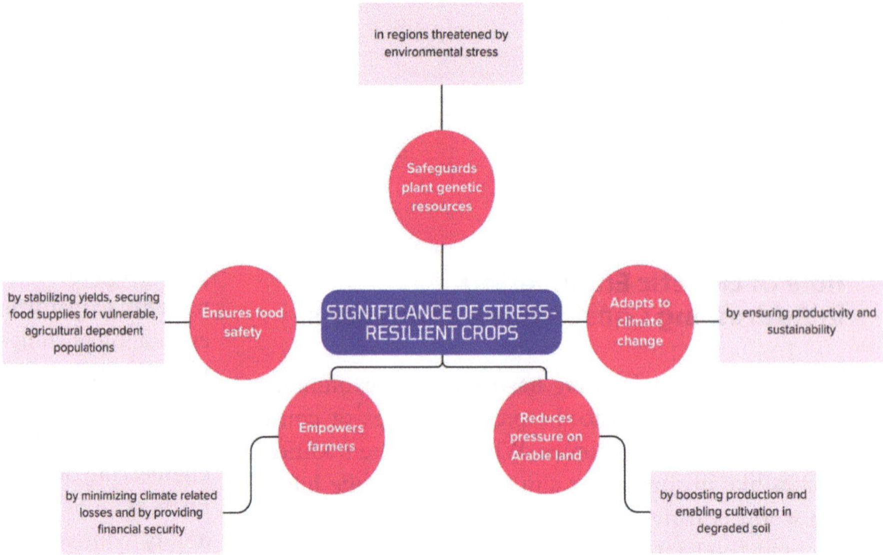

Fig. 6 Importance of stress-resilient crops

parts of South Asia, and drought-resilient rice varieties have been developed by incorporating drought-tolerant genes from wild rice relatives, enabling their cultivation in rain-fed and drought-prone regions. These drought-tolerant crops are transforming agriculture by enhancing yields and helping farmers adapt to increasingly unpredictable rainfall patterns.

7.2.2 Flood-Tolerant Crops

Flooding has become more frequent and intense due to climate change, particularly in low-lying areas and river basins where rice is a staple crop. Traditional rice varieties are highly sensitive to submergence, and extended flooding can destroy entire crops. Genetic engineering has tackled this challenge by introducing flood-tolerance genes, enabling rice plants to survive and recover after prolonged periods of submergence. These genes permit plants to Flood-tolerant crops can temporarily lower metabolic activity, enabling them to survive in oxygen-deficient, waterlogged conditions and Flood-tolerant crops can quickly recover and resume growth once the water recedes. For example, Swarna-Sub1 rice, developed through genetic engineering, carries the **SUB1A** gene for submergence tolerance, allowing it to survive up to 14 days underwater. Widely adopted in flood-prone regions of South Asia, including Bangladesh and India, this variety helps farmers mitigate crop losses, enhance food security, and achieve economic stability amid recurring monsoons (Salgotra et al. 2019).

7.2.3 Salt-Tolerant Crops

Soil salinity is another major challenge exacerbated by climate change, particularly in coastal regions and areas affected by poor irrigation practices. Salinity reduces the availability of water to plants and can cause toxic ion accumulation, severely impacting crop yields. Genetic engineering allows scientists to introduce salt-tolerance traits into crops, enabling them to survive and produce yields in saline-affected soils. Salt-tolerant genes are often derived from halophytes and wild crop relatives. These genes enable plants to use exclusion mechanisms to prevent sodium ion buildup in tissues, enhance salt tolerance, regulate water and ion uptake, maintain cellular balance under saline conditions, and produce protective proteins that shield plant cells from salinity-induced stress. Genetically engineered salt-tolerant rice genes incorporated from wild species like *Oryza coarctata* thrives in saline-affected fields, while genetically modified salt-resilient wheat, Kharchia 65, thrives in high-salinity soils, enabling cultivation in degraded and coastal lands. These salt-tolerant crops ensure that previously uncultivable saline soils can be utilized for food production, helping to secure agricultural productivity in affected regions (Legesse 2020). Figure 6 depicts the significance of stress-resilient crops.

7.3 Enhancing Nutritional and Agronomic Trait

Genetic engineering has revolutionized agriculture by way of enhancing the dietary satisfactory and agronomic traits of plants. These advancements are vital for addressing global challenges along with malnutrition, food insecurity, and the need to sustain agricultural productivity under developing population pressures and environmental stresses. By precisely modifying crop genomes, genetic engineering guarantees that plants meet human nutritional needs even while maintaining genetic diversity and improving yields without further encroachment into natural ecosystems (Shukla and Shukla 2024).

8 Nutrient-Enriched Crops

Genetic engineering combats micronutrient deficiencies by developing biofortified crops enriched with essential vitamins and minerals like Golden Rice (engineered with beta-carotene to address vitamin A deficiency, preventing blindness and immune disorders in vulnerable populations), iron and zinc-enriched crops (genetically modified rice, wheat, and beans with higher iron and zinc levels help combat anemia and stunted growth), and protein-enriched crops (enhanced maize and potatoes provide improved protein content, addressing malnutrition in protein-deficient regions). Nutrient-enriched crops are an innovative solution to malnutrition, offering a cost-effective, scalable, and sustainable approach to improving health outcomes in communities with limited access to diverse and nutritious diets.

Higher Yields to Ensure Food Security

Genetic engineering boosts crop productivity to meet rising food demand without expanding farmland, enhancing yield, resilience, and sustainability while preserving biodiversity.

Improved Photosynthesis Efficiency

Photosynthesis, the process by means of which plants convert sunlight into energy, is inefficient in many plants. Genetic engineering enables scientists to enhance photosynthetic efficiency through modifying genes that adjust mild absorption and carbon fixation. For instance, studies have shown that introducing genes to optimize the RuBisCO enzyme—a key player in photosynthesis—can appreciably improve crop yields. Improved photosynthesis results in faster growth, greater biomass manufacturing, and higher yields, specifically for crops like rice and wheat.

Disease and Pest Resistance

Genetic engineering reduces crop losses by introducing resistance genes that protect plants from diseases, pests, and pathogens. This minimizes the need for chemical insecticides, which can harm ecosystems and human health. Bt crops: Crops like Bt cotton and Bt maize are genetically engineered to produce a protein from the bacterium *Bacillus thuringiensis*, which is toxic to specific pests but safe for humans and the environment. These crops notably lessen pest damage and enhance yields. Disease-resistant types, which includes wheat proof against stem rust or potatoes proof against late blight, make sure strong production in areas vulnerable to crop illnesses.

Stress Tolerance for Higher Yields

Crops engineered to face up to environmental stresses, consisting of drought, salinity, and flooding, hold productivity underneath challenging conditions. For example: Drought-tolerant maize sorts make sure stable yields in water-scarce regions. Salt-tolerant rice lets in cultivation in saline soils, which can be growing due to sea-degree upward thrust and terrible irrigation practices. By incorporating pressure-resilient trends, genetically engineered vegetation improve productivity on existing agricultural land, reducing the pressure to clear natural habitats for farming.

9 Conclusion

The conservation and sustainable utilization of plant genetic resources are crucial for addressing the interconnected challenges of biodiversity loss, global food security, and climate change. While conventional in situ and ex situ conservation strategies remain indispensable, genetic engineering has emerged as a pivotal tool in preventing genetic erosion, improving crop resilience, and restoring endangered species. Genetic engineering complements traditional conservation efforts by preserving genetic trait for improving yields and nutritional value. Genetic engineering allows scientists to preserve important genetic traits from landraces, wild relatives, or traditional varieties, and it also reduces genetic erosion by broadening the genetic base of crops and incorporating diverse traits, by mitigating the loss of biodiversity caused by monoculture farming. For instance, traits from wild rice species have been used to enhance the resilience of cultivated rice varieties, ensuring that critical genetic material is not lost but instead leveraged for modern agriculture. Successful global case studies highlight the potential of integrating these approaches to preserve plant diversity and promote sustainable agriculture. However, this progress must be supported by strong policy frameworks, ethical considerations, and equitable access to technology to ensure global food security and environmental sustainability. Preserving the genetic legacy of plant species is crucial for building a resilient future for agriculture and ecosystems, even in the face of unprecedented challenges.

References

Abiha U, Phutela S, Shukla S (2021) Biodiversity conservation: an imperial need in combatting pandemic and healthcare emergencies. In: The impact of the COVID-19 pandemic on green societies: environmental sustainability, pp 323–347

Aubry S (2023) Genebanking plant genetic resources in the postgenomic era. Agric Human Values 40(3):961–71

Das S, Shukla S, Debnath S, Manjula IK (2024) Synergistic role of plant tissue culture and plant growth-promoting micro-organisms as conservation strategies for Himalayan medicinal plants amidst climate change. In: Mishra AP, Kumar A, Chandra N, Singh G, Pande CB (eds) Threatened medicinal plants in the Indian Himalayan region. World Sustainability Series. Springer, Cham. https://doi.org/10.1007/978-3-031-73687-2_8

Ebert AW, Engels JMM (2020) Plant biodiversity and genetic resources matter! Plants 9:1706. https://doi.org/10.3390/plants9121706

Engels JMM, Ebert AW (2021) A critical review of the current global ex situ conservation system for plant agrobiodiversity. I. history of the development of the global system in the context of the political/legal framework and its major conservation components. Plants 10:1557. https://doi.org/10.3390/plants10081557

Guzzon F, Gianella M, Giovannini P, Payne TS (2022) Conserving wheat genetic resources. In: Wheat improvement: food security in a changing climate. Springer International Publishing, pp 299–318. https://doi.org/10.1007/978-3-030-90673-3_16

Henry RJ (2014) Sequencing of wild crop relatives to support the conservation and utilization of plant genetic resources. Plant Genet Resour Characterisation Util 12(SUPPL. 1):S1-4. https://doi.org/10.1017/S1479262114000033

Hufford MB, Berny Mier Y Teran JC, Gepts P (2019) Crop biodiversity: an unfinished magnum opus of nature. Annu Rev Plant Biol 70:727–751. https://doi.org/10.1146/annurev-arplant-042817-040240

Legesse A (2020) Assessment of coffee (*Coffea arabica* L.) genetic erosion and genetic resources management in Ethiopia. Int J Agric Ext 7(3):223–229

Louwaars N (2020) Increasing or limiting genetic diversity of crops through plant biotechnology: a review

Meyer A, Barton N (2019) Botanic gardens are important contributors to crop wild relative preservation.Crop Sci 59(6):2404–2412. https://doi.org/10.2135/cropsci2019.01.0033

Miladinovic D, Antunes D, Yildirim K, Bakhsh A, Cvejić S, Kondić-Špika A et al (2021) Targeted plant improvement through genome editing: from laboratory to field. Plant Cell Rep 40:935–951. https://doi.org/10.1007/s00299-021-02689-0

Ogwu MC, Osawaru ME, Ahana CM (2014) Challenges in conserving and utilizing plant genetic resources (PGR). Int J Genet Mol Biol 6(2):16–23. https://doi.org/10.5897/IJGMB2014.0093

Olorode O (2004) Conservation of plant genetic resources 1:1–10. Available from: www.africanethnomedicines.net

Pathirana R, Carimi F (2022) Management and utilization of plant genetic resources for a sustainable agriculture. Plants 11:2760. https://doi.org/10.3390/plants11202760

Priyanka V, Kumar R, Dhaliwal I, Kaushik P (2021) Germplasm conservation: instrumental in agricultural biodiversity-a review. Available from: https://www.preprints.org/manuscript/202104.0461/v1

Rajpal VR, Sehgal D, Valluru R, Singh S (2023) Editorial: current advances in genomics and gene editing tools for crop improvement in a changing climate scenario. Front Genet 14:1176016. https://doi.org/10.3389/fgene.2023.1176016

Rao VR, Hodgkin T (2002) Genetic diversity and conservation and utilization of plant genetic resources. Plant Cell Tissue Organ Cult 68(1):1–19. https://doi.org/10.1023/A:1013359015812

Ritambhara, Shukla SK, Shukla S (2021) Automation, modern tools and technique for sustainable agriculture–an important parameter toward advance plant biotechnology. In: Green technological innovation for sustainable smart societies: post pandemic era, pp 281–300

Salgotra RK, Chauhan BS (2023) Genetic diversity, conservation, and utilization of plant genetic resources. Genes 14:205. https://doi.org/10.3390/genes14010205

Salgotra RK, Sharma M, Pandotra P (2019) Biotechnological interventions for sustainable conservation of plant genetic resources in the scenario of climate change. Nat Resour Conserv Res 1(4):46–55. https://doi.org/10.26599/NRCR.2019.9040004

Schmitz S, Barrios R, Dempewolf H, Guarino L, Lusty C, Muir J (2023) Crop diversity, its conservation and use for better food systems. In: Science and innovations for food systems transformation. Springer International Publishing, pp 545–52. https://doi.org/10.1007/978-3-031-15730-5_59

Schneider H (2023) Integrating genomics and conservation to safeguard plant diversity. Integr Conserv 2:10–18. https://doi.org/10.1111/icv.12001

Shruti, Tripathi A, Shukla S (2023) Methods of genetic transformation: major emphasis to crop plants. J Microbiol Biotechnol Food Sci 13(4):1–12. https://doi.org/10.55251/jmbfs.10276

Shukla S, Shukla SK (2024) Micropropagation for crop improvement and its commercialization potential, IN: the potential of microbes for a circular economy, Paperback ISBN: 9780443159244 eBookISBN: 9780443159251

Shukla S, Patra D, Sinha A, Saha R, Tripathi Y, Borah NK Shukla SK (2025) Medicinal and aromatic plant cultivation, improvement and conservation for sustainable development. In: Kumar N (ed) Industrial crops improvement. Sustainable landscape planning and natural resources management. Springer, Cham, pp 183–204

Singh R, Shukla S, Shukla SK (2023) Impact of sterilant on in vitro seed germination of *Citrus macroptera* montr.: an important measure. Indian J Agric Biochem 36(2):183–188

Singh R, Shukla S, Shukla SK, Katare DP (2024) Chemical profiling of endangered *Citrus macroptera* leaf extracts and evaluation of its cytotoxic activity. Plant Sci Today 11(2):233–241. https://doi.org/10.14719/pst.3044

Tripathi A, Rathore M, Shukla S, Das A, Debnath S (2024) *Agrobacterium* and biolistic mediated genetic transformation of mungbean cultivar Samrat using embryogenic explant. Plant Cell Tiss Organ Cult 157:72. https://doi.org/10.1007/s11240-024-02780-y

Yadav B, Pandit DL, Banjade D, Mehata DK, Bhattarai S, Bhandari S et al (2024) Insights into the germplasm conservation and utilization: Implications for sustainable agriculture and future crop improvement. Arch Agric Environ Sci 9(1):180–193. https://doi.org/10.26832/24566632.2024.0901019

Yadesa L (2022) Overview on heritability concept, application and its importance in quantitative genetics. Int J Res Agron 5(2):12–17. https://doi.org/10.33545/2618060X.2022.v5.i2a.95

An Integrated Approaches of Conversation and Improvement of Medicinal Plants

Meenambiga Setti Sudharsan, Lakshmipriya Selvam, Jayashree Pandurangan, V. Keerthi, Vivek Pazhamalai, S. Ivo Romauld, and Sowmya Hari

Abstract

Medicinal plants have been indispensable to global healthcare systems, offering therapeutic, economic, and cultural benefits. However, biodiversity loss of medicinal plants is caused by factors such as overharvesting, deforestation, climate change, and natural disasters threatening their survival. The importance of sustainable practices while examining integrated approaches for the preservation and enhancement of medicinal plants. Conservation methods include in situ strategies, such as preserving plants in their natural environment, and ex situ techniques, like botanical gardens, seed gene banks, and cryopreservation. Long-term preservation heavily relies on advanced biotechnological techniques, such as synthetic seed technology and plant tissue culture. Effective conservation is hampered by factors like resource scarcity, research limitations, and biopiracy. Public awareness campaigns, strong regulatory frameworks, and sustainable funding mechanisms are all necessary to address these problems. This chapter also explores bioprospecting as a method for finding novel therapeutic chemicals while advancing conservation. In addition to ecological preservation, medicinal plant conservation promotes medication development, cultural heritage preservation, and economic growth. The sustainability of medicinal plants for future generations is the aim of these integrated initiatives, which involve cooperation between governments, researchers, local communities, and industries. This chapter emphasizes the potential of medicinal plants to improve the healthcare sector worldwide and underscores the necessity of balancing conservation with their sustainable utilization. Combining both traditional knowledge and contemporary scientific discoveries is essential to protect and utilize their full potential for healthcare development.

Keywords

Medicinal plants · Biodiversity · Sustainable use · Conservation · Healthcare

1 Introduction

The term "medical plants" covers an extensive variety of plants that are utilized in herbalism, some of which possess therapeutic properties. In less industrialized nations, medicinal botanicals are the "backbone" of conventional medicine, with over 3.3 billion individuals engaging in their regular use. These therapeutic plants are considered a plethora of components that can be employed in the synthetic and development of novel drugs. Also, these organisms are indispensable to the development of human societies on a global scale. Numerous plant species in a variety of environments may be found in the Indian subcontinent. About 8000 of the roughly 17,000 species of higher plants are thought to be therapeutic and are utilized by village populations, especially tribal tribes, or in traditional medical systems like Ayurveda. UNESCO, 1996, has noted that traditional medicine and medicinal plants are used in the majority of developing nations as a foundation for maintaining good health. Furthermore, the extraction and development of various medications and chemotherapeutics from these plants, as well as from traditionally used herbal treatments in rural areas, have been linked to an increased dependence on the usage of medicinal plants in industrialized countries (Fig. 1). Traditional medical methods have gained international attention in the last ten years. According to current estimates, a sizable section of the populace in many developing nations primarily depends

M. S. Sudharsan (✉) · S. Lakshmipriya · J. Pandurangan · V. Keerthi · P. Vivek · S. Ivo Romauld · S. Hari
Department of Bioengineering, School of Engineering, Vels Institute of Science, Technology and Advanced Studies (VISTAS), Pallavaram, Chennai, India
e-mail: meenambiga.se@velsuniv.ac.in

N. Kumar (ed.), *Biotechnology in Plant Conservation*, Sustainable Landscape Planning and Natural Resources Management,
https://doi.org/10.1007/978-3-032-02605-7_12

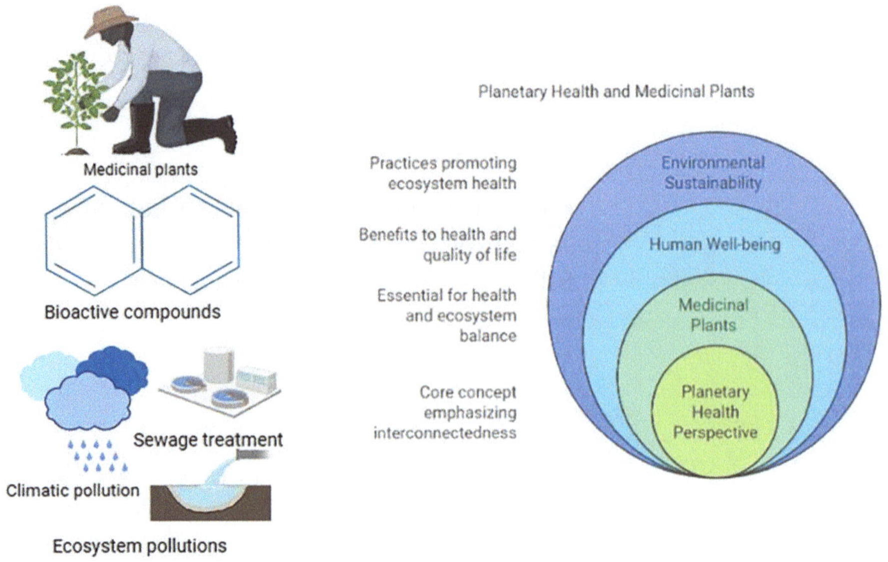

Fig. 1 A planetary health perspective of medicinal plants

on traditional healers and medical herbs to cover their basic medical needs. These nations may have access to modern medicine, but historically and culturally, herbal remedies, or phytomedicines, have frequently remained popular. Medicinal plants are commonly utilized as raw materials to extract active chemicals that are then used to make various medications. Plant-based compounds are included in blood thinners, antibiotics, antimalarial drugs, and laxatives. Additionally, the active components of morphine, vincristine, and taxol were separated from opium poppies, yew, periwinkle, and foxglove, respectively. In many developing countries, medicine remains the bedrock of healthcare, which uses the local customs and beliefs. The World Health Organization (WHO) defines health as a comprehensive state of physical, mental, and social well-being, rather than merely the absence of sickness or disability. The World Health Organization's objective of ensuring that all individuals worldwide lead sustainable, socioeconomic, and productive lifestyles by the year 2000 can be significantly furthered by medicinal plants. The growth and commercialization of bioindustries based on medicinal plants in developing nations are contingent upon the availability of facilities and knowledge regarding the industrial potential of medicinal plants, including upstream and downstream bioprocessing, extraction, purification, and marketing. Moreover, the dependence of lower-income urban and rural communities on traditional medicinal plants and herbs as supplemental aids to standard pharmaceutical market products is exacerbated by the absence of updated socioeconomic and public healthcare systems. Approximately 9000 plants have been documented as having therapeutic properties in various cultures and countries, according to recent estimates (Singh et al. 2015).

2 Importance of Medicinal Plants

Plants have been used for healthcare from the earliest days of civilization. Direct use of plants for therapeutic purposes is not just an ancient practice. Across the world, many rural and ethnic communities still employ medicinal plants for their medical needs. Approximately 80% of people worldwide, as the WHO estimates, rely mostly on traditional medicines, many of which are derived from herbs. Available locally and restricted access to modern medicines are the primary causes of the extensive use of plant resources in healthcare (Pattanayak et al. 2021). Alternative medicine is the use of medicinal plants for therapeutic purposes. Almost every culture uses alternative medicine, but Asian and Western cultures use it the most. However, the majority of people still think that the only reliable and efficient medication is one that comes in a dosage form, such as tablets, capsules, etc. Aspirin, digoxin, paclitaxel, and many more plant-based substances are among the numerous pills and capsules that are used on a regular basis. The earliest humans used plants and herbs to flavor and preserve food, soothe pain, cure headaches, and even fend against diseases like epidemics. Throughout the ages, human groups have shared knowledge of their therapeutic qualities with one another (Singh et al. 2015). The biological characteristics of plant species that are employed for a variety of reasons worldwide, including the treatment of infectious diseases, are typically attributed to active chemicals generated during secondary metabolism (Munita et al. 2016). People are currently being cautioned by numerous studies about the dangers of pathogenic germs that have developed resistance to newly discovered antimicrobials. Several studies are required to describe the chemical makeup of these plant antimicrobials and the mechanisms involved in microbial growth inhibition, either independently or in conjunction with conventional antimicrobials. This is because data on the antimicrobial activity of many plants, which were previously thought to be empirical, have been scientifically confirmed (Ur Rehman et al. 2021).

2.1 Healthcare Importance

According to ethno-medical traditions, India's medicinal plant includes a valuable bioresource of the richest plant. The use of therapeutic plants has been documented in several ancient medical systems, including Ayurveda, Unani, and Siddha, as well as in the Indian pharmaceutical keycodex and the British and American pharmacopoeias. One of the first Himalayan states, Himachal Pradesh, is home to a wealth of therapeutic plants. The Himalayan Significance of Medicinal Plants in Human Life 7 state is home to a wealth of traditional medical knowledge and medicinal plants. The current population of medicinal plants is declining as a result of irresponsible wild harvesting and habitat loss, which has caused plant populations to dwindle and, in some cases, even become extinct. Such a collection of endangered animals is illegal under the law. *Picrorhiza kurroa* and *Aconitum heterophyllum* were highlighted in the technical sessions of a workshop on Himachal Pradesh endangered medicinal plants species because of their elite populations, propagule availability, high-quality production, value addition, and guaranteed markets. Despite being one of the best for high elevations, *Swertia chirata* was not allowed to be grown right away because of its lengthy gestation time and sensitive field management. The Red Data Book has a lengthy list of several endangered medicinal plants which gives the mountain areas and villages a significant source of revenue. The issue of depletion has also escalated due to the huge demand for these plants, since many of them are primarily harvested from the wild (Srivastava et al. 2018).

2.2 Economic Importance

Nations all over the world are currently at risk from the transition of acute to chronic illnesses, which directly affects the supply and demand for pharmaceuticals. The aforementioned circumstance is thought to mostly affect the expanding populations of developing nations. Essential medications are not readily available, and the distribution of medications for both infectious and noncommunicable diseases does not satisfy the needs of a sizable portion of the populace in developing nations. Even though countries have invested a lot of money in the creation of pharmaceutical items, there has been some evidence of a slight impact on the health of the populace. It was believed that the experience was severe and that a cost-effective methodology was used in the balanced approach to the production, distribution, and use of medication. This strategy has significantly decreased the cost of producing medicines in developing nations and may have helped their circumstances. However, several nations, particularly those in Southeast Asia, were unable to put the strategy into practice and were unable to lower their excessive pharmaceutical spending.

2.3 Cultural Importance

As far back as time began, healing plants have made a big difference in people's health. India has a lot of different kinds of medical plants. Tribal people use these plants to treat a wide range of illnesses, and they play an important role in both healthcare and the local economy. They need these medicinal plants for their religious work, ceremonies, rites, and festivals, but most of them do not know that these plants really help them in their daily lives. In the Indian area, traditional medicine is still an important part of the healthcare system. Researchers have found that medicinal plants from Central India can fix life-threatening illnesses like coronary artery disease, diabetes, cancer, inflammatory conditions, viral diseases, neurological problems, and kidney problems. As a result, the demand for medicinal plants around the world has been slowly growing. Ayurveda, Unani, and Siddha are some of the traditional medicine systems used in India to make most of the dosage forms. A new generation should learn about economic medicinal plants. This will help protect medicinal plants and create jobs, which will improve the health of the indigenous and rural people in India (Neupane et al. 2022; Yadav et al. 2022).

3 Factors Lead to Biodiversity Loss of Medicinal Plants

Since the beginning of time, medicinal and aromatic herbs have been essential in reducing human suffering from illnesses, afflictions, and disorders. Natural and man-made disasters, such as fast climate change, urbanization, industrialization, pollution from overpopulation, declining forest cover, habitat loss, overharvesting, destructive harvesting, and floods, are the main reasons for the reduction of wild medicinal plant species. Like all other living things in the biosphere, medicinal and aromatic plants are susceptible to the effects of climate change. As a result, some medicinal plants that are indigenous to particular ecosystems or geographic locations may eventually be endangered, and they are especially susceptible to the effects of climate change.

As human needs, populations, and economic trade increase, so does the demand for a diverse range of wild species. Several organizations are advocating the introduction of wild species into agriculture systems due to the growing awareness that certain wild species are being overfished.

However, it is important to comprehend that cultivation might likewise have an impact on conservation. For instance, the cultivation of medicinal plants might lessen the amount of wild populations that are harvested, but it can also result in environmental damage, genetic diversity loss, and a reduction in the incentives to preserve wild populations.

4 Medicinal Plant Conservation Techniques

Conservation aims to promote sustainable development by securing and utilizing biological resources in ways that do not deplete the diversity of species and genes found worldwide or devastate significant habitats and ecosystems. Activities including collecting, propagation, characterization, evaluation, identifying and eradicating diseases, storage, and dissemination are generally included. Plant genetic resource conservation has long been seen as a crucial component of biodiversity conservation (Table 1).

4.1 In Situ *Conservation*

The main goal of the in situ conservation method is to protect wild genetic variety in its natural environment. In India, protected areas like national parks, fauna protected areas, and ecological reserves help to keep forests thriving. This kind of conservation is accomplished by designating certain places as wild nurseries and nature reserves as well as by making sure that the greatest number of wild species can continue to exist in managed environments, such as plantation woods and farms (Kadam et al. 2020). The government ought to evaluate how well the nation's medicinal plants are covered by the protected areas system. Re-introducing species that have been severely reduced by over-collection into their natural habitats is a good idea.

4.2 Ex Situ Conservation

Ex situ preservation is not always clearly different from in situ conservation, but it works well with it when it does. This is especially true for medical plants that are being overused and are in danger of going extinct because they grow slowly, are scarce, and are easily infected when they are planted again. Ex situ conservation tries to grow and reintroduce endangered species into their natural habitats to make sure they stay alive. Sometimes this is done to get a lot of plant material that can be used to make drugs, and it is often done right away to protect medical plant resources. A lot of species of medical plants that used to live in the wild can still be grown in gardens that are far from where they normally live. These plants can also have their seeds collected and saved for later planting.

4.3 Bioprospecting and Its Role in Conservation of Medicinal Plant

Medical and pharmacological fields are expanding thanks to medicinal bioprospecting. A person is generally bioprospecting if they are seeking for food or other biological value in their surroundings. The multidisciplinary endeavor of medicinal bioprospecting, which involves a range of researchers and logical subjects, is to search for compounds for medications in the natural environment. Drug bioprospecting could yield fresh leads against pharmaceutical targets like malaria, human immunodeficiency virus (HIV)/acquired immune deficiency syndrome (AIDS), cancer, and Alzheimer's disease, among others. Broad revisions of natural products made from plants (herbal treatments) are currently required, and these items require further research and study of medications that have been approved by the Food and Drug Administration (FDA) (Juan et al. 2017). According to a recent review of the literature, over 70% of medications used to treat anti-infectious disorders and 60% of medications used for oncotherapy come from natural sources. These recommendations date back to 2002. Drug compounds generated from plants increase the likelihood of creating more potent medications through laboratory synthesis and aid in determining the relationship between medications and their target location. To supply the available natural source for treating health concerns, bioprospection is required. It has frequently been discovered that the compounds extracted from medicinal plants may not be used as the finished product, but they may contain lead compounds for the creation of new medications. The aspects of floristic composition, forest stand structure, species distribution, species diversity, soil physicochemical qualities, etc., are all covered in ecological studies on medicinal plants. The management and maintenance of source plant species, as well as the development of new drugs, depend heavily on the findings of species ecology studies. The scientific community's interest in ecological research on medicinal plants for global resource management has grown (Dwivedy et al. 2019).

Table 1 Various medicinal plant conservation techniques

Conservation technique	Description	Method/strategy	Examples/applications	References
In situ conservation (on-site)	Conserving medicinal plants in their natural habitat	• Protecting natural habitats • Sustainable harvesting • Habitat restoration	• National parks • Wildlife sanctuaries • Biosphere reserves	Khan et al. (2019)
Protected areas	Designated regions where plants are protected from human exploitation	• Establishing conservation areas • Banning or regulating resource extraction	• Forest reserves • National parks	Das et al. (2016)
Sustainable harvesting	Managing the collection of medicinal plants to ensure long-term survival	• Harvesting only mature plants • Limiting the amount collected • Leaving seeds or roots for regeneration	• Regulated collection of endangered herbs	Khasim et al. (2020)
Habitat restoration	Restoring or enhancing the natural environment of medicinal plants	• Reforestation • Soil and water conservation • Removing invasive species	• Restoring degraded ecosystems	Gowthami et al. (2021)
Ex situ conservation (off-site)	Preserving plant species outside their natural habitats	• Gene banks • Seed banks • Botanical gardens	• Seed storage in gene banks • Cultivating plants in botanical gardens	Yadav et al. (2016)
Seed banks	Storing seeds for long-term preservation and future reintroduction	• Collecting and storing seeds in controlled conditions	• Millennium seed bank	Gupta et al. (2022)
Botanical gardens	Growing medicinal plants in gardens to preserve their species	• Cultivating plants in protected garden spaces • Educating the public	• Royal Botanic Gardens, Kew	Pandey and Savita (2017)
Tissue culture	Propagating plants through tissue culture techniques to ensure genetic diversity	• Cloning plants in laboratories using tissue culture	• Commercial propagation of rare plants	Yadav et al. (2016)
Gene banks	Storing genetic material of plants for future research and restoration	• Cryopreservation of plant genetic material	• Cryopreserved seeds of medicinal plants	Rajeswara Rao et al. (2016)
Propagation and cultivation	Growing medicinal plants on farms or controlled environments	• Cultivation in agroforestry systems • Organic farming techniques	• Domesticated cultivation of Ginseng	Gaurav et al., 2018
Agroforestry practices	Integrating medicinal plants within agricultural systems	• Intercropping medicinal plants with other crops	• Medicinal plants grown alongside food crops	Bijalwan et al. (2019)
Domestication and cultivation	Developing commercial cultivation systems for medicinal plants	• Selecting plants for cultivation • Ensuring optimal growth conditions	• Cultivation of Echinacea	Ramawat et al. (2021)
Research and education	Conducting research and educating the public about medicinal plant conservation	• Ethnobotanical studies • Training and outreach programs	• Workshops for communities • Botanical research on plant properties	Samal et al. (2016)
Ethnobotanical research	Studying the traditional uses and significance of medicinal plants	• Documenting indigenous knowledge • Studying plant–medicine relationships	• Research into traditional healing practices	Kichu et al. (2015)
Policy and legal support	Implementing regulations and policies to protect medicinal plants	• National laws • International agreements	• Convention on International Trade in Endangered Species (CITES)	Gowthami et al. (2021)
National regulations	Legal frameworks to protect endangered species of medicinal plants	• Banning or regulating harvesting of endangered species	• Endangered species protection laws	Kharel et al. (2021)

4.4 Increasing Botanical Gardens for Medicinal Plant Conservation

Nowadays, there are several botanic gardens that focus primarily on the research and cultivation of therapeutic plants. Furthermore, a lot of different botanical parks either have a dedicated medical plant garden or keep unique collections of medicinal plants. Botanic gardens should and can play a larger role in the conservation of plant resources, of which traditional plant conservation is but one example. The majority of gardens and arboreta, particularly in North America, have focused the most of their efforts on their traditional functions, such as display, education, and horticultural research, but many individual botanic gardens have been involved in conservation initiatives for a long period of time. Growing plants successfully in botanic gardens is nothing special!

The fact that many of these gardens now prioritize conservation as an extra activity is a recent trend. In fact, several gardens in North America have included native species conservation in their charters and accessions policies, and comparable initiatives have occurred in other nations to differing degrees. For instance, in collaboration with the Museum National d'Histoire Naturelle, plant conservatoriums have been established in Brest, Porquerolles, and Nancy, France, with the goal of preserving and reviving regionally endangered species. The Conservatoire du Mascarin, which is being built on the Indian Ocean island of Reunion, was listed as a fourth in 1987. Similar to this, the Macaronesian region's endangered species are cultivated and propagated in the Jardin botánico Viera y Clavijo on the Canary Island of Las Palmas. The following succinctly describes the justification for viewing botanic gardens as the "sleeping constituency," or natural collaborators of the conservation movement. It is acknowledged that farmed plant populations and their propagation must become essential components of our worldwide strategy for the preservation of the genetic variety of plant species given the ongoing and growing destruction and change of the world's plant life—the World Conservation Strategy confirms (Dwivedy et al. 2019).

4.5 Medicinal Plant Seed Gene Bank Conservation Techniques

Plant genetic wealth is preserved in a gene bank. Seeds and vegetative plants preserve the rich legacy of plants that provide food and support for people.

DNA, gametes or cells, embryos, tissue culture, propagules, etc. In addition to conventional seeds, refractory species and vegetatively propagated clonal material are kept in field, cryo, and in vitro settings.

Gene banks have been around for a while in many parts of the world, and their significance has long been acknowledged. Just a few years ago, the biggest and most advanced gene bank in India opened its doors in New Delhi. The main purpose of this bank is to store agricultural species and their wild cousins.

The establishment of a gene bank for forest germplasm has not yet begun, despite the topic having been brought up in recent debates. Given the speed at which forests are being destroyed globally, especially on the Indian subcontinent, it is imperative that the genetic material of Indian forests be preserved.

In order to support in situ conservation activities, the gene bank intends to do the following:

- Plant genetic resources should be kept safe for a long time using methods like ex situ protection for seeds, in vitro growth, and cryopreservation.
- As a role of the global system, the gene bank serves as a regional depository for identical collections and also as a repository for gathered, threatened, and high-quality material.
- Monitoring and maintaining the current collections, which makes it easier to plan the regeneration program.
- Ensuring that both native and exotic germplasm are available for assessment, use, and conservation by periodically increasing the number of seeds.
- Researching germplasm conservation in the medium and long term.
- Creating and running a forest genetic resource database and information network system.
- Help plan and support graduate programs and short- to medium-term education programs on how to work with forest genetics at the national, regional, and global levels.

With resistant seeds and vegetatively propagated plants, the most popular ex situ technique for preserving genetic resources is the field gene bank (FGB). By transferring and preserving samples of a species, subspecies, or variety as live collections, this conservation technique preserves genetic variation away from its natural habitat. By creating a FGB, collections are guaranteed to continue to exist for a long time and to provide authentic materials or propagule for both use and research. A practical benefit of this approach is the availability of the germplasm and ongoing plant growth assessment. The conserved material includes both wild and cultivated species that can be used for research, reintroduction, breeding, and other uses as needed.

4.6 Advanced Techniques for Long-Term Conservation

Around 50,000–80,000 species of therapeutic plants are known to exist, according to the World Wide Fund for Nature (WWF) and International Union for Conservation of Nature (IUCN). From there, overharvesting and habitat loss have already put about 15,000 species in danger. Additionally, an additional 20% of these plant resources are used by the population increase. Many recommendations have been produced based on the thorough study on how to protect medical plants and use them in a way that does not harm them. For their preservation and sustainable usage, only some of the suggestions—such as in situ conservation in botanic gardens and other natural reserves—have been implemented. Progression has been made in plant tissue growth, and solicitation methods, as well as the vast collection of natural products made possible by biotechnological breakthroughs, are ultimately leading to the effective synthesis of desired bioactive chemicals. Plant cell and genetically edited hairy root culture are a good option to making rare and expensive secondary metabolites that are used in medicine. You can use micropropagation and put tissue or organs inside fake seeds to make them grow. It is now possible to achieve higher in vitro regeneration rates as well as an appropriate technique for the transportation and storage of germplasms.

Furthermore, in the event that normal seed production is insufficient, synthetic seed technology offers a feasible alternative for propagation. However, cryopreservation remains the only technology that can preserve problem species germplasm for an extended period of time (Gantait et al. 2017).

4.6.1 Cryopreservation for Future Use of Medicinal Plants

Some parts of plants, like regular seeds and frost-hardy quiescent buds, do not have much water, so they can be frozen right away without being prepared first. However, most of the things that are frozen for experiments—like cell samples, calli, shoot tips, and embryos—have a lot of water in them, which makes them more likely to get damaged by freezing because they are not naturally freeze-tolerant. Therefore, artificial dehydration is required to shield cells from the harm that results from intracellular water crystallizing into ice. Classical and novel cryopreservation methods differ in the methods used and the physical mechanisms they are based on (Kundu et al. 2018).

Traditional methods use freeze-induced dehydration, but more recent methods rely on vitrification—the direct conversion of liquid water into an amorphous phase or glass without the creation of crystalline ice.

Slow cooling to a predetermined pre-freezing temperature and then quick immersion in liquid nitrogen are the steps of traditional cryopreservation methods. Due to the need for expensive and sophisticated programmed freezers, they are typically complicated to operate. By using a home or laboratory freezer to complete the freezing stage, their use can occasionally be avoided.

4.6.2 Clonal Germplasm Repositories

Clonal repositories are similar to field gene banks, but they only contain genetically identical material or clonal material. In contrast, field plantings that emphasize intraspecific variability are referred to as FGB. Clonal repositories were the main focus of nearly all previous horticultural crop research. But there are now more FGBs to protect diversity even in annual gardening crops, like tree crops. The twin function of FGBs is to conserve intraspecific and interspecific variation. The field gene bank's (FGB's) primary goal is to maximize variety while minimizing sample size and quantity. Both random and nonrandom sampling techniques may be used for gathering the samples. Only samples whose shape shows biological stress, like disease protection and other physiological traits, will be chosen by nonrandom selection.

Clonal crops are typically preserved in FGBs as an adjunct to cryo-banks and in vitro conservation methods. Because the germplasm in FGBs is susceptible to disorders, testing, checking in, and management methods are necessary to preserve the accessions. FGBs are the only feasible and economical method of conserving clonal crops, even though it is impossible to claim that they are the most secure type of conservation. Careful preparation and field management are necessary to reduce the dangers (Fig. 2).

5 Challenges for Conservation of Medicinal Plants

Medicinal plant conservation faces several critical challenges that hinder effective efforts to protect and sustain these valuable resources. A major issue is the lack of people don't know enough about the biological, cultural, and economic value of medicinal plants, so protection efforts don't get enough support. To get people to care about and fight for their protection, it is important to make people more aware of them through education programs, public campaigns, and community involvement. Additionally, funding and resource constraints significantly limit progress. Conservation efforts require financial, human, and infrastructural resources for research, habitat restoration, and outreach, but these are often inadequate. Increased investment, public–private partnerships, and sustainable financing mechanisms are crucial for addressing these gaps (Table 2).

As the economy and means of existence of human civilizations in emerging nations rely heavily on forest products,

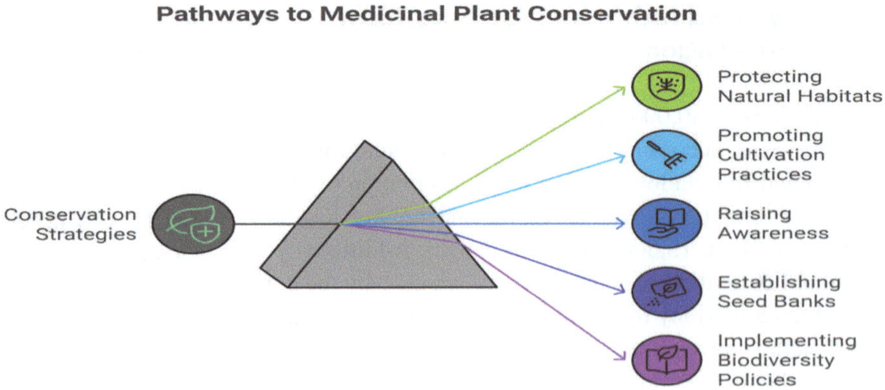

Fig. 2 Conservation strategies of medicinal plants

Table 2 The challenges for the conservation of medicinal plants, along with descriptions and potential solutions

Challenge	Description	Potential solutions	Reference
Overharvesting	Excessive collection of medicinal plants for commercial use or local consumption can lead to depletion of wild populations	• Implement sustainable harvesting methods • Encourage the cultivation of medicinal plants in gardens or farms • Create harvest quotas or licenses to regulate collection	Pandey et al. (2017)
Habitat loss and degradation	Destruction of natural habitats due to deforestation, urbanization, agriculture, and mining threatens plant populations	• Protect critical habitats through conservation laws and policies • Restore degraded ecosystems • Promote land-use practices that are compatible with plant conservation	Yousaf and Toor (2021)
Climate change	Changes in temperature, rainfall patterns, and extreme weather events can alter the growing conditions of medicinal plants, potentially threatening their survival	• Develop climate-resilient cultivation methods • Monitor plant species' responses to climate change and adapt conservation strategies accordingly	Das et al. (2017)
Invasive species	Invasive plant species can compete with native medicinal plants, affecting their growth and survival	• Implement programs to control and remove invasive species • Restore native plant species and ecosystems • Educate communities about the risks posed by invasive species	Dar et al. (2019)
Genetic erosion	Overharvesting and habitat loss can reduce the genetic diversity of medicinal plants, making them more susceptible to diseases and climate change	• Establish seed banks and botanical gardens for ex situ conservation • Promote in situ conservation in protected areas • Support breeding programs to maintain genetic diversity	Singh (2015)
Unsustainable trade	The global trade in medicinal plants can lead to overexploitation, especially when plant species are not properly regulated	• Establish sustainable trade guidelines and certifications (e.g., fair trade and organic certification) • Implement monitoring systems for trade and collection of endangered species	Raju et al. (2024)
Loss of traditional knowledge	The decline in indigenous knowledge of medicinal plants can lead to the loss of valuable cultural practices and sustainable use techniques	• Document traditional knowledge through ethnobotanical research • Involve indigenous communities in conservation and knowledge preservation • Promote intergenerational knowledge transfer	Juan (2017)

the ongoing growth of the human population is one of the reasons for concern over the ability to meet daily needs for food and medicine. Because of this phenomenon, forests and their products are continuously being eroded, making it more difficult to meet demands and preserve valuable bioresources. The requirements for their purity and accurate identification, however, are not keeping up with the growth. The market pricing for medicinal plants and their derivatives do not give us exact information about supply and demand, profits, or how the market functions.

Another key challenge is there is not enough study and data on medicinal plant types, including their ecology, distribution, and threats, which hampers effective conservation planning. Increased research and collaborations among the researchers, local communities, and traditional physicians are vital to bridging this knowledge gap. Lastly, conservation prioritization and tradeoffs add complexity, as decisions must consider species' conservation status, they have monetary worth, cultural importance, and environmentally friendly roles, and they need to be balanced with tradeoffs like protection and safe use. To make sure that medicinal plants are protected for the foreseeable future, authorities, conservation groups, researchers, communities, and the business sector must all work together to overcome these challenges.

5.1 Biopiracy

"Biopiracy" means using natural resources and traditional knowledge in a way that is inappropriate or against the law, especially that of indigenous groups, without proper acknowledgment or payment. The appropriation of therapeutic plants like neem, turmeric, and basmati rice are notable examples that show how susceptible indigenous knowledge systems are to being exploited by international enterprises.

As part of global endeavors to make sure that gains from using biological assets are shared fairly, the International Convention on Biological Diversity (CBD) and its follow-up, the Nagoya Protocol, have been created. These programs stress that access and benefit-sharing (ABS) processes, terms that are agreed upon by both parties, and informed consent are necessary to stop biopiracy. However, their usefulness has been restricted due to enforcement issues and important nations' lack of engagement.

The intricacy of biopiracy is demonstrated by the neem tree case. Indian groups had been using traditional knowledge for centuries, and Western firms tried to patent it on the grounds of innovation. A significant win against biopiracy was achieved when these patents were eventually revoked as a result of opposition from groups in India and Europe as well as campaigners like Vandana Shiva. The commercialization of the hoodia plant, which the San people have long used as an appetite suppressant, in South Africa highlights the difficulties and advancements made in combating biopiracy. Despite the exclusion of indigenous groups from early accords, agitation led to a modest profit-sharing arrangement, setting the stage for fair cooperation.

National measures like Brazil's Biodiversity Law, which forbids biopiracy and aims to guarantee equitable benefit-sharing with indigenous communities, have been influenced by the CBD and Nagoya Protocol. Administrative obstacles continue to impede implementation despite legislative progress (Huggan and Tiffin 2015). Biopiracy is getting more acute because pharmaceutical sales are going through unprecedented levels. It has been estimated that the global medicinal plant market, which was discovered by following the advice of local and indigenous tribes, is worth $43 billion. For this reason, the knowledge of indigenous peoples is so valuable: "Discovering traditional information makes the method of looking for medicinal plants 400% more effective." When indigenous groups lack official legal protection, biopiracy frequently serves as a quick route to enormous profits without having to fairly compensate the information's original sources.

5.2 Research Restrictions

Restrictions that unintentionally hinder research have resulted from the controversy around bioprospecting and intellectual property rights. After the Convention on Biological Diversity (CBD), strict regulations were put in place to protect indigenous knowledge and biodiversity. These limitations do, however, occasionally impede scientific study that is unconnected to biopiracy. As an example: Access to plant samples is hampered by bureaucratic obstacles, which deters real scientific research. Because of their fear of exploitation, some nations have implemented unduly restrictive laws that prevent or postpone crucial conservation research.

Overly stringent laws designed to stop biopiracy frequently fall short of striking a balance between sustainable research methods and medicinal plant conservation. The issue is made worse by misunderstandings over the extent of bioprospecting and irrational hopes for financial gain. These overreactions limit prospects for cooperation and the creation of solutions to protect medicinal plants. Fragmented conservation efforts are the result of inconsistent research licenses worldwide. Other countries lag behind, enforcing inconsistent prohibitions that sabotage international cooperation, while China and other countries are developing frameworks to standardize medicinal plant research through laws and quotas (Chen et al. 2016).

There are claims that certain nations and territories have overreacted to concerns about intellectual resource theft and biopiracy by imposing harsh limits on research that could have major negative effects to sustainable development and conservation. It is very likely that these purported overreactions have been mostly caused by problems with medicinal plants in particular. It is likely that there are frequent misunderstandings regarding the scope of bioprospecting, the potential financial rewards, and the relative importance of bioprospecting research in comparison with research with other goals.

5.3 Overharvesting

Growing demands associated with population increase and the corresponding expansion of agricultural land into forests and woods, habitat destruction, and land degradation are all directly linked to the overexploitation of plants for medicinal purposes and other biotic resources.

Destructive harvesting typically leads to resource depletion and potentially species extinction for medicinal plants with sluggish growth and limited availability.

As a result, it is important to think about the sustainable usage of medicinal herbs and develop appropriate harvesting techniques. Medicinal plants (such as herbs, shrubs, and trees) are more harmed by root and whole-plant harvesting than by leaf, flower, or bud collection. Herbal medicines made from whole plants or roots can be safely swapped over to medicines made from plant parts.

Controlled growing conditions can make sure that production stays stable and boost the production levels of active chemicals, which are usually metabolites that are secondary in nature. In order to obtain more of the things that people want, farming methods are designed to give the plants an adequate amount of water, nutrients, supplements that people can choose to add, and natural factors like light, temperature, and humid. Increasing cultivation also helps restore their wild resources, brings down their selling prices, and drops the amount of medicinal plants that can be harvested.

5.4 Deforestation

Overexploitation, careless collection, unchecked deforestation, and habitat damage can all make a species less common, but they cannot tell you by themselves how sensitive or resistant a species is to harvest pressure. The risk of extinction is linked to many biological features, such as ecological specialization, dispersion spread, size of population, species diversity, development rate, and the system of reproduction.

5.5 Natural Threats

In response to destructive gathering, medicinal plants are becoming infrequent and hard to find. Many species have lost their genetic stock and will never be able to get it back because they were taken from their natural habitats to be used as medicine. Medicinal plants are in danger because their habitats are being lost or broken up. Natural ecosystems are quickly being changed or killed by growing farming, cities, exploitation, and building new roads and bridges. Medicinal plants often need certain types of ecosystems to survive. When these habitats are destroyed or gone, the populations of those plants become more fragile. Fragmentation stops the flow of genes, lowers population numbers, and raises the chance of local mass extinctions. To protect medicinal species of plants and make sure they live on, conservation initiatives ought to concentrate on keeping and restoring good environments. Medicinal plants are in danger when they are harvested without following the rules and in a way that does not support their communities. As the need for herbal remedies grows, people may collect plants in ways that are not viable, which can hurt wild populations and cause overexploitation. Some medical plants grow slowly or cannot reproduce well, which makes them more likely to be harvested too much. Harvesting ways that hurt plants or kill whole individuals without letting them grow back make the problem even worse. Sustainable harvesting methods, like encouraging farming and accreditation initiatives, can help wild species by reducing the amount of pressure on them.

Changes in the Climate and Environmental Factors

The changing climate is a big problem for plants that are used for medicine. Changing trends of rainfall, increasing temperatures, and catastrophic storms can throw off the intricate equilibrium of the environment that they need to survive. Temperature, moisture, and fluctuations in humidity can have an effect on medicinal plants, making them unable to grow, reproduce, or stay healthy in general. Changes in climate can also affect where species live, causing their ranges to grow or shrink. To respond to and lessen the effects of climate change, conservation approaches should include finding species that can survive in changing climates and protecting a variety of habitats that act as ecological barriers (Shackleton et al. 2015).

To deal with these dangers, we need an integrated strategy that includes protecting habitats, using environmentally friendly havesting methods, and adapting to climate change. Protected areas and nature parks can help keep important places for plants that heal safe. Agroforestry practices or community-based projects that promote the growing of healing plants can help populations of wild plants by easing the stress on them. For long-term preservation, it is also important to teach locals, traditional medicine specialists, and customers about ecological harvesting methods and how important it is to protect medicinal plants (Kharel et al. 2021). To sum up, the problems that medicinal plants face, such as ecosystem loss, overharvesting, and climate change, are big problems that make it hard for them to stay alive. To deal with these dangers, we need to take protection actions that protect habitats, use sustainable methods, and come up with ways for animals to adapt. By reducing these risks, we can make sure that medical plants will be around for subsequent generations and keep the many health and well-being benefits they offer.

6 Potential Benefits of Medicinal Plant Conservation

The preservation of healing plants can be helped through the integration of traditional medicine into modern healthcare systems. A lot of traditional medicine is based on healing plants. Including them in national healthcare guidelines and procedures can help them stay in use and protect them. Using scientific study and evidence-based methods to backup traditional healing practices can help people accept them more and make it easier for traditional healers and modern healthcare providers to work together. This kind of mixing can lead to the creation of healthcare practices that are respectful of different cultures and last a long time, using the wide range of healing plants (Shukla et al. 2023). The pharmaceutical industry understands that medicinal plants have an enormous amount of potential as an origin of new compounds. Many medications that save lives, such as antibiotics, anticancer drugs, and medicines for heart disease and blood vessels, were made from substances found in plants. It is very important to protect medicinal plants so that there are still a lot of different chemical compounds that can be tested and used in drug research. Medicinal plants are also very important in traditional systems of medicine like Ayurveda, traditional Chinese medicine, and Indian healing techniques. These methods take an in-depth look at healthcare and have been praised for how well they work. It is important to protect medicinal plants so that traditional health practices can continue and be accepted. To sum up, therapeutic plants are important in terms of history, culture, and the economy. They have deep roots in cultural practices, help local economies, and are very important to both conventional medical systems and pharmaceutical firms. Preserving medicinal plants protects cultural history, helps the economy grow in a way that does not harm the environment, and keeps these plants' medical benefits hidden for future use (Chen et al. 2016). Medicinal plants are especially important for protection because of the reverence that people have for them. These values are about how medicinal plants can help with healthcare, making money, preserving national identity, and making sure people can make a living. People think that these effects can sometimes be so big that focusing on medicinal plants can help with a lot more than just protecting endangered communities of medicinal plants or making sure that medicines are made in a way that does not harm the environment (though these are both good goals on their own). It would be helpful to look into the link across where different types of plants grow and how valuable medicinal plants are to people (if this could be measured). It is thought that medicinal plants often have the best relative values in places with a lot of different kinds of plants. If this link is proven to be real, then projects at many plant variety centers that aim to protect all living things could often gain from relying on medicinal plants.

7 Conclusion

In summary, a multifaceted approach to protecting and improving medical plants is needed to keep ecological diversity high, encourage long-term use, and make healthcare better. With this method, medicinal plants will be around for future generations because it uses many strategies, such as protecting habitats, improving farming methods, using biotechnology, and getting people involved. Nature reserves need to be protected and excessive exploitation stopped, and healthy approaches to harvesting should be pushed at the same time. New farming methods, like selective breeding, genetic editing, and tissue culture, can be used along with traditional knowledge to make medicinal plants better. The quality, yield, and healing qualities of these plants are all improved by this harmonious integration. Keeping plants for medicinal purposes alive is a difficult and multifaceted task that needs people to work together, use long-lasting methods, and combine science and traditional wisdom. We can make sure that medicinal plants will be available and useful for generations to come by dealing with threats, taking advantage of chances, and absorbing information from past successes. It is important to protect these plants because they are part of nature and help people's health and well-being all over the world. It is up to all of us to make sure that medicinal plants continue to grow and help both people and wildlife in the future. Important to the achievement of these activities is cooperation between investigators, neighborhoods, policymakers, and businesses. To make practices that are good for the environment and traditional heritage last, we need to work together. This will also help make medicinal plants more useful for medicine and drugs. When everything is put together, it not only protects and improves medical plants, but it also helps reach bigger goals like protecting the environment along with rendering people healthier.

References

Bijalwan A, Verma P, Dobriyal MJ, Patil AK, Thakur TK, Sharma CM (2019) Trends and insights of agroforestry practices in Madhya Pradesh, India. Current Sci 117(4):597–605

Chen SL, Yu H, Luo HM, Wu Q, Li CF, Steinmetz A (2016) Conservation and sustainable use of medicinal plants: problems, progress, and prospects. Chinese Med 11:1

Dar JA, Subashree K, Sundarapandian S, Saikia P, Kumar A, Khare PK, Dayanandan S, Khan ML (2019) Invasive species and their impact on tropical forests of Central India: a review. In: Tropical ecosystems: structure, functions and challenges in the face of global change, pp 69–109

Das M, Jain V, Malhotra SK (2016) Impact of climate change on medicinal and aromatic plants. Indian J Agric Sci 86(11):1375–1382

Dwivedy AK, Singh VK, Das S, Chaudhari AK, Upadhyay N, Singh A, Singh A, Dubey NK (2019) Biodiversity bioprospection with respect to medicinal plants. In: Ethnopharmacology and biodiversity of medicinal plants. Apple Academic Press, pp 3–30

Gantait S, Kundu S, Yeasmin L, Ali MN (2017) Impact of differential levels of sodium alginate, calcium chloride and basal media on germination frequency of genetically true artificial seeds of *Rauvolfia serpentina* (L.) Benth. ex Kurz. J Appl Res Med Aromatic Plants 4:75–81

Gaurav N, Juyal P, Tyagi M, Chauhan N, Kumar A (2018) A review on in vitro propagation of medicinal plants. J Pharmacognosy Phytochemistry 7(6):2228–2231

Gowthami R, Sharma N, Pandey R, Agrawal A (2021) Status and consolidated list of threatened medicinal plants of India. Genet Resour Crop Evol 68(6):2235–2263

Gupta R, Nigam A, Kapila R (2022) Cultivation and conservation of underutilized medicinal and agricultural plants in India. Proc Natl Acad Sci India Section B Biol Sci 92(4):741–745

Huggan G, Tiffin H (2015) Postcolonial ecocriticism: literature, animals, environment. Routledge

Juan B (2017) Bioprospecting and drug development, parameters for a rational search and validation of biodiversity. J Microb Biochem Technol 9:e128

Kadam ST, Pawar AD (2020) Conservation of medicinal plants: a review. Int Ayurvedic Med J 8:3890–3895

Kharel M, Ghimire YN, Timsina KP, Adhikari SP, Subedi S, Poudel HK (2021) Economics of production and marketing of wheat in Rupandehi district of Nepal. J Agric Natural Resources 4(2):238–245

Khasim SM, Long C, Thammasiri K, Lutken H (eds) (2020) Medicinal plants: biodiversity, sustainable utilization and conservation. Springer, Singapore

Kichu M, Malewska T, Akter K, Imchen I, Harrington D, Kohen J, Vemulpad SR, Jamie JF (2015) An ethnobotanical study of medicinal plants of Chungtia village, Nagaland, India. J Ethnopharmacol 26(166):5–17

Kundu S, Salma U, Gantait S (2018) Cryopreservation of medicinal herbs: major breakthroughs, hurdles and future. In: Biotechnological approaches for medicinal and aromatic plants: conservation, genetic improvement and utilization, pp 353–381

Munita JM, Arias CA (2016) Mechanisms of antibiotic resistance. Virulence Mech Bacterial Pathogens 22:481–511

Neupane NP, Yadav E, Verma A (2022) Cultural, practical, and socioeconomic importance of edible medicinal plants native to Central India. In: Edible plants in health and diseases: volume 1: cultural, practical and economic value. Springer Nature Singapore, Singapore, pp 181–207

Pandey A, Savita R (2017) Harvesting and post-harvest processing of medicinal plants: problems and prospects. Pharma Innov J 6(12):229–235

Pattanayak S (2021) Plants in healthcare: past, present and future. Exploratory Animal Med Res 11(2)

Rajeswara Rao BR (2016) Genetic diversity, genetic erosion, conservation of genetic resources, and cultivation of medicinal plants. In: Genetic diversity and erosion in plants: case histories, pp 357–407

Ramawat KG, Arora J (2021) Medicinal plants domestication, cultivation, improvement, and alternative technologies for the production of high value therapeutics: an overview. In: Medicinal plants: domestication, biotechnology and regional importance, pp 1–29

Samal J (2016) Medicinal plants and related developments in India: a peep into 5-year plans of India. Indian J Health Sci Biomed Res KLEU 9(1):14–19

Shackleton CM, Ticktin T, Pandey AK (2015) Introduction: the need to understand the ecological sustainability of non-timber forest products harvesting systems. In: Ecological sustainability for non-timber forest products. Routledge, pp 3–11

Shukla SK (2023) Conservation of medicinal plants: challenges and opportunities. J Med Bot 7:5–10

Singh R (2015) Medicinal plants: a review. J Plant Sci 8(2):50–55

Srivastava AK (2018) Significance of medicinal plants in human life. In: Synthesis of medicinal agents from plants. Elsevier, pp 1–24

Ur Rehman F, Kalsoom M, Adnan M, Fazeli-Nasab B, Naz N, Ilahi H, Ali MF, Ilyas M, Yousaf G, Toor MD (2021) Importance of medicinal plants in human and plant pathology: a review. Int J Pharm Biomed Res 8:1–1

Yadav N (2016) Conservation of some endangered and economically important medicinal plants of India: a review. J Integr Sci Technol 4(2):59–62

Applications of DNA Banks in Plant Genetic Resource Conservation

Kaushik Kumar Panigrahi, Ayesha Mohanty, Laxmipreeya Behera, Madhusmita Pradhan, Darshana Patra, Rojalin Pradhan, Suman Dutta, and Sourav Ranjan Mohapatra

Abstract

The conservation of plant genetic resources (PGRs) has been recognized as an essential strategy to sustainable agriculture and enhance biodiversity. Notably, the application of DNA banks to support conservation is a new area in PGR conservation that is evolving but is proven to be extremely useful and practical. In the unlikely scenario that seed storage or in vitro methods are ineffective, DNA banking has been claimed to represent genetic control, which includes rebreeding. It preserves the actual gene pool for several years. This approach has a greater implication in conserving recalcitrant-seeded species that cannot be preserved using conventional methods, enhancing genetic resources conserved in particular. By the use of molecular techniques like DNA barcoding and genotyping, the effectiveness of DNA banks is enhanced to facilitate proper identification and determination of genetic diversity within and between species. Currently, new advances in science and technology will make it easier for us to recognize what is most important to the improvement of crops, there being many more challenges fomented by climate change and ever-increasing importance of ensuring a food supply. Furthermore, global approaches supporting the preservation of crop wild relatives and underused species in normal conservation plans, usually forgotten, could be enhanced by DNA banks. In brief, the strategic utilization of DNA banks is an innovative strategy to promote genetic resource conservation in order to ensure that future generations can rely on the genetic diversity which is vital to sustainable agriculture.

Keywords

Biodiversity conservation · Crop wild relatives · DNA banking · Ex situ conservation · Plant genetic resources (PGRs)

K. K. Panigrahi (✉)
AICRN on Potential crops, Odisha University of Agriculture & Technology (OUAT), Bhubaneswar, Odisha, India
e-mail: kaushikouat@gmail.com

A. Mohanty
AINP on Soil Biodiversity Biofertilizers, Odisha University of Agriculture and Technology, Bhubaneswar, Odisha, India

L. Behera
Department of Agricultural Biotechnology, Odisha University of Agriculture and Technology, Bhubaneswar, Odisha, India

M. Pradhan
School of Biotechnology, Centurion University of Technology and Management, Bhubaneswar, Odisha, India

D. Patra
Department of Genetics and Plant Breeding, Siksha 'O' Anusandhan Deemed to be University, Bhubaneswar, Odisha, India

R. Pradhan
Department of Vegetable Science, Siksha 'O' Anusandhan Deemed to be University, Bhubaneswar, Odisha, India

S. Dutta
Department of Genetics and Plant Breeding, Ramakrishna Mission Vivekananda Educational and Research Institute, Kolkata, West Bengal, India

S. R. Mohapatra
Department of Forest Biology and Tree Improvement, Odisha University of Agriculture and Technology, Bhubaneswar, Odisha, India
e-mail: srmohapatra.forestry@ouat.ac.in

1 Introduction

The conservation of plant genetic resources is crucial for ensuring food security and enhancing global biodiversity. The intricate connection between food security and biodiversity emphasizes the vital need for integrated conservation strategies that emphasize the preservation of crop genetic diversity while also acknowledging the importance of safeguarding the larger ecosystems which nurture these crops. The expanding global population, along with the rising threats

N. Kumar (ed.), *Biotechnology in Plant Conservation*, Sustainable Landscape Planning and Natural Resources Management,
https://doi.org/10.1007/978-3-032-02605-7_13

from climate change, loss of habitat, and other environmental problems rendering the preservation of plant genetic resources becoming more important. Policymakers, researchers, and local communities have to work together to prioritize conservation measures. By doing so, we are ensuring that future generations have access to a diverse array of crops, which are essential for fulfilling their dietary needs and conserving the fragile equilibrium of our natural ecosystem.

Moreover, strengthening conservation strategies may encourage stronger resilience in food systems, enabling them to adapt to novel shifts and challenges. This comprehensive strategy acknowledges that crop health has a significant relationship to the sustainability of ecosystems, thereby ensuring sustainable development in the future. The joint efforts of every stakeholder will be crucial to preserving both the genetic diversity of plants and the well-being of the environment and its inhabitants.

1.1 Significance of Plant Genetic Resource Conservation

The conservation of plant genetic resources (PGRs) is paramount to ensure food security and biodiversity conservation. It is critical, considering the mounting pressures that are being faced by the agricultural systems currently. Factors responsible are climate change, habitat loss, and increases in the global population. The PGR plays an essential role in food security where this resource is the base for all breeding programs to develop other fortified crops that could be able to withstand environmental challenges and meet increasing demands for food by more people. As it stands, approximately 7.4 million germplasm accessions which are equivalent of over 16,500 plant species are located worldwide in 1750 gene banks, which shows the great strides made to preserve these very important resources (Fu 2017). Plant breeders may access the genetic diversity stored in gene banks and put it to work in crop improvement that creates new cultivars to face different climate and pest pressures. Higher food security is at stake because genetic erosion has been made more acute by the rise of monocultures and high-yielding varieties, which thus becomes a significant threat. Such circumstances may an increased risk of crop failures and food scarcity, for agricultural systems also lose their potential diversity under natural resources that counteract disease and insect threats.

It is not the only reason but also will have deep implications for ecosystem stability and resilience. The resilience of ecosystems breaks down to be able to provide its essential services: pollination, cycling of nutrients, and soil formation, all of which are critical to sustainable agricultural production (Yadav et al. 2024). Therefore, nobody wants to lose them. The preservation of PGR means not only preserving genetic diversity but also the strategy toward ecological balance necessary for sustainable food system (Table 1). The different types of conservation for plant genetic resource (PGR) strategies are depicted in Fig. 1.

1.1.1 Importance for Food Security

Food security is directly related to access to a wide range of plant genetic resources. The genetic diversity inherent in PGR is crucial for the creation of crops that can survive under a wide range of environmental conditions and can resist biotic and abiotic stresses (Lee et al. 2024). Crop wild relatives (CWRs) are increasingly well known for their capacity to enhance the genetic makeup of cultivated plants, with benefits such as yielding increased qualities of disease resistance and nutritional content. Their significant contribution to agricultural output emphasizes the necessity for the conservation of these wild relatives, valued at over $116 billion annually in the United States.

The International Treaty on Plant Genetic Resources for Food and Agriculture (ITPGRFA) emphasizes the importance of conserving plant genetic resources for food security, recognizing that they are indispensable for sustainable agriculture, and because they are so fundamental for the resilience of food systems. Therefore, the treaty establishes international cooperation on the conservation and sustainable utilization of PGR with the provision for encouraging the sharing of such genetic resources between the countries to add to the global food security.

In addition, the incorporation of local practices and traditional knowledge into PGR conservation can promote food security through access to a range of crop diversity that is adapted to their individual environments by the local communities. Local conservation methods, as evident in different regions, have been effective in securing agricultural biodiversity as well as maintaining food availability (Priyanka et al. 2021).

1.1.2 Biodiversity Loss and Its Implications

The consequences of losing biodiversity go far beyond food safety to influence the ecological integrity and well-being of the planet as a whole. Plant genetic loss degrades ecosystems' resilience to render them increasingly vulnerable to climate change and other anthropogenic impacts. About 22% of plant species are projected to be threatened with extinction, and this necessitates the immediate need for conservation to counteract these losses (Yadav et al. 2024).

Genetic erosion leads to loss of genetic diversity that may cost the crop plant its adaptability to altered environmental conditions, especially with regard to climate change.

Table 1 Types of conservation for plant genetic resource (PGR) management

Conservation type	Subtype	Description	Key technologies/methods	Challenges	Applications
In situ	Protected areas	Conservation within natural ecosystems (e.g., forests and wetlands)	Global Positioning System (GPS) mapping, ecological monitoring, and habitat restoration	Climate change, habitat fragmentation, and invasive species	Wild relatives of crops and endemic species
	On-farm conservation	Maintenance of traditional crop varieties by farmers in agricultural landscapes	Participatory plant breeding and seed exchange networks	Genetic erosion due to commercial hybrids and loss of traditional knowledge	Landraces, heirloom crops (e.g., Andean potatoes)
	Community-based conservation	Conservation led by local communities, integrating cultural and traditional practices	Indigenous governance systems and ethnobotanical documentation	Marginalization of traditional knowledge and policy neglect	Sacred groves and traditional medicinal plants
Ex situ	Seed banks	Long-term storage of orthodox seeds at −20 °C (e.g., Svalbard Global Seed Vault)	Cryopreservation, vacuum sealing, and moisture control	Recalcitrant seeds (e.g., cocoa and mango) and viability decline over time	Global crop diversity (e.g., wheat and rice)
	Field gene banks	Living collections of plants in controlled environments (e.g., orchards)	Clonal propagation, grafting, and pest management	High costs, susceptibility to pests/diseases, and space limitations	Perennial crops (e.g., apples and bananas)
	In vitro conservation	Tissue culture under sterile, slow-growth conditions	Micropropagation, nutrient media optimization, and cold storage	Somaclonal variation and contamination risks	Endangered species (e.g., orchids and medicinal plants)
	Cryopreservation	Ultra-low temperature storage (−196 °C) of seeds, embryos, or meristems	Vitrification and encapsulation–dehydration	Technical complexity and species-specific protocols	Recalcitrant seeds and vegetatively propagated species
Complementary strategies	DNA banking	Storage of genomic material for future analysis and synthetic biology	DNA extraction, silica drying, and lyophilization	Ethical concerns (e.g., biopiracy) and incomplete genetic representation	Gene cloning, phylogenomics, and de-extinction efforts
	Pollen banks	Preservation of pollen for hybridization and genetic diversity	Freeze-drying and liquid nitrogen storage	Short viability periods and species-specific viability	Breeding programs (e.g., maize and Brassica)

Increasing temperatures together with erratic weather patterns can mean that crops without genetic diversity cannot readily cope with new threats, leading to lower agricultural productivity along with increasing food insecurity. Therefore, the next generation must have an improved store of genetic material for future food production when the world changes (Yadav et al. 2024).

Aside from the immediate effects on food security, loss of biodiversity can have ripple effects on the ecosystem that are crucial for human existence. Ecosystems in good health offer a variety of services, such as water filtration, soil quality, and climate regulation, all of which are necessary for sustainable agriculture (Yadav et al. 2024). The decline of these services as a result of biodiversity loss can result in higher costs of agricultural production and increased dependence on chemical inputs, further contributing to environmental degradation (Priyanka et al. 2021).

1.2 Evolution of Conservation Methods

The traditional approaches for the conservation of plant genetic resources have evolved over time to give room for the use of the more sophisticated procedures for conservation. Unlike the old, more advanced molecular techniques have become necessary and adopted especially in the last decade of the twentieth century owing to the rising demand to improve strategies for conservation of biological resources as a response to the enhanced global concerns; also, it was a response to the changes in ecology as well as anthropogenic pressures. Although the traditional methods have managed to conserve genetic diversity, molecular biology is more preferred due to its accuracy, sustainability, and reliability over time, which can give protection against the loss of genetic diversity. The conservation of plant genetic resources has so far been attributed to the application of these two types

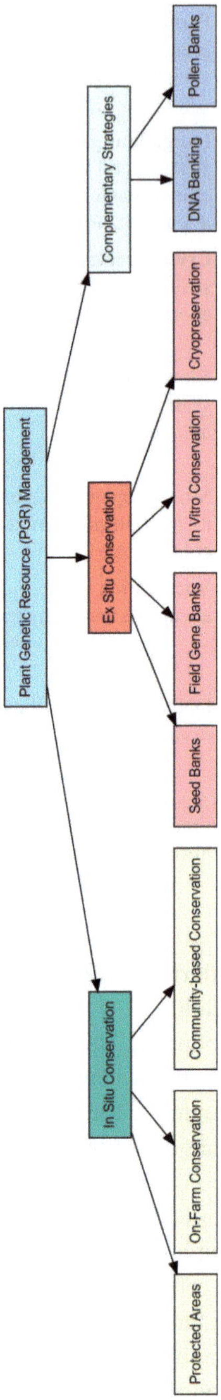

Fig. 1 Types of conservation for plant genetic resource (PGR) management

of methods. The following parts will delve into the detailed inventories and descriptions of the methods and heritage regarding the conservation of plant genetic resources.

1.2.1 Traditional Methods (Seed Banks and Field Gene Banks)

Seed Banks

- Seed banks have been the mainstay of plant genetic conservation, storing seeds under controlled conditions to preserve their viability over extended periods. These are holding centers for genetic material, and they ensure that plant species can be conserved and hopefully reintroduced as and when required.
- Seed banking has progressed from mere storage to a sophisticated process of generating genetically similar duplicate copies, improving the security and integrity of such collections (Aribi 2024; Coelho et al. 2020).

Field Gene Banks

- Field gene banks involve the cultivation of plant species in particular areas to preserve genetic resources in a viable condition. This method is especially advantageous for species that fail to produce viable seeds or possess recalcitrant seeds undesirable for storage in seed banks (Aribi 2024).
- They are vital in ensuring genetic diversity and enabling research and breeding programs through access to live planting material (Yadav et al. 2024).

1.2.2 Transition to Molecular Conservation

Molecular Techniques

- Molecular conservation has emerged as one of the essential tools for assessment and protection of germplasm in plants. This sophisticated process, such as DNA fingerprinting, marker-assisted selection (MAS), and even whole-genome sequencing of the genome of chosen plants, allows for precise identification and monitoring of genetic diversity (Shukla et al. 2022).
- These methods enable the conservation of genetic material at the molecular level, eliminating environmental variables and providing a more comprehensive understanding of genetic variation (Shukla et al. 2022).

Cryopreservation and In Vitro Culture

- Cryopreservation involves the storage of plant tissues at ultra-low temperatures, thereby interrupting metabolic processes and conserving genetic material indefinitely. This technique is especially beneficial for species with seeds that are difficult to store using traditional techniques.
- In vitro culture techniques, such as micropropagation and slow-growth storage, offer alternative methods for the conservation of plant genetic resources, especially for clonally propagated and endangered species (Agrawal et al. 2024).

While traditional methods like seed and field gene banks remain significant, the inclusion of molecular tools offers an important breakthrough in conservation approaches. These latest methods not only enhance the accuracy of genetic preservation but also open fresh avenues for breeding and research with the objective of boosting sustainable agriculture and food security. However, their application poses concerns regarding accessibility, expertise, and infrastructural requirements, which may limit their use in situations with limited resources. The integration of traditional and molecular methodologies may provide a comprehensive strategy for the conservation of plant genetic resources in diverse environments.

1.3 Role of DNA Banks in Modern Conservation Strategies

DNA banks are key not just in modern conservation efforts, but also in preserving the gene pool which is essential for the preservation of biodiversity and also in support of all conservation efforts. DNA banks lend themselves as reservoirs that are available to increase the genetic diversity in endangered populations, as well as to yield ground for the establishment and propelling of innovations which develop conservation science. The specific objectives and focus are likely to be influenced by how DNA banks can be interrelated into conservation strategies and the support of these strategies for biodiversity preservation with the progress in genomics.

1.3.1 Complementing Existing Conservation Techniques

Plant DNA banks are a key component of contemporary conservation by supplementing the conventional preservation techniques like in situ conservation and seed banking. In situ conservation tries to conserve plants in their natural habitats, whereas DNA banks offer a critical ex situ conservation technique that guarantees the conservation of genetic diversity outside their ecosystem or habitat. The integration of strategies is necessary, particularly when habitat destruction and global warming impact the majority of plant species. DNA banks store the genetic material to prevent the loss of genetic diversity due to environmental factors (Yadav et al. 2024; Coelho et al. 2020).

Furthermore, DNA banks enable the implementation of advanced biotechnological approaches, including cryopreservation and tissue culture, which can improve the effectiveness of conservation efforts (Coelho et al. 2020). These methods facilitate the everlasting preservation of genetic material, ensuring that even if a species becomes extinct in its natural habitat, its genetic information can be utilized for future ecological restoration efforts. The inclusion of molecular techniques in conservation strategies enables the identification of genetic variations crucial to breeding programs intended to enhance resistance to biotic and abiotic stresses (Priyanka et al. 2021).

In addition, DNA banks can facilitate the restoration of species into their natural habitats by providing genetically diverse material, thus improving the likelihood of successful establishment and adaptability (Shukla et al. 2022). This is especially important for endangered species, since preserving genetic diversity is essential for their long-term survival. The genetic material stored in DNA banks can be employed to create new populations or strengthen existing ones, thereby improving the overall genetic health of the species (González-Toral and Cires 2022).

1.3.2 Advancing Research and Innovation

The creation of plant DNA banks greatly contributes to plant conservation and genetic research and innovation. Through access to a large number of genetic resources, DNA banks allow researchers to carry out research that can help identify new traits and genetic adaptations important for the survival of plants. For instance, molecular methods like DNA barcoding and genome sequencing can be used for the genetic material in DNA banks to identify genetic diversity and evolutionary relationships between plant species.

In addition, the information obtained from DNA banks can guide breeding programs to create new crop varieties that are more resistant to climate change and pests. This is especially relevant to global food security, where there is a growing need for crops that can endure shifting environmental conditions. By incorporating genetic data from DNA banks into breeding plans, scientists are able to strengthen the genetic foundation of crops under cultivation, which results in higher yields and sustainability.

Aside from its agricultural uses, DNA banks also assist in ecological studies through information about plant reaction to environmental stresses. Conservation of genetic information enables researchers to investigate how plants evolve to resist different stress factors, including drought or soil erosion, which is vital for designing successful conservation practices (Coelho et al. 2020). This study can be used to inform policymaking and conservation actions intended to safeguard biodiversity in the presence of prevailing environmental pressures.

1.4 Aims and Scope of the Chapter

This chapter discusses the many aspects in which plant DNA banks contribute to conservation in current conservation efforts, their significance in enhancing the current conservation methods and also in advancing the field of research and technological developments. The chapter aims to explore the roles of DNA banks by assessing how such facilities are helping in preserving plant genetic resources and promoting biodiversity in any ecosystems. This chapter includes an in-depth analysis of the methodologies utilized in DNA banking, such as cryopreservation and molecular techniques, along with the implications of these methods for conservation practices. The chapter will also examine cases illustrating the successful utilization of DNA banks in conservation efforts and demonstrating their capacity to strengthen the capacity for resilience of plant species and ecosystems.

Additionally, the chapter will discuss the drawbacks and limitations in DNA banking such as problems connected with accessibility, financing, and cooperation among conversationalists, scientists, and policymakers. By analyzing these challenges, the chapter shall put forward recommendations to enhance the effectiveness of DNA banks in helping conservation efforts and ensuring long-term existence of plant species by using cryopreservation. The chapter will highlight the significance of plant DNA banks in contemporary conservation approaches, calling for their incorporation into larger conservation programs for natural improvement in the conservation of plant genetic resources and biodiversity. By this examination, the chapter hopes to contribute to the current debate on conservation approaches and the need to conserve genetic diversity against global environmental changes.

2 Understanding DNA Banks

2.1 Concept and Definition

DNA banks are specialized, categorized stores for storing genetic materials from extensive plant species and are critical in maintaining genetic variation, more importantly for the conservation of plant genetic resources (PGRs). Many plant species face several anthropogenic challenges to their existence, considering habitat loss, climate change, and the like. Therefore, genome resource banks (GRBs) and DNA banking become vital to safeguarding genetic information critical to biodiversity (Ballou et al. 2022). Unlike traditional seed banks that deal exclusively with the preservation of seeds, DNA banks go further to include nuclear DNA, chloroplast DNA, and other molecular parts. The construction of DNA banks is greatly important concerning ex situ conservation approaches aiming toward protection and

Table 2 Types of DNA bank conservation strategies

DNA bank type	Preservation method	Target material	Key technologies	Applications	Limitations
Genomic DNA banks	Silica-dried tissue	Leaves, stems, and roots	CTAB extraction, spectrophotometry, and nanodrop quantification	PCR, SSR markers, and whole-genome sequencing	Degradation in low-quality samples and inhibitor contamination
	Lyophilized DNA	Purified DNA	Freeze-drying and vacuum-sealed storage	Long-term storage (decades) and synthetic biology	Requires high initial DNA quality
Functional DNA banks	Plasmid/Cosmid Libraries	Cloned genes in bacterial vectors	Restriction digestion, ligation, and transformation	Gene editing (CRISPR), trait stacking	Vector instability, limited insert size (~40–50 kb)
	cDNA Banks	Reverse-transcribed mRNA	Reverse transcription (RT)-PCR and normalization libraries	Transcriptomics and stress-responsive gene analysis	Bias in mRNA representation and lack of non-coding regions
Spatial DNA banks	Metagenomic banks	Environmental DNA (eDNA) from soil and water	Metabarcoding and shotgun sequencing	Biodiversity monitoring and microbiome studies	Low DNA yield, contamination from host organisms
	Herbarium genomics	Historical specimens (century-old samples)	Ancient DNA protocols and hybridization capture	Evolutionary studies and extinct species analysis	DNA fragmentation and post-mortem damage
Synthetic DNA banks	Synthetic oligo archives	Artificially synthesized genes/constructs	Gene synthesis and error correction algorithms	Synthetic biology and metabolic pathway engineering	High cost and ethical risks (e.g., synthetic pathogens)
Cryo-DNA banks	Cryopreserved cells	Cell suspensions, meristems in liquid nitrogen	Cryoprotectants (e.g., DMSO) and controlled-rate freezing	Regeneration of whole plants and cell line preservation	Requires specialized infrastructure and viability testing

preservation of genetic resources outside their natural settings (Table 2). This enables long-term storage of genetic materials which promotes future research and breeding attempts as well as restoration activities (González-Toral and Cires 2022). Remember that the stability of genetic material in plants increases the chances of studying genetic variability, evolutionary interrelationships, and suitable changes toward different environments.

2.2 Major Functions and Objectives

2.2.1 Preservation of Genetic Material

DNA banks serve several primary functions, including the storage of genetic material from a diversity of plant species. This conservation is crucial for genetic diversity, which is key to the health and adaptability of a species (Ballou et al., 2022). Genetic information is protected against loss through extinction or genetic erosion by storing DNA samples. When high-quality DNA had been extracted from high-quality udder milk and frozen using advanced cryopreservation techniques, researchers found that they were able to store the DNA for long periods without it degrading significantly, and hence still viable (Coelho et al., 2020).

2.2.2 Supporting Genomic Research

DNA banks support genomic research—but preservation is only one part of that story. They are excellent resources for scientists who study plant genetics, evolution, and breeding. Having well-characterized DNA samples allows researchers to perform high-throughput genomic analyses, which can identify valuable traits relevant for crop improvement and conservation (González-Toral and Cires 2022). Additionally, DNA banks aid in the creation of molecular markers and genetic maps that are key tools in plant breeding programs to improve crop resilience and productivity (Ballou et al. 2022).

2.3 Types of Genetic Material Stored

2.3.1 Nuclear DNA Versus Chloroplast DNA

Typically, DNA banks store nuclear DNA, but some have chloroplast DNA as well. Nuclear DNA is actually the genetic material encoded by cells that makes up most of the genetic information for an organism's existence and an important factor to understand genetic timbers of a species. On the other hand, chloroplast DNA is passed down maternally, helps inform evolutionary history, and reveals phylogenetic relationships between plant species. For genetic studies, preservation of both plays an important role, as they provide different insights into genetic diversity and evolution.

2.3.2 RNA and Other Molecules

Besides DNA, other DNA banks also preserve RNA and other molecular constituents. RNA preservation is especially crucial for the analysis of gene expression and regulatory processes in plants (Coelho et al. 2020). The preservation of other molecules, including proteins and metabolites, can give a more comprehensive picture of the genetic and biochemical pathways in plants, improving the understanding of plant biology and responses to environmental stressors (Ballou et al. 2022).

2.4 Key Differences Between DNA Banks and Seed Banks

2.4.1 Storage Methods

The storage techniques used in DNA banks are quite different from the storage techniques used in seed banks. DNA banks use cryopreservation as a technique to store genetic material in ultra-low temperatures, mostly liquid nitrogen (Coelho et al. 2020). With this process, degradation is avoided, and viability of the DNA is retained for long periods. Conversely, seed banks are mainly concerned with drying and cooling seeds to extend their lifespan, which works for orthodox seeds but is perhaps unsuitable for recalcitrant seeds that fail to endure desiccation.

2.4.2 Accessibility and Usage

The other significant difference is in the availability and use of the stored genetic material. DNA banks tend to make it possible for scientists to access more diverse genetic material, such as rare and endangered species, that may not be accessible in seed banks (González-Toral and Cires 2022). Such an access is necessary to further study plant genetics resources conservation because through this, it becomes possible to conduct research in genetic diversity as well as learn methods of the conservation and restoration of endangered plants (Ballou et al. 2022). In addition, DNA bank-derived molecular data can be incorporated into breeding programs to expand the genetic pool of cultivated plants and make them more resistant to environmental stressors (González-Toral and Cires 2022).

To summarize, DNA banks are considered a very important feature for conservation of genetic resources in plants. It comprises genetic diversity and genomic research. DNA banks provide an excellent opportunity in the field of plant biology and conservation through their storage of nuclear and chloroplast DNA, along with RNA or other molecule forms. DNA banks versus seed banks increased awareness of diversified strategies for plant genetic resource conservation, which takes on new meanings with the passing of time in the face of ever-growing environmental challenges.

3 Global Overview of DNA Banks

DNA banks are essential for the preservation of genetic materials for research purposes, conservation, and agriculture. They preserve genetic information to conserve biodiversity, assist in genetic research, and tackle problems at local and global levels. Some prominent DNA banks in these areas include the Millennium Seed Bank Partnership (MSBP) in the United Kingdom, Svalbard Global Seed Vault, Global Genome Biodiversity Network, International Rice Research Institute (IRRI) Gene Bank, Vavilov Institute of Plant Genetic Resources (VIR), and the National Centre for Genetic Resources Preservation in the United States. These institutions provide crucial support to biodiversity hotspots, research in genetics on a worldwide scale, and collaborations on the regional level (Table 3).

3.1 Millennium Seed Bank (UK)

Established in the year 2000, the Millennium Seed Bank Partnership (MSBP) is one of the largest seed banks in the world and is managed under the aegis of Royal Botanic Gardens, Kew. MSBP has a unique role in conserving the genetic diversity of plants. It plans to collect and preserve seeds from at least 25% of the world's flora, but mainly from wild plant species that are endangered due to habitat loss and climate change, among other human-induced factors. The partnership involves collaboration with more than 188 organizations from 97 countries, making it a leader at global level in ex situ conservation efforts (Raven et al. 2023).

The Millennium Seed Bank, situated in the Royal Botanic Gardens, Kew, is among the largest and most extensive seed banks in the world. It includes over 13,598 species, representing a diverse array of beneficial plants, including those with medicinal, ecological, and nutritional significance. The bank's collection covers all continents and features species from 34 biodiversity hotspots, underscoring its global importance in the conservation of plant diversity.

The primary aim of MSBP is to collect, conserve, and utilize seeds of a large number of plant species, especially those under threat. The Bank has a very stringent protocol for seed collection to adhere to international standards and maintain genetic diversity. The seeds are preserved at ultra-low temperatures to increase their longevity and will ultimately serve the purposes of conservation and restoration and research activities in the future (Liu et al. 2018).

Table 3 Global plant gene/DNA banks: overview and references

Gene bank	Location	Year established	Focus area	Storage methods	Key contributions to research
Millennium Seed Bank (Kew)	United Kingdom	2000	Wild species, crop relatives, and seed longevity	Seeds at −20 °C; genomic libraries	• Drought-tolerant alleles • Seed longevity studies
Global Genome Biodiversity Network (GGBN)	Global Network	2011	Standardized DNA/tissue storage and biodiversity genomics	DNA/tissue samples at −80 °C; linked biorepositories	• Genomic data sharing for conservation • Global DNA storage standards
Svalbard Global Seed Vault	Norway (Svalbard)	2008	Global crop backup and climate resilience	1.2 million seed duplicates in permafrost (−18 °C)	• Post-conflict crop restoration • Seed viability in Arctic conditions
IRRI Gene Bank	Philippines	1971	Rice diversity and salinity/submergence tolerance	130,000+ rice accessions; DNA markers for *Saltol* and *Sub1A* genes	• *Sub1A* gene for flood tolerance • Salinity-tolerant rice breeding
Vavilov Institute (VIR)	Russia	1921	Historic crop diversity and cold tolerance	330,000+ seed samples; focus on cereals, legumes, and wild relatives	• Cold-tolerant wheat alleles • Conservation of landraces for climate adaptation
USDA-ARS National Centre for Genetic Resources Preservation (NCGRP)	United States (Colorado)	1958	Cryopreservation, and agricultural diversity	Seeds, tissues, and DNA at −196 °C (liquid nitrogen)	• Salt-tolerant germplasm • Cryopreservation protocols for long-term storage
Australian Plant DNA Bank	Australia	1994	Native flora, extremophiles, and medicinal plants	50,000+ DNA samples at −80 °C; desert-adapted species	• Heat-shock proteins in *Acacia* • Salinity tolerance in *Eucalyptus*
Embrapa Cenargen	Brazil	1984	Amazonian biodiversity, tropical crops, and CRISPR-edited varieties	DNA and tissues for cassava, sugarcane, and wild rice	• Flood-tolerant sugarcane via CRISPR • Drought-resistant cassava
ICRISAT Gene Bank	India	1979	Drought-tolerant crops (sorghum, millet, and chickpea)	131,000+ seed accessions; genomic datasets	• *PgPIP1* gene in pearl millet • Heat tolerance in chickpea
Nordic Genetic Resource Center (NordGen)	Sweden	1979	Nordic crops (barley and oats) and cold tolerance	DNA and seeds at −18 °C; boreal plant collections	• Frost-resistant barley alleles • Climate adaptation of Nordic crops

Apart from seed banking, the MSB conducts research not limited to seed biology but also germination and some ecological requirements of different plant species. The research also leads to conservation strategies as well as improving the requirements for changing environments of plant populations (Hill et al. 2021). The MSBP also has outreach and educational programs to raise awareness about the importance of plant conservation-related seed banks in the preservation of biodiversity.

The MSBP also aims to solve some global challenges of biodiversity loss and food security. The conservation of seeds from various plant species is integrated in the MSBP, which acts as a significant resource that will call on agricultural and ecological restoration efforts in the future. For example, seeds stored at the MSBP may be used to reintroduce native species into their natural habitats.

In addition, the MSBP supports research on the genetic diversity of crops and wild relatives necessary for breeding programs directed toward developing new varieties that can withstand environmental stresses. This research plays a vital role in the context of climate change, as it identifies traits that can improve the adaptability of plant species to varying conditions (Liu et al. 2018).

3.2 Global Genome Biodiversity Network (GGBN)

The Global Genome Biodiversity Network (GGBN) is a worldwide effort aimed at the conservation and accessibility of genomic biodiversity data, including plant genetic resources. The GGBN connects institutions for the collection, preservation, and distribution of genomic specimens, thereby enhancing our knowledge of genetic diversity and its application in agriculture and conservation. The GGBN promotes the use of advanced genomic technologies to catalogue and characterize plant species, which is essential for discovering favorable genetic traits for plant improvement. The network's focus on biodiversity genomics becomes especially relevant considering habitat loss and climate change, as it seeks to make genetic resources available for future research and breeding initiatives.

3.3 Global Seed Vault (Svalbard Global Seed Vault)

The Svalbard Global Seed Vault, commonly referred to as the "Doomsday Vault," serves as an essential repository for the preservation of plant genetic resources. Situated in the Svalbard archipelago of Norway, it serves as a global backup repository for seed samples, preserving biodiversity against natural disasters, climate change, and human conflict. The vault currently houses over one million seed samples, representing various crops for food security and agricultural sustainability. The establishment of the Seed Vault emerged from the recognition of an urgent necessity for a global conservation strategy, an idea that has reflected since the 1970s through various international agreements and initiatives aimed at preserving agricultural biodiversity. The Svalbard Global Seed Vault is a crucial component of the global ex situ conservation framework that ensures genetic diversity for future generations and can be utilized in crop enhancement initiatives.

3.4 International Rice Research Institute (IRRI) Gene Bank

The International Rice Research Institute (IRRI) Gene Bank is an essential repository for the preservation of rice genetic diversity. The IRRI Gene Bank, situated in the Philippines, contains more than 132,000 accessions of rice germplasm, constituting it one of the largest repositories of rice genetic resources worldwide (Jamora and Ramaiah 2022). The gene bank is essential for global food security, providing breeders and researchers access to a variety of rice cultivars that can be utilized to create new strains with improved characteristics, including disease resistance, drought tolerance, and nutritional quality. The IRRI Gene Bank conducts comprehensive research to characterize and assess the genetic materials it preserves, ensuring their effective application in breeding programs. The IRRI Gene Bank significantly contributes to global efforts to protect agricultural biodiversity and improve food production systems through its initiatives.

3.5 Vavilov Institute of Plant Genetic Resources (VIR)

Russia's Vavilov Institute of Plant Genetic Resources (VIR) is known for having the most significant number of plant genetic resources, specifically crop-related ones with economic value. Founded early in the twentieth century, the VIR stores more than 330,000 accessions such as seeds, tubers, and other plant materials. The institute is committed to conserving, investigating, and exploiting plant genetic resources, with particular emphasis on crop genetic diversity vital to food security. The research endeavors of VIR cover characterization and testing of genetic materials, which form the cornerstone for breeding for increased crop tolerance and adaptation to shifting environmental parameters. The historical importance and continuing role in conservation of agricultural biodiversity at the international level make it a significant part of the worldwide network of gene banks.

3.6 National Center for Genetic Resources Preservation (NCGRP)

The National Center for Genetic Resources Preservation (NCGRP) of the United States is a central organization devoted to the preservation of plant genetic resources. It comes under the patronage of the US Department of Agriculture and is tasked with the conservation of a vast array of plant germplasm, such as seeds, tissues, and DNA samples. The NCGRP was previously referred to as the National Seed Storage Laboratory (NSSL) before it was renamed in 2001 to accommodate an enlarged mission. Its Plant Genetic Resources Preservation Program (PGRPP) takes care of the National Plant Germplasm System (NPGS) base collection in which duplicate units of the germplasm maintained in the working collections are stored in long-term storage. The NCGRP uses new biotechnological techniques, for example, cryopreservation and in vitro culture, to sustain the viability of genetic material over long periods. This center not only aims at the conservation of genetic diversity but also at the realization of the need to use these resources for research and breeding programs to enhance crop resilience and productivity. The NCGRP's work is in line with international efforts to counter genetic erosion and achieve food security in the context of environmental adversity.

4 Technologies and Methods in DNA Banking

Conserving the genetic resources of plants is crucial to food security and the maintenance of biodiversity. DNA banking or gene banking has emerged as a significant strategy for the preservation of genetic material in plants, especially against the threats of climate change, habitat destruction, and other human-related pressures. This article would present a holistic view of the advanced DNA extraction techniques, conservation methods, storage and cataloguing practices, and trends toward automation and artificial intelligence in the context of DNA banking for plant genetic resources (Table 4). The methods which are used in DNA banking is illustrated in Fig. 2.

4.1 Advanced DNA Extraction Techniques in Plants

4.1.1 CTAB and Phenol–Chloroform Methods

The extraction of high-quality DNA is a fundamental step in establishing DNA banks. Among the various methods available, the Cetyltrimethylammonium bromide (CTAB) method is widely recognized for its effectiveness in isolating DNA from plant tissues, particularly those with high polysaccharide content. This method employs a cationic detergent to lyse cells and precipitate nucleic acids, resulting in high yields of intact DNA suitable for molecular analyses.

Contrary to that, though it is really laborious, the phenol–chloroform method of extraction is famous because it takes away proteins and other contaminants quite effectively. This method uses organic solvents to separate DNA from proteins and lipids, producing pure DNA. The methods have advantages and disadvantages, so the selection of extraction drilling often depends on the unique needs of downstream applications such as polymerase chain reaction (PCR) amplification or sequencing.

4.1.2 Modern Automation in Extraction

In recent times, there have been advancements in automation profoundly changing DNA extraction processes with increased throughput and reproducibility. Automated systems have the advantage of carrying out DNA extraction on multiple samples at the same time, with a subsequent lowering of contamination and human error. Robotic platforms and special reagents streamline the extraction process, making it possible to study large-scale projects involving the conservation of various plant species in gene banks. Automated extraction techniques can also be implemented in conjunction with real-time monitoring systems, thereby providing a greater degree of control over extraction conditions and, in turn, improving the integrity of DNA being extracted.

4.2 Preservation Methods: Cryopreservation and Others

4.2.1 Comparison of Longevity and Cost

Preservation methods are the backbone of the shelf life of the stored genetic material. Cryopreservation is among the best methods to keep plant genetic resources alive for long periods, involving the preservation of biological samples at ultra-low temperatures. It has been used for a broad spectrum of plants with respect to those with recalcitrant seeds that cannot be stored under conventional seed banking techniques (Nguyen and Norton 2020).

By comparison, lyophilization or freeze-drying represents another preservation mechanism that has acquired increased popularity because it is cost-friendly and easy. Freeze-drying entails dehydration of biological material, thus avoiding microbial growth and enzyme action that will result in damage (Whitehouse et al. 2020). Although storage at ambient conditions is possible using lyophilized samples, durability may not reach the level in cryopreserves, especially for delicate genes.

Thus, the decision between cryopreservation and lyophilization frequently depends upon the particular needs of the preserved genetic material, such as sensitivity to temperature variations and storage duration.

4.2.2 Recent Innovations in Preservation Media

Innovative preservation media have also considerably improved the efficacy of DNA banking. Cryoprotectants have recently been investigated in studies aimed at increasing the survival rates of plant tissues during cryopreservation. Specifically, dimethyl sulfoxide (DMSO) and glycerol were found to have high potential for improving survival rates. These prevent ice crystal formation that destroys cellular structures and compromises DNA quality. Furthermore, improvements in lyophilization medium formulations further improve the stability of the nucleic acids during freeze-drying and thus the quality of preserved DNA (Li and Pritchard 2009).

4.3 Storage and Cataloguing of Genetic Material

4.3.1 Digital Documentation Practices

Proper storage and cataloguing of genetic material are vital for efficient DNA bank management. Digital documentation practices have developed as a primary aspect of DNA banking

Table 4 Technologies and methods in DNA banking

Technology/method	Purpose/application	Key features	Advantages	Limitations	Plant-specific use cases	Conservation goals
DNA extraction (CTAB method)	Isolate high-quality DNA from plant tissues (leaves and seeds) with high polysaccharide content	Cetyltrimethylammonium bromide (CTAB) buffers to remove polyphenols and polysaccharides	Effective for recalcitrant plants (e.g., conifers and tropical species)	Time-consuming; requires optimization for specific species	Conservation of endangered plants and seed banks	Preserve genetic diversity of rare/endemic species
Cryopreservation	Long-term storage of plant germplasm (seeds, embryos, and shoot tips) in liquid nitrogen	Vitrification or slow-freezing to avoid ice crystal damage	Preserves viability for centuries; critical for clonal crops (e.g., banana)	Requires specialized protocols for recalcitrant seeds (e.g., cocoa and mango)	Storing orthodox and recalcitrant seeds at global seed vaults (e.g., Svalbard)	Safeguard crop wild relatives and endangered flora
Lyophilization	Stabilize plant DNA/RNA from leaves or seeds for room-temperature storage	Freeze-drying tissues or extracted nucleic acids	Reduces degradation risks in field conditions; portable	May reduce DNA yield in lignified tissues	Field collections in biodiversity hotspots (e.g., Amazon rainforest)	Enable long-term storage in resource-limited settings
FTA cards	Preserve plant DNA from leaf punches or pollen for field collections	Chemically treated cards lyse cells and inhibit nucleases	Rapid, low-cost; ideal for remote fieldwork	Not suitable for large-scale genomic DNA extraction	Documenting genetic diversity of wild crop relatives (e.g., wheat and rice landraces)	Facilitate in situ conservation monitoring
Seed banking	Store orthodox seeds at low humidity and temperature (−20 °C)	Desiccation-tolerant seeds dried to 3–7% moisture	Cost-effective; preserves thousands of species	Fails for recalcitrant seeds (e.g., avocado and oak)	Millennium Seed Bank Partnership (Kew Gardens); crop seed repositories	Protect agrobiodiversity and wild plant species
In vitro conservation	Maintain plant tissues (meristems and callus) in sterile culture	Slow-growth cultures or synthetic seed technology	Preserves clonal crops; avoids genetic drift	Risk of somaclonal variation; high maintenance	Conservation of vegetatively propagated crops (e.g., potato and yam)	Maintain genetic integrity of clonal species
Automated DNA barcoding	Identify plant species using standardized genetic markers (e.g., *rbcL* and *matK*)	High-throughput sequencing of chloroplast/mitochondrial genes	Accelerates species identification	Limited resolution in closely related taxa	Tracking illegal timber trade (e.g., rosewood) or invasive species	Combat biopiracy; monitor endangered species trade
Whole-genome sequencing	Catalogue complete genomes of crop wild relatives and endangered plants	Illumina, PacBio, or Nanopore platforms for high-resolution data	Enables allele mining for climate resilience	High cost; requires bioinformatics infrastructure	Genomic resource banks for crops (e.g., IRRI rice gene bank)	Identify adaptive traits for breeding programs
Ethnobotanical databases	Link genetic data with traditional knowledge of plant uses	Integrates genomic metadata with indigenous community inputs	Enhances equitable benefit-sharing	Risk of misappropriation without ethical frameworks	Documenting medicinal plants (e.g., neem and turmeric) with local communities	Promote biocultural conservation

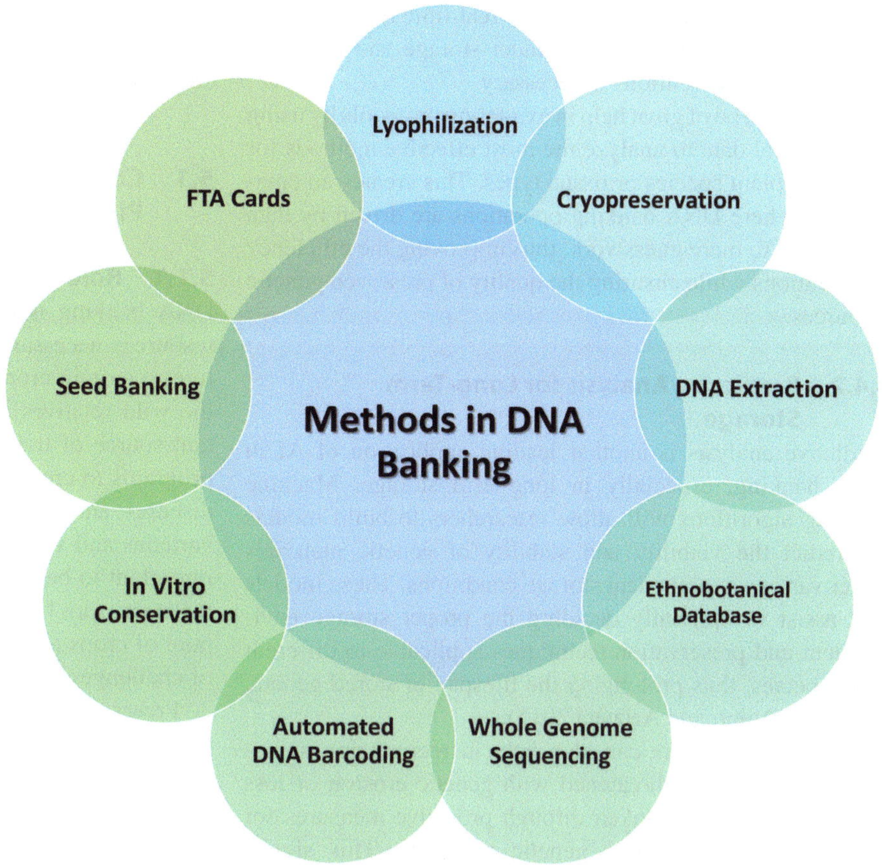

Fig. 2 Technologies and methods in DNA banking

in today's era, making it more straightforward to organize and retrieve genetic materials. Extensive databases with complete metadata on every accession, including details of its source, morphological traits, and genetic data, make it possible for researchers to retrieve and work with genetic material more efficiently.

Additionally, the use of Geographic Information Systems (GISs) in digital documentation processes enables spatial mapping of genetic resources, yielding important information regarding the distribution and diversity of plant species. In addition to facilitating greater access to genetic materials, this method also aids in conservation by pinpointing regions of high genetic diversity that can be prioritized for protection (González-Toral and Cires 2022).

4.3.2 Use of Barcoding for Efficient Retrieval

DNA barcoding is a powerful way to identify and recover any genetic material from DNA banks. In short, using a sequence of DNA contains valuable information that allows researchers to identify an organism or species and accession more reliably than cases where morphological character descriptions are too ambiguous to identify it. This methodology has facilitated assessments in plant genetic resources, including rapid assessments of genetic diversity and identification of unique accessions with potential appeal in breeding programs.

The combination of DNA barcoding with digital documentation improves retrieval efficiency, enabling researchers to quickly and accurately locate specific genetic materials. The amalgamation of molecular techniques with digital technologies signifies a substantial progression in the administration of plant genetic resources.

4.4 Emerging Trends: Automation and Artificial Intelligence in DNA Banking

4.4.1 AI-Based Quality Control

Artificial intelligence (AI) application in DNA banking is an infant field that has a great promise of assuring quality control over genetic materials. AI algorithms can analyze large datasets generated during extraction and preservation of DNA, thereby picking specific patterns or anomalies that may signify possible threats to the integrity of samples (Gupta et al. 2024). Such capability, when folded into the operation of the DNA bank, could establish near-real-time quality monitoring of the genetic materials under storage and therefore allow timely interventions if necessary.

AI can also be of great help in extracting protocols by using the historical data to analyze the most effective methods for particular plant species or tissue types. This creates an environment where DNA banking operations are driven by data as opposed to mere guesswork, thus improving the efficiency of operations while ensuring the quality of preserved genetic resources.

4.4.2 Predictive Analysis for Long-Term Storage

Predictive analysis is another leading application of AI in DNA banking, especially in long-term storage. Machine learning algorithms will allow researchers to build models to predict the viability and stability of genetic materials under various hypothetical storage conditions. These models will assist in optimally deciding the proper storage environment and preservation techniques applicable to different plant species, thus prolonging the lifespan of stored genetic resources (Niino and Arizaga 2015).

Predictive analysis can also help in recognizing accessions that could be threatened with genetic erosion or loss and thereby possibly taken through proactive measures for securing these invaluable genetic materials. This significant step into the future with the introduction of AI-driven predictive analysis into DNA banking practice offers researchers advanced techniques to better conserve plant genetic resources.

DNA banking for plant genetic resources has developed rapidly owing to advances in extraction methodologies, preservation techniques, and digital documentation. Automation and AI applications in DNA banking operations hold great prospects for streamlining genetic resource conservation. With climate change and biodiversity loss becoming even more pressing, strong DNA banking practices will increasingly ensure sustainable agriculture and food security through access for future generations to this genetic diversity.

5 Applications of DNA Bank in Plant Genetic Resource Conservation

The conservation of plant genetic resources is crucial for ensuring food security, biodiversity, and ecological balance. DNA banking has emerged as a pivotal strategy in this context, providing a means to preserve genetic material from both wild and cultivated plant species. This comprehensive overview will delve into the applications of DNA banking in the conservation of plant genetic resources, focusing on its role in maintaining crop diversity, protecting rare and endangered species, facilitating crop improvement programs, and contributing to ecosystem and habitat restoration.

5.1 Conservation of Wild and Cultivated Plant Species

5.1.1 Role in Maintaining Crop Diversity

DNA banking is essential for the conservation of genetic resources necessary for breeding programs and for agriculture to sustain crop diversity. Genetic variability existing in the wild relatives of cultivated crops is, in fact, an important source of traits that could be introduced into breeding programs to create cultivars that can cope better with pests, diseases, and/or environmental changes. Integration of local varieties and wild ancestors into breeding programs is very important to become a source of genes for developing pest-resistant crops (Milczarek et al. 2020). It keeps all the genetic pool of crops and prepares future generations for new drafts of challenges.

Preservation of different genetic materials in DNA banks explores the new genetic traits for crop improvements that have never been tapped into previously. Free access to various genetic diversities is important for breeding programs that will improve yield as well as nutritional quality and stress tolerance. DNA banking thus makes the sustainable agricultural foundation by establishing the genetic diversities required for resilient crop variety development.

5.1.2 Protection of Rare and Endangered Species

In addition to the cultivated crops, this DNA banking is of practical importance for the conservation of rare and endangered plants. A number of plant species are constantly being threatened by habitat destruction, climate change, and introduced species invasiveness, causing a lowered population. DNA banks thus conserve the genetic material of these species against the loss during their habitat loss over time, so that much of their genetic makeup will be available for future restoration. Just like that, the preserved genetic material can be used under re-introduction programs of endangered species which are again conserved and reintroduced to their original habitat.

Moreover, conserving genetic materials from endangered species can be beneficial to research on their genetics, allowing scientists to understand the factors leading to their decline and formulating ideas for their conservation. DNA banks, by maintaining genetic resource repositories, are critical in biodiversity conservation and for the protection of species facing extinction.

5.2 Use in Crop Improvement Programs

5.2.1 Breeding for Pest and Disease Resistance

One of the significant application areas of DNA banking and gene banking is the crop improvement programs, especially in breeding for pest and disease resistance. Genetic diversity preserved in the DNA bank can be afforded a good opportunity to identify and introgress resistance genes into the cultivated varieties. For example, genes in the wild relatives that grant resistance against specific pests and diseases can now be used within breeding programs to develop more resistant cultivars (Souza and Long 2018). This method has been implemented successfully in breeding potatoes; resistance genes from wild Solanum species have been introgressed into cultivated varieties for the control of cyst nematodes (Milczarek et al. 2020).

By incorporating molecular markers and genomic tools with DNA banking, the precise breeding strategies could be made more accurate. Marker-assisted selection allows for the identification and selection of desirable characters from the plants at faster rates than traditional breeding procedures (Shcherban 2023). Such new molecular techniques combined with old breeding have the potential to make crop improvement programs much more effective and to speed up the development of varieties. These types can survive all the peculiarities of the environment into which they are released, are resistant to pests, and develop well as required.

5.2.2 Enhancing Climate Resilience in Crops

Enhancing crop resilience to abiotic stresses, particularly drought, salinity, and extreme temperatures, is urgent in the face of climate change. DNA banks are facilities meant to preserve genetic resources related to climate-resilience traits. For instance, among the genetic diversity in wild relatives of crops exist valuable alleles for such qualities as drought tolerance and heat resistance, both of which are essential for maintaining agricultural productivity during climate change. Research shows that such genetic resources can be integrated into breeding programs for generating climate-resilient crop varieties. It is the margins retained in DNAs that are expected to yield better economic conditions in developing cultivars that withstand future climate changes while assuring food security for future generations.

6 Plant DNA Banks in Biotechnology and Genomics

6.1 Marker-Assisted Breeding and Genetic Mapping in Plants

Biotechnology and genomics with plant DNA banks have now changed everything in plant breeding. These banks are repositories for genetic materials by which it is possible to effect the desired genetic changes through marker-assisted breeding, genetic mapping, and many more applications. Furthermore, this section describes the milestone achievements of marker-assisted breeding, development of high-throughput markers, and the application of these technologies in improving crop traits.

6.1.1 Development of High-Throughput Markers in Plants

High-throughput markers are crucial to effective genetic mapping and selection during breeding programs. The generation of these markers has been made easier by improvements in genomic technologies to enable quick determination of genetic diversity in a variety of plant species. Next-generation sequencing (NGS) technology has transformed identification of single nucleotide polymorphisms (SNPs) and microsatellites (SSRs), which form the basis for genetic mapping as well as marker-assisted selection (MAS).

The integration of bioinformatics tools has further increased the efficiency of the marker development process. Some of these may be potential markers for desirable traits, which can be identified by researchers through the analysis of large genomic datasets, thereby facilitating the downstream breeding process. In addition, this approach enables not only the rapid identification of high-throughput markers but also their relevance to specific traits of interest for different plant species.

6.1.2 Microsatellites (SSRs) and Single Nucleotide Polymorphisms (SNPs)

Microsatellites, or simple sequence repeats (SSRs), are short, tandemly repetitive DNA sequences that are highly polymorphic and serve as genetic markers in plant breeding. Abundant and variable, they are good markers for genetic diversity studies and genetic map construction. Conversely, single nucleotide polymorphisms (SNPs) are the most common type of genetic variation in plants, and their use is increasing in the context of marker-assisted breeding due to very high genome density.

The combination of SSRs and SNPs allows for a comprehensive understanding of genetic diversity within plant populations. Utilizing these markers, researchers can identify and select for traits like disease resistance, drought tolerance, and yield improvements (Ahmad et al. 2018). The advent of extensive databases harboring information on such markers enhances their accessibility and utility in breeding programs and will thus facilitate the development of improved crop varieties.

6.1.3 Genotyping by Sequencing (GBS) for Large-Scale Applications

Genotyping by sequencing (GBS) is powerful technique that allows the simultaneous genotyping of thousands of individuals at a very low scale price compared to those employed in traditional methods. Basically, genomic DNA is digested with restriction enzymes, and sequencing of the resulting fragments occurs. GBS became popular in plant breeding because it generates large volumes of data quickly and at low costs (He et al. 2014).

GBS technologies are thus applied in plant DNA banks for rapid assessment of genetic diversity as well as the identification of markers for trait association. This method has significance for breeding programs of complex-genome crops as it provides an insight into the genetic architecture underlying comparatively important agronomic traits (Nybom and Lācis 2021). GBS can additionally be integrated with other genomic technologies to fine-tune breeding, thus resulting in more resilient and productive crop varieties.

6.1.4 Bioinformatics Tools for Marker Discovery

The significance of bioinformatics in marker discovery is paramount. The prompt expansion of genomic data has made bioinformatics technology essential for the analysis and interpretation of sophisticated information. This software facilitates the identification of genetic markers linked to specific traits, enabling researchers to make accurate choices regarding breeding programs.

Bioinformatics tools enable the integration of different types of data, such as genomic, transcriptomic, and phenotypic data, to reveal associations between genetic variations and traits of interest. Using sophisticated statistical approaches and machine learning algorithms, scientists can increase the precision of marker discovery and enhance the effectiveness of marker-assisted selection in plant breeding. This combination of bioinformatics with plant DNA banks greatly enhances the breeding process and increases the scope for improving crops.

6.2 Applications in Molecular Breeding Programs in Plants

Marker-assisted selection and the other methods that are affiliated with molecular breeding are utilizing the genetic resources deposited in plant DNA banks to improve crop attributes. Molecular breeding applications related to the identification and selection of drought resistance, disease resistance, yield improvements for cereals, and reduction of breeding cycle duration in perennial cropping systems are discussed.

6.2.1 Marker-Assisted Selection (MAS) for Drought Resistance

Drought has proved to be one of the critical challenges facing global agriculture in developing water-resilient crop varieties. Marker-assisted selection (MAS) has emerged as a strong tool for the identification and selection of drought resistance traits in plants. Breeders can thus efficiently select individuals for drought resistance traits using genetic markers associated with drought tolerance, thereby speeding up the breeding process.

The incorporation of molecular markers into breeding programs has thus far proven successful for a number of crops, including maize, wheat, and rice, in the development of drought-resistant varieties. Such advancements will therefore not only enhance crop stability but also contribute toward food security of drought-prone areas. Exploration of the genetic diversity still available in plant DNA banks would enhance further chances for obtaining drought-resistant crops.

6.2.2 Disease-Resistant Trait Identification

Another important area of application for molecular breeding programs is the characterization of traits involved in disease resistance. Plant diseases can severely hinder agricultural productivity, so developing resistant varieties is key to sustainable crop production. Breeders would be able to use any available genetic markers for disease resistance so as to allow for rapid selection of the trait in breeding populations.

Plant DNA banks are a very important resource to identify genetic variations associated with disease resistance. The exploration of wild relatives and landraces stored in these banks has led to the discovery of new resistance genes that can be introgressed into cultivated varieties. Such ventures are important in creating more genetic diversity into the breeding programs and improving the overall resilience of crops to a variety of pathogens.

6.2.3 Crossbreeding for Yield Improvement in Cereals

Crossbreeding remains the traditional breeding method and is also a fundamental means for improving yield. It gets more sophisticated when molecular markers are used in cross-breeding programs for precise selection of desired traits, thereby increasing the yield of cereal crops (Chaudhary et al. 2023). The diversity of traits available to breeders can be exploited from genetic resources contained in plant DNA banks.

Improvements with marker-assisted selections in cross-breeding programs have brought tremendous advancement in cereal production. For instance, the identification of markers for grain size, plant height, and disease resistance helped breeders develop high-yielding wheat varieties. There are possibilities for yield improvement in cereal crops from further harnessing genetic diversity stored in plant DNA banks.

6.2.4 Reducing the Breeding Cycle Time in Perennial Crops

Perennial crops pose unique challenges for breeding because of their longer breeding cycles compared to annual crops. According to Samantara et al. (2022), integrating molecular markers into breeding programs for perennial crops can substantially reduce breeding cycle time. With the speed of marker-assisted selection, breeders can identify and select traits that are more desirable sooner into the cycle, thereby accelerating the process of developing improved varieties.

With the exploration of genetic resources present in plant DNA banks, breeding for perennial crops has advantages in that the genetic resources offer a range of traits that may not be present in cultivated varieties.

7 Role in Gene Editing Technologies (Clustered Regularly Interspaced Short Palindromic Repeats (CRISPR)–CRISPR-Associated Protein 9 (Cas9))

The advent of gene editing technologies, especially CRISPR-Cas9, has radically transformed plant biotechnology by being able to effect precise changes in plant genomes. The role of plant DNA banks in making gene editing applications possible is therefore critical, since they access a mind-boggling genetic biodiversity for many purposes, including identification of target genes or construction of pest-resistant crop varieties.

7.1 Identification of Target Genes

The identification of target genes is a fundamental step in the application of gene editing technologies. Plant DNA banks serve as repositories for genetic material, allowing researchers to access conserved DNA sequences that can be utilized for gene editing. This section explores the various approaches to identifying target genes in plant DNA banks.

7.1.1 Use of Conserved DNA Sequences from Plant DNA Banks

Conserved DNA sequences are critical for identifying target genes associated with specific traits. The exploration of genetic resources stored in plant DNA banks enables researchers to access conserved sequences that can be utilized for gene editing applications. By analyzing these sequences, researchers can identify genes that play essential roles in plant development, stress responses, and other vital traits.

The utilization of conserved DNA sequences from plant DNA banks enhances the precision of gene editing efforts, allowing for targeted modifications that can improve crop performance. This approach not only accelerates the development of improved varieties but also contributes to a better understanding of the genetic basis of essential traits in plants.

7.1.2 Annotation of Functional Genes in Wild Relatives

The annotation of functional genes in wild relatives stored in plant DNA banks is another critical aspect of gene editing applications. Wild relatives often harbor unique genetic variations that can be harnessed for crop improvement. By annotating these genes, researchers can identify potential targets for gene editing that may confer desirable traits, such as disease resistance or abiotic stress tolerance.

The exploration of functional genes in wild relatives enhances the potential for developing crops with improved traits. By utilizing gene editing technologies in conjunction with the genetic resources stored in plant DNA banks, researchers can create varieties that are better adapted to changing environmental conditions.

7.1.3 Discovery of Alleles Responsible for Abiotic Stress Tolerance

Abiotic stress, including drought, salinity, and extreme temperatures, poses significant challenges to crop production. The identification of alleles responsible for abiotic stress tolerance is essential for developing resilient crop varieties. Plant DNA banks provide a valuable resource for discovering these alleles, as they contain genetic material from diverse plant species that may possess unique adaptations to abiotic stress.

By leveraging the genetic diversity stored in plant DNA banks, researchers can identify alleles associated with stress tolerance and utilize gene editing technologies to introduce these traits into cultivated varieties. This approach not only enhances crop resilience but also contributes to food security in regions vulnerable to climate change.

7.1.4 Enhancing Precision in Crop Modification

The integration of gene editing technologies with plant DNA banks enhances the precision of crop modification efforts. By utilizing genetic resources stored in these banks, researchers can target specific genes for modification, leading to more predictable outcomes in breeding programs. This precision is particularly valuable in developing crops with improved traits, as it minimizes the risk of unintended consequences associated with traditional breeding methods (Aziz et al. 2022).

The application of gene editing technologies in conjunction with plant DNA banks represents a significant advancement in crop improvement. By harnessing the genetic diversity stored in these banks, researchers can create crops that are better suited to meet the challenges of modern agriculture.

7.2 Development of Pest-Resistant Crop Varieties

The development of pest-resistant crop varieties is a critical application of gene editing technologies. Plant DNA banks provide access to a wealth of genetic diversity that can be utilized to enhance pest resistance in crops. This section explores the various approaches to developing pest-resistant varieties using genetic resources stored in plant DNA banks.

7.2.1 Improving Nutrient Use Efficiency in Plants

Nutrient use efficiency is a crucial factor in sustainable agriculture, as it directly impacts crop yields and environmental sustainability. The exploration of genetic resources stored in plant DNA banks allows researchers to identify genetic variations associated with improved nutrient use efficiency. By utilizing gene editing technologies, researchers can introduce these traits into cultivated varieties, enhancing their ability to utilize nutrients effectively.

The development of crop varieties with improved nutrient use efficiency not only contributes to higher yields but also reduces the environmental impact of agricultural practices. By harnessing the genetic diversity stored in plant DNA banks, researchers can create crops that are better adapted to nutrient-limited conditions (Abbas et al. 2023).

7.2.2 Editing Metabolic Pathways for Secondary Metabolites

Secondary metabolites play essential roles in plant defense against pests and diseases. The editing of metabolic pathways responsible for the production of these compounds is a promising approach to enhancing pest resistance in crops. Plant DNA banks provide access to genetic resources that can be utilized to identify and modify genes involved in secondary metabolite biosynthesis.

By leveraging the genetic diversity stored in plant DNA banks, researchers can develop crop varieties with enhanced levels of secondary metabolites, improving their resistance to pests and diseases. This approach not only enhances crop resilience but also contributes to the overall sustainability of agricultural systems (Şimşek et al. 2024).

7.2.3 Elimination of Anti-Nutritional Factors in Staple Crops

Anti-nutritional factors in staple crops can negatively impact human health and nutrition. The elimination of these factors through gene editing technologies is a critical application of plant DNA banks. By utilizing genetic resources stored in these banks, researchers can identify genes associated with anti-nutritional factors and employ gene editing techniques to eliminate or reduce their expression.

The development of staple crops with reduced anti-nutritional factors not only improves their nutritional value but also enhances their overall acceptance and consumption. By harnessing the genetic diversity stored in plant DNA banks, researchers can create healthier food options for populations reliant on staple crops.

8 Synthetic Biology and Novel Crop Development

Synthetic biology represents a transformative approach to crop development, enabling the design and construction of novel biological systems. Plant DNA banks play a crucial role in facilitating synthetic biology applications by providing access to diverse genetic resources that can be utilized for various purposes, including the creation of synthetic genomes and the engineering of crops for enhanced traits.

8.1 Creation of Synthetic Genomes

The creation of synthetic genomes involves the design and assembly of complete genomes with specific traits. Plant DNA banks provide a valuable resource for this process, as they contain genetic material from a wide range of plant species (Coelho et al. 2020). This section explores the various approaches to creating synthetic genomes using genetic resources stored in plant DNA banks.

8.1.1 Computational Design of Plant Genome Scaffolds

Computational tools play a critical role in the design of plant genome scaffolds. By utilizing genetic resources stored in plant DNA banks, researchers can create computational models that simulate the structure and function of plant genomes. This approach allows for the identification of key genes and regulatory elements that can be targeted for modification in synthetic genome construction. The integration of computational design with plant DNA banks enhances the precision of synthetic genome creation, enabling researchers to develop crops with specific traits tailored to meet agricultural challenges. This approach represents a significant advancement in the field of synthetic biology and crop improvement (Jayakodi et al. 2021).

8.1.2 Assembly of Minimal Genomes for Targeted Traits

The assembly of minimal genomes involves the construction of simplified genomes that retain essential functions while eliminating nonessential elements. Plant DNA banks provide access to genetic resources that can be utilized to identify the core components necessary for plant growth and development. By assembling minimal genomes, researchers can create crops with targeted traits that are optimized for specific agricultural applications.

The development of crops with minimal genomes not only enhances their efficiency in resource utilization but also contributes to the overall sustainability of agricultural practices. By leveraging the genetic diversity stored in plant DNA banks, researchers can create crops that are better suited to meet the demands of modern agriculture.

8.1.3 Challenges in Synthesizing Plant-Specific Organelles like Chloroplasts

The synthesis of plant-specific organelles, such as chloroplasts, presents unique challenges in synthetic biology. Chloroplasts play a critical role in photosynthesis and energy production, making them essential for plant growth. The exploration of genetic resources stored in plant DNA banks can provide insights into the genetic basis of chloroplast function and development.

However, the complexity of chloroplast structure and function poses significant challenges for their synthesis. Researchers must navigate the intricate interactions between nuclear and organellar genomes to achieve successful synthesis. Despite these challenges, the potential for creating crops with enhanced photosynthetic efficiency through synthetic biology remains a promising avenue for future research.

8.2 Applications in Food and Biofuel Production

The applications of synthetic biology in food and biofuel production are vast, with the potential to revolutionize agricultural practices. Plant DNA banks provide access to genetic resources that can be utilized to engineer crops for enhanced traits, contributing to improved food security and sustainable biofuel production.

8.2.1 Designing Crops with Enhanced Photosynthesis Efficiency

Enhancing photosynthesis efficiency is a critical goal in crop improvement, as it directly impacts yield and resource utilization. The exploration of genetic resources stored in plant DNA banks allows researchers to identify genetic variations associated with improved photosynthetic traits. By utilizing synthetic biology approaches, researchers can engineer crops with enhanced photosynthesis efficiency, ultimately leading to higher yields and more sustainable agricultural practices. The development of crops with improved photosynthetic efficiency not only contributes to food security but also reduces the environmental impact of agriculture. By harnessing the genetic diversity stored in plant DNA banks, researchers can create crops that are better adapted to meet the challenges of climate change.

8.2.2 Production of Bioplastics from Genetically Modified Plants

The production of bioplastics from genetically modified plants represents a promising application of synthetic biology. By engineering plants to produce bioplastics, researchers can create sustainable alternatives to traditional petroleum-based plastics (Latour-Paczka and Luciński 2024). Plant DNA banks provide access to genetic resources that can be utilized to identify and modify genes involved in bioplastic production.

The development of crops capable of producing bioplastics not only contributes to reducing plastic waste but also enhances the sustainability of agricultural practices. By leveraging the genetic diversity stored in plant DNA banks, researchers can create crops that are better suited for bioplastic production.

8.2.3 Role in Algae-Based Biofuel Development

Algae-based biofuels represent a sustainable alternative to fossil fuels, with the potential to reduce greenhouse gas emissions and dependence on nonrenewable resources. The exploration of genetic resources stored in plant DNA banks can provide insights into the genetic basis of algal growth and lipid production. By utilizing synthetic biology approaches, researchers can engineer algae for enhanced biofuel production (Radakovits et al. 2010).

The development of algae-based biofuels not only contributes to energy sustainability but also promotes the utilization of agricultural by-products and waste materials. By harnessing the genetic diversity stored in plant DNA banks, researchers can create algal strains that are better suited for biofuel production.

8.2.4 Engineering Plants for Higher Biomass Yield

The engineering of plants for higher biomass yield is a critical goal in sustainable agriculture. By utilizing genetic resources stored in plant DNA banks, researchers can identify genetic variations associated with increased biomass production. Through synthetic biology approaches, researchers can engineer crops to optimize growth and resource utilization, ultimately leading to higher biomass yields.

The development of crops with enhanced biomass yield not only contributes to food security but also supports the production of renewable resources for bioenergy and bioproducts. By leveraging the genetic diversity stored in plant DNA banks, researchers can create crops that are better suited to meet the demands of modern agriculture.

The plant DNA banks play a pivotal role in advancing biotechnology and genomics in agriculture. Through the development of high-throughput markers, gene editing technologies, and synthetic biology applications, these banks provide access to a wealth of genetic resources that can be harnessed for crop improvement. The integration of molecular breeding, marker-assisted selection, and synthetic biology approaches enhances the potential for developing resilient and productive crop varieties that can meet the challenges of modern agriculture. As the global population continues to grow and environmental challenges escalate, the role of plant DNA banks in supporting sustainable agricultural practices will become increasingly critical.

9 Conclusion

Crop DNA banks act in a very important way to support genetic diversity, aiding crop improvement and ecosystem restoration. These serve as repositories for genetic material, ensuring that traits of value are conserved for the sake of future generations. Also, the work of DNA banks is important in research directed toward understanding the genetic basis of traits crucial for agricultural productivity and resilience. Providing access to alternative genetic resources, crop DNA banks enhance the breeding process of variety development suitable for environmental stresses and the food production industry. Moreover, crop DNA banks are also a help in conservation activities that work to restore ecosystems and protect endangered species. The preservation of genetic diversity fostered by these banks, in turn, increases the resilience of plant populations along with ecosystem health. Thus, collaborative efforts among crop DNA banks, science researchers, and international organizations are important to answer the critical challenges posed by biodiversity loss and climate change.

References

Abbas MT, Batool Z, Ahmad CA, Khan HAA, Akhter A (2023) Application of modern biotechnology and bioinformatics approaches in agricultural sciences; a systematic review. World J Biol Biotechnol 8(2):13. https://doi.org/10.33865/wjb.008.02.0906

Agrawal A, Gowthami R, Chander S, Srivastava V, Shankar M, Deepak DA (2024) Biotechnological tools for conservation of plant genetic diversity. In: Sustainable development and biodiversity, pp 711–752. https://doi.org/10.1007/978-981-99-5245-8_21

Ahmad A, Wang J, Pan Y, Sharif R, Gao S (2018) Development and use of simple sequence repeats (SSRS) markers for sugarcane breeding and genetic studies. Agronomy 8(11):260. https://doi.org/10.3390/agronomy8110260

Aribi MM (2024) Plant gene banks: conservation of genetic resources. In: Sustainable development and biodiversity, pp 753–775. https://doi.org/10.1007/978-981-99-5245-8_22

Aziz MA, Brini F, Rouached H, Masmoudi K (2022) Genetically engineered crops for sustainably enhanced food production systems. Frontiers Plant Sci 13. https://doi.org/10.3389/fpls.2022.1027828

Ballou JD, Lacy RC, Traylor-Holzer K, Bauman K, Ivy JA, Asa C (2022) Strategies for establishing and using genome resource banks to protect genetic diversity in conservation breeding programs. Zoo Biol 42(2):175–184. https://doi.org/10.1002/zoo.21741

Chaudhary N, Salgotra RK, Chauhan BS (2023) Genetic enhancement of cereals using genomic resources for nutritional food security. Genes 14(9):1770. https://doi.org/10.3390/genes14091770

Coelho N, Gonçalves S, Romano A (2020) Endemic plant species conservation: biotechnological approaches. Plants 9(3):345. https://doi.org/10.3390/plants9030345

Fu YB (2017) The vulnerability of plant genetic resources conserved ex situ. Crop Sci 57(5):2314-2328. https://doi.org/10.2135/cropsci2017.01.0014

González-Toral C, Cires E (2022) Relevance of DNA preservation for future botany and ecology. Mol Ecol 31(20):5125–5131. https://doi.org/10.1111/mec.16652

Gupta DK, Pagani A, Zamboni P, Singh AK (2024) AI-powered revolution in plant sciences: advancements, applications, and challenges for sustainable agriculture and food security. Explor Foods Foodomics 2:443–459. https://doi.org/10.37349/eff.2024.00045

He J, Zhao X, Laroche A, Lu ZX, Liu H, Li Z (2014) Genotyping-by-sequencing (GBS), an ultimate marker-assisted selection tool to accelerate plant breeding. Front Plant Sci 5:484. https://doi.org/10.3389/fpls.2014.00484

Hill R, Llewellyn T, Downes E, Oddy J, MacIntosh C, Kallow S, Gaya E et al (2021) Seed banks as incidental fungi banks: fungal endophyte diversity in stored seeds of banana wild relatives. Frontiers Microbiol 12. https://doi.org/10.3389/fmicb.2021.643731

Jayakodi M, Schreiber M, Stein N, Mascher M (2021) Building pan-genome infrastructures for crop plants and their use in association genetics. DNA Res 28(1):dsaa030. https://doi.org/10.1093/dnares/dsaa030

Jamora N, Ramaiah V (2022) Global demand for rice genetic resources. CABI Agric Biosci 3(1). https://doi.org/10.1186/s43170-022-00095-6

Latour-Paczka K, Luciński R (2024) Artificial biopolymers derived from transgenic plants: applications and properties—a review. Int J Mol Sci 25(24):13628. https://doi.org/10.3390/ijms252413628

Lee SA, Seo H, Han S, Jeong JS, Jung TW, Lee KJ (2024) Exploring genetic diversity of *Centella asiatica* (L.) Urb. across six Korean island regions: implications for conservation and utilization. Plant Genetic Resources 22(1):53–55. https://doi.org/10.1017/s1479262123001090

Li D, Pritchard HW (2009) The science and economics of ex situ plant conservation. Trends Plant Sci 14(11):614–621. https://doi.org/10.1016/j.tplants.2009.09.005

Liu U, Cossu TA, Davies RM, Forest F, Dickie JB, Breman E (2020) Conserving orthodox seeds of globally threatened plants ex situ in the Millennium Seed Bank, Royal Botanic Gardens, Kew, UK: the status of seed collections. Biodivers Conserv 29(9–10):2901–2949. https://doi.org/10.1007/s10531-020-02005-6

Milczarek D, Tatarowska B, Plich J, Przetakiewicz A, Flis B (2020) Solanum gourlayi—a source of cyst nematode resistance in potato breeding. Potato Res 63(4):589–595. https://doi.org/10.1007/s11540-020-09459-9

Nguyen GN, Norton SL (2020) Genebank phenomics: a strategic approach to enhance value and utilization of crop germplasm. Plants 9(7):817. https://doi.org/10.3390/plants9070817

Niino T, Arizaga MV (2015) Cryopreservation for preservation of potato genetic resources. Breed Sci 65(1):41–52. https://doi.org/10.1270/jsbbs.65.41

Nybom H, Lācis G (2021) Recent large-scale genotyping and phenotyping of plant genetic resources of vegetatively propagated crops. Plants (Basel, Switzerland) 10(2):415. https://doi.org/10.3390/plants10020415

Priyanka V, Kumar R, Dhaliwal I, Kaushik P (2021) Germplasm conservation: instrumental in agricultural biodiversity—a review. Sustainability 13(12):6743. https://doi.org/10.3390/su13126743

Radakovits R, Jinkerson RE, Darzins A, Posewitz MC (2010) Genetic engineering of algae for enhanced biofuel production. Eukaryot Cell 9(4):486–501. https://doi.org/10.1128/ec.00364-09

Raven PH, Ouyang Z, Smith P, Wackernagel M (2023) The Shenzhen congress and plant conservation: what have we accomplished in the 6 years since? J Syst Evol 61(6):949–956. https://doi.org/10.1111/jse.13030

Samantara K, Bohra A, Mohapatra SR, Prihatini R, Asibe F, Singh L, Reyes VP, Tiwari A, Maurya AK, Croser JS, Wani SH, Siddique KHM, Varshney RK (2022) Breeding more crops in Less time: a perspective on speed breeding. Biology 11(2):275. https://doi.org/10.3390/biology11020275

Shcherban S, Shcherban NI, Abitayev F, Hryhorchuk N, Sotsioner A (2023) Evaluation of sunflower breeding material for resistance to broomrape (*orobanche cumana wallr.*) and herbicides under artificial climate conditions to accelerate the breeding process. Sci Horizons 26(11):111–122. https://doi.org/10.48077/scihor11.2023.111

Shukla S, Singh R, Bhutani R, Tripathi A, Shukla SK (2022) Restoration and conservation of plant genetic resources via molecular techniques: an important measure for sustainable agriculture. In: Agro-biodiversity and agri-ecosystem management. https://doi.org/10.1007/978-981-19-0928-3_12

Şimşek Ö, Isak MA, Dönmez D, Şekerci AD, İzgü T, Kaçar YA (2024) Advanced biotechnological interventions in mitigating drought stress in plants. Plants 13(5):717. https://doi.org/10.3390/plants13050717

Souza APD, Long SP (2018) Toward improving photosynthesis in cassava: characterizing photosynthetic limitations in four current African cultivars. Food Energy Security 7(2). https://doi.org/10.1002/fes3.130

Whitehouse KJ, Hay FR, Lusty C (2020) Why seed physiology is important for genebanking. Plants 9(5):584. https://doi.org/10.3390/plants9050584

Yadav B, Pandit DL, Banjade D, Mehata DK, Bhattarai S, Bhandari S, Ghimire NP, Yadav P, Paudel P (2024) Insights into the germplasm conservation and utilization: implications for sustainable agriculture and future crop improvement. Arch Agric Environ Sci 9(1):180–193. https://doi.org/10.26832/24566632.2024.0901026

MIX
Papier aus verantwortungsvollen Quellen
Paper from responsible sources
FSC® C105338

If you have any concerns about our products,
you can contact us on
ProductSafety@springernature.com

In case Publisher is established outside the EU,
the EU authorized representative is:
Springer Nature Customer Service Center GmbH
Europaplatz 3, 69115 Heidelberg, Germany

Printed by Libri Plureos GmbH
in Hamburg, Germany